Applications of Picosecond Spectroscopy to Chemistry

NATO ASI Series

Advanced Science Institutes Series

A series presenting the results of activities sponsored by the NATO Science Committee, which aims at the dissemination of advanced scientific and technological knowledge, with a view to strengthening links between scientific communities.

The series is published by an international board of publishers in conjunction with the NATO Scientific Affairs Division

A	Life Sciences	Plenum Publishing Corporation
B	Physics	London and New York
C	Mathematical and Physical Sciences	D. Reidel Publishing Company Dordrecht, Boston and Lancaster
D	Behavioural and Social Sciences	Martinus Nijhoff Publishers
E	Engineering and Materials Sciences	The Hague, Boston and Lancaster
F	Computer and Systems Sciences	Springer-Verlag
G	Ecological Sciences	Berlin, Heidelberg, New York and Tokyo

Series C: Mathematical and Physical Sciences Vol. 127

Applications of Picosecond Spectroscopy to Chemistry

edited by

Kenneth B. Eisenthal

Department of Chemistry, Columbia University, New York, U.S.A.

D. Reidel Publishing Company

Dordrecht / Boston / Lancaster

Published in cooperation with NATO Scientific Affairs Division

Proceedings of the NATO Advanced Research Workshop on
Applications of Picosecond Spectroscopy to Chemistry
Acquafredda di Maratea, Italy
June 6-10, 1983

Library of Congress Cataloging in Publication Data

NATO Advanced Research Workshop on Applications of Picosecond Spectroscopy to Chemistry (1983) : Acquafredda di Maratea, Italy)
Applications of picosecond spectroscopy to chemistry.

(NATO ASI series. Series C, Mathematical and physical sciences; vol. 127)
"Published in cooperation with NATO Scientific Affairs Division."
"Proceedings of the NATO Advanced Research Workshop on Applications of Picosecond Spectroscopy to Chemistry, held at Acquafredda di Maratea, Italy, June 6–10, 1983"—CIP t.p. verso.
Includes index.
1. Laser spectroscopy—Congresses. 2. Picosecond pulses—Congresses. I. Eisenthal, K. B. II. Title. III. North Atlantic Treaty Organization. Scientific Affairs Division. IV. Series: NATO ASI series. Series C, Mathematical and physical sciences; Vol. 127.
QD96.L3N38 1983 543'.0858 84-6992
ISBN 90-277-1788-5

Published by D. Reidel Publishing Company
P.O. Box 17, 3300 AA Dordrecht, Holland

Sold and distributed in the U.S.A. and Canada
by Kluwer Academic Publishers,
190 Old Derby Street, Hingham, MA 02043, U.S.A.

In all other countries, sold and distributed
by Kluwer Academic Publishers Group,
P.O. Box 322, 3300 AH Dordrecht, Holland

D. Reidel Publishing Company is a member of the Kluwer Academic Publishers Group

Printed in The Netherlands.

TABLE OF CONTENTS

PREFACE

With the development of lasers that can generate light pulses ranging from 10^{-11} - 10^{-14} sec duration, and capable of peak powers in excess of 10^{13} watts scientists have been able to investigate the interactions of light with matter in a time and power domain not previously possible. These ultrashort laser pulses provide a powerful tool for the study of chemical phenomena at the most fundamental level. Many of the elementary processes of importance in chemistry including energy dissipation, molecular motions, structural and chemical changes occur on a very short time scale and thus require special approaches. The use of ultrashort laser pulses to perturb and to probe systems of interest affords a direct approach to the time resolution of very rapid chemical phenomena.

It was recognition of the impact of these relatively new approaches to chemical phenomena that motivated NATO to sponsor a meeting on the applications of picosecond spectroscopy in chemistry. The primary aim of the NATO workshop was to gain some perspective on the status of the field in terms of present research activities, technological developments and if possible the difficult task of sensing future directions. The way we decided to approach these issues was to gather together the main contributors to the field, fortunately many of whom were able to attend, to present their work and to participate in what turned out to be lively discussions of the field.

It was generally agreed that major factors in the expanding applications of ultrashort laser pulses to chemistry have been the development of new techniques, the improvement of existing ones, and the commercial availability of sophisticated lasers. New areas of research have opened up with the generation of ultrashort laser pulses which extend from the infra-red ($\sim$1000 cm^{-1}) to the extreme ultraviolet (below 100 Å) and which can time resolve events in the 10^{-12} sec time domain, and recently down to 10^{-14} sec at certain visible and UV frequencies. The refinement of tunable high repetition rate lasers ($\sim$100 MHz) is emerging as a powerful tool for studying molecular phenomena which require high fields or the deposition of large amounts of energy. The new developments and advances in techniques such as the combination of ultrashort laser pulses with supersonic molecular beams, transient grating methods, photoacoustic detection, molecular spectroscopy and the combination of a colliding pulse laser with pulse compression methods to generate pulses of the order of 10^{-14} sec have produced exciting results and once again opened up new directions of research in the studies of energy relaxation, coherence effects, molecular motions in liquids and thermochemistry.

The impact in chemistry resulting from the continuing interplay between technology and the development of new techniques is seen clearly in the more sophisticated studies of complex chemical phenomena. This exciting development marks the further growth, though not yet matured, aspect of this young field. At this meeting we saw advances in our understanding of important yet elusive liquid state phenomena, such as effects of solute-solvent interactions on structural changes, on molecular motions and vibrational and electronic energy relaxation, and the relation of these processes to solvation, charge transfer and chemical reactions. The range of systems amenable to study extended from the weakly bound van der Waals complexes to cold, isolated molecules in supersonic jets to the complexity of photosynthesis which is the outstanding single problem in photochemistry, to the strongly coupled molecules in liquids and crystals. Energy migration and relaxation, dephasing processes, and the dynamics of the melting of solids induced by femtosecond laser pulses were among the highlights of elegant studies of rapid solid state phenomena.

A closing comment on this workshop is the recognition that an important aspect of this rapidly growing research area is the fundamental nature of fast processes as reflected in the commonality of approach extending across the scientific and engineering boundaries as defined by chemistry, physics, biology, quantum electronics, and electrical engineering. The close contact between researchers in these diverse fields most assuredly contributes to its impressive growth rate. As the field continues to expand this positive feature of cutting across disciplines will certainly decline. Nonetheless at the moment it is a source of great energy and progress.

Kenneth B. Eisenthal

ACKNOWLEDGEMENTS

I think it is clear that a workshop of this sort would not have been developed nor succeeded without the contributions of a number of people. It is in this spirit that I acknowledge with great pleasure the NATO Scientific Committee, sponsors of the meeting. Their help and useful advice contributed to the success and enjoyment of this conference. In particular, I thank Dr. Craig Sinclair, Director of the Advanced Study Institute Program for his valuable assistance provided in numerous discussions, and Mr. John A.S. Walker, Deputy Assistant Secretary General for Scientific Affairs, for his participation in the meeting and for his useful discussions with us of the many aspects and objectives of the NATO scientific program.
I am also pleased to acknowledge the many contributions of Dr. E.V. Sitzmann to the preparation of this book.

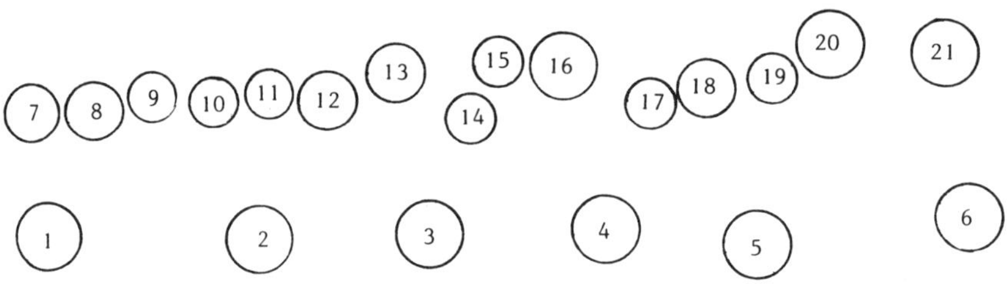

1. C. Shank
2. G. Porter
3. K.B. Eisenthal
4. G. Kenney-Wallace
5. N. Mataga
6. A. Zewail
7. K. Yoshihara
8. A. Laubereau
9. J. Walker
10. V. Sundstrom
11. P. Mataloni
12. G. Fleming
13. K. Peters
14. J. Langelaar
15. A.E. Siegman
16. R. Hochstrasser
17. F. Dorr
18. J. Faure
19. C. Flytzanis
20. D. Wiersma
21. M. Topp

PART I PHOTOCHEMISTRY

THE SUB-NANOSECOND REGIME IN PHOTOSYNTHESIS

George Porter

Davy Faraday Laboratory, The Royal Institution,
21 Albemarle Street, London W1X 4BS, U.K.

ABSTRACT

Photosynthesis is not only the most important of all photochemical and, indeed, chemical, reactions but its primary photochemical processes all occur on the sub-nanosecond time scale. The two principal phenomena are light harvesting of the absorbed energy in the antenna pigments of the photosynthetic unit and electron transfer chemistry in the reaction centre to which this energy is transferred. Picosecond studies of the former are mainly by time resolved fluorescence measurements on algae and the chloroplasts of green plants, and in the latter by absorption measurements on the isolated reaction centres from photosynthetic bacteria. In the early experiments, high intensities led to unexpected results but these are now well understood and the effect can either be utilised or eliminated. The subject is developing very rapidly and further progress is more dependent on the availability of well characterised biological or model systems than on further developments of laser technology.

Photosynthesis begins with very fast reactions of the electronically excited state and ends with the growth of the whole plant, which may occupy several thousand years. Only the former very fast processes are photochemical in the sense that they involve the excited electronic state; subsequent processes are dark reactions driven by the chemical potential accumulated by light absorption.

The earliest events are mainly of two forms: light harvesting by electronic energy transfer from the antennae to the reaction centre and electron transfer within the reaction centre. Both occur in times less than a nanosecond. These processes

K. B. Eisenthal (ed.), Applications of Picosecond Spectroscopy to Chemistry, 3–19.

present one of the greatest of all challenges to photochemical kineticists; photosynthesis is the most important of all photochemical and, indeed chemical, processes; being initiated by light it must of necessity be studied by light initiation methods, and it is one of the few important chemical processes whose primary intermolecular reactions occur on a picosecond time scale.

The primary event of photosynthesis following optical absorption is excitation of chlorophyll, and it is well established that it is the singlet state which undergoes chemical reaction before intersystem crossing to the triplet state occurs. The singlet fluorescence emission, which is always more easily studied than absorption, has therefore provided the easiest experimental approach to study of the primary photochemical processes. Its lifetime usually lies in the range between a few picoseconds and one or two nanoseconds.

Early measurements, with nanosecond spark or Q switched laser excitation, gave little more than upper limits to the rates of these events, while phase delay methods gave only an average delay time, assumed to be exponential. Mode-locked laser pulses in the picosecond region opened up a new regime of kinetic study in photosynthesis as it did in many other areas. For the most precise measurements, but limited to times longer than about 30 picoseconds, single-photon-counting time-resolved fluorescence is by far the most precise method and has the advantage that the pulse energy is small, although there may be a cumulative build-up of long lived products. Most work with time resolution better than 100 picoseconds has utilised single pulses from mode-locked Nd lasers and streak camera detection. This has two disadvantages; first the dynamic range, and therefore the precision of the measurements, is rather low and, second, the pulse energy necessary to obtain measurable fluorescence intensities for single pulse measurement is of the order of 10^{13} photone cm^{-2}/pulse which, as we shall see, is dangerously near the intensity where two-photon absorption becomes significant.

The newer picosecond techniques now available will undoubtedly overcome some of these problems. They include synchroscan streak cameras which allow signal averaging and lower pulse intensities (but introduce cumulative pulse to pulse effects), and fluorescence up-conversion techniques which use the same low pulse intensities and have a time resolution which is limited only by the pulse duration.

Absorption techniques have been less used than fluorescence in picosecond studies of whole cells, although they have been employed very effectively to investigate isolated reaction centres from photosynthetic bacteria. For the more complex whole cells a wide spectral range and precise difference spectra must be recorded. This is now possible by picosecond flash photolysis and spectroscopy where the transient absorption spectrum initiated by the photolysis flash is probed by a white spectroscopic probe flash, created by self-phase-modulation of a high-intensity picosecond pulse. This technique should be particularly useful for studies of photosynthesis in vivo, where simultaneous recording of the whole spectrum of transient change is necessary because of the different behaviour of dark-adapted and light-adapted specimens.

Photosynthesis is an area of picosecond research where the preparation and characterisation of the system to be studied is less advanced than the picosecond techniques themselves. What has been achieved so far, and will now be described, is rather modest but it is an encouraging harbinger of what is to come.

First, let us recall the overall pattern of natural photosynthesis. It consists of a chain of electron transfer reactions which, in green plants and algae, result, at the electron donor end, in the oxidation of water to oxygen, and, at the electron acceptor end, in reduction processes such as the conversion of carbon dioxide to carbohydrate, of nitrogen to ammonia or of protons to hydrogen. These involve two photochemical steps and two separate, but linked, photochemical systems. Photosystem II oxidises water and produces a reduced intermediate (e.g. a hydroquinone). Photosystem I oxidises the hydroquinone and produces the reducing equivalents to make carbohydrate etc. Each photosystem requires at least one photon for each electron transfer so that two photons are utilised for one complete electron transfer from the oxidising to the reducing side. Since four electron transfers are necessary to liberate one molecule of oxygen, a minimum of eight photons are required per oxygen molecule liberated in the overall photosynthesis. In bacterial photosynthesis, the oxidant is some other substance than water (e.g. H_2S), no oxygen is liberated and the overall process, which is essentially PS I, requires only four photons but stores little energy.

Each of the two photosystems consists of two important parts (see Figure 1). First, there is an array of pigment molecules, in the light harvesting antenna, whose purpose is to collect the light energy and transfer it to the reaction centre. Typically, this consists of a few hundred chlorophyll and other pigment molecules, closely coupled so that electronic energy transfer between them is efficient. The reaction centre, to which this energy is transferred, consists of a special pigment molecule, which becomes electronically excited by energy transfer from the antenna, and several redox systems coupled in such a way that the excited pigment receptor can bring about electron transfers sequentially along the redox chain.

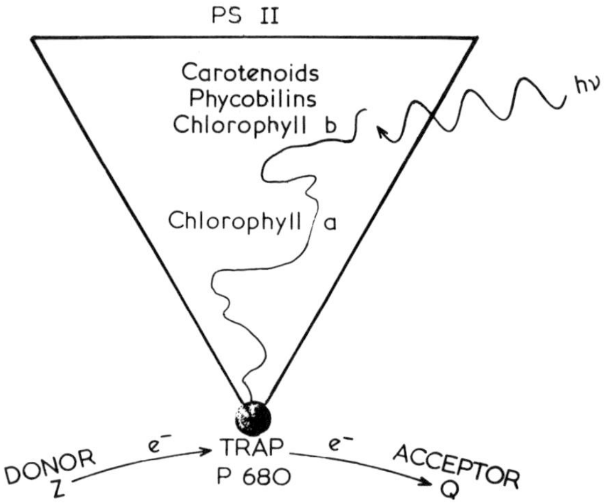

Figure 1. Scheme of antenna and reaction centre in a photosynthetic unit

Picosecond studies of photosynthesis so far fall mainly into two distinct categories. (a) Fluorescence studies of energy transfer in the light harvesting antenna, mainly on algal whole cells or the chloroplasts of green plants, and (b) transient absorption spectral studies of electron transfer processes in isolated reaction centres obtained almost exclusively from photosynthetic bacteria.

ENERGY TRANSFER IN THE ANTENNA

Since there are several hundred times as many pigment molecules in the antenna of the whole plant as in the reaction centre, the antenna absorbs most of the incident light. It turns out that the time required for energy transfer within the antenna is usually longer than that for the first electron transfer from the excited state pigment in the reaction centre, so that nearly all the fluorescence arises from the antenna pigments. It is also found that there is more fluorescence from PS II than PS I and fluorescence studies of the whole plant, in vivo, are therefore mainly concerned with mechanisms of energy transfer within the antenna of PS II.

The pioneering work of Brody and Rabinowitch (1) using nanosecond spark excitation had given lifetimes which were comparable with the time resolution of the method but phase shift methods gave apparently reliable lifetimes (assuming single exponential decay), for chlorophyll fluorescence in chlorella, of 600 ps (dark adapted) and 1700 ps (light adapted) (2,3). These times were also in good agreement with those calculated from fluorescent yields under the same conditions. It was therefore surprising when the first direct picosecond laser measurements, from a number of laboratories, all gave much shorter lifetimes.

These measurements used pulses from a frequency doubled (530 nm) mode-locked Nd laser and the first detection was by electro-optical shutter which gave very irreproducible results (4,5). Reproducibility was improved by using streak camera detection (6), especially when the photographic recording was replaced by silicon-target vidicon multichannel detection (7). All gave fluorescence lifetimes much shorter than the phase shift methods (e.g. our lifetime for chlorella was 108 ± 10 ps). Measurements of fluorescence yield under these conditions gave low values corresponding to the short lifetimes (8,9).

It was immediately appreciated that the short lifetimes might be a result of the high intensities used since, under such intensities, the possibility of excited state-excited state annihilation arose for singlet-singlet and triplet-singlet interactions in the same way as had long been known for triplet-triplet annihilation (10). Many experiments could be, and were, chosen to confirm this hypothesis and, as sometimes happens, the first ones happened to be untypical and showed no intensity effect on lifetime. For example, we chose the red alga previously studied (Porphyridium cruentum) with a lifetime of fluorescence of 92 ± 10 ps. We found no change in this lifetime when the intensity was reduced by a factor of ten, and concluded that exciton annihilation was not occurring. This was a correct lifetime and conclusion for this alga.

Other measurements on chlorella also found no intensity dependence in lifetime of chlorella fluorescence, a result which was, in same cases, due to the fact that at very high intensities, in the region of 10^{16} photons/cm^2/pulse, the annihilation effect saturates and, in other cases, due to the fact that pulse trains were used and earlier pulses had built up a high concentration of triplet states. Breton and Giacintov (11) showed that there was a cumulative decrease in the quantum yield of fluorescence over several pulses of a laser pulse train. Porter, Synowiec and Tredwell (12) measured yields and lifetimes as a function of single pulse intensity as well as the number of the pulse in the train and showed that there were both short time (single pulse) and long time (pulse train) annihilation effects and that, when single pulse intensities below 3×10^{13} photons/cm^2 were used, annihilation effects were eliminated and lifetimes became comparable with those obtained by phase shift and other low intensity methods.

These high intensities are, of course, irrelevant to the normal conditions of photosynthesis and most subsequent work has been at low intensities where exciton annihilation is not important. But high intensity exciton-exciton annihilation may be used as an experimental tool to investigate the size and structure of the light harvesting antenna, an approach which has been effectively exploited(13).

The fall off in fluorescence yield and lifetime will begin at exciting intensities which give more than one photon hit/ photosynthetic unit. Making certain assumptions about the arrangement of the pigment molecules and their intermolecular transfer rate, the number of pigment molecules in the average light harvesting antenna can be estimated. Studies of the intensity/fluorescence-yield curves showed that, in chlorella, 2000 chlorophyll molecules are interconnected in the exciton transfer process. Since, from other evidence, there are only 300 antennae molecules to each reaction centre in photosystem 2, this would imply some interconnection between the photosynthetic units. The intensity threshold where exciton annihilation begins is between 10^{13} and 10^{14} photons/cm^2/pulse in chlorella. In the smaller units, of red algae and bacteria, the threshold occurs at higher intensities and antennae units containing only 8 pigment molecules have been studied (14). For this reason, and the more rapid transfer to the reaction centre, the red alga, Porphyridium cruentum, does not show an intensity dependence of fluorescence lifetime below 10^{15} photons/cm^2 and can therefore be safely studied using a single-pulse solid-state laser and streak camera. On the other hand, chloroplasts of green plants, and green algae such as chlorella must be studied with intensities a hundredfold lower, where the sensitivity of the streak camera method is poor. Fortunately these large antennae also have relatively long fluorescence lifetimes so that excellent time resolution as well as sensitivity can be attained by the use of mode-locked pulse trains, single photon counting and signal averaging. Examples of each of these systems will now be described.

ENERGY TRANSFER IN A MULTIPIGMENT ARRAY

The red and blue-green algae have antennae which are most elegantly composed of three or more pigments which are arranged, both morphologically and in order of excitation energy, so as to channel the energy most efficiently to the reaction centre.

The arrangement of the four pigments in one of these algae, Porphyridium cruentum, is shown in Figure 2. Three pigment protein complexes are arranged in the phycobilisome like the shells of an onion with the shortest wavelength absorber, phycoerythrin, on the outside, then phycocyanin, then allophycocyanin.

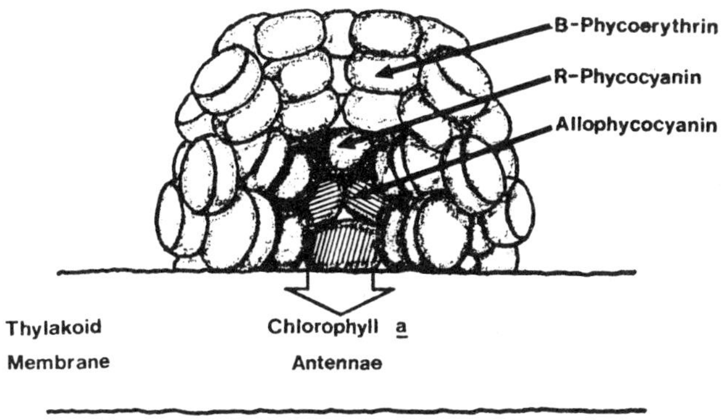

Figure 2. Arrangement of pigments in a phycobilisome

The last pigment, to which the energy is finally transferred is chlorophyll which is in the membrane to which the phycobilisome is attached. In our experiment, the 530 nm pulse from a frequency doubled Nd laser was used to excite the phycoerythrin, being relatively little absorbed by the other pigments. All four pigments fluoresce and the fluorescences can be fairly well separated by interference filters transmitting 10 nm wide bands centred at 576, 640, 661 and 685 nm.

The results of these experiments (15,16,17) are summarised in Figure 3 and in Table 1. The sequential rise and decay of the fluorescence at the four wavelengths is clearly shown and the times were independent of pulse intensity over a tenfold intensity range. When the phycobilisomes were removed from the membrane, the lifetime of the allophycocyanin was increased from 118 ps to 4 ns since the chlorophyll acceptor was no longer present, but the kinetics of the fluorescence of the other two pigments were unaffected, showing that the energy transfer was sequential and direct transfer from the outer pigments to chlorophyll was not important. Pulse intensity studies on the detached phycobilisomes showed that exciton annihilation did occur in the long lived allophycocyanin under these conditions but not in the other pigments, and this was also reflected in the intensity dependence of fluorescence yields.

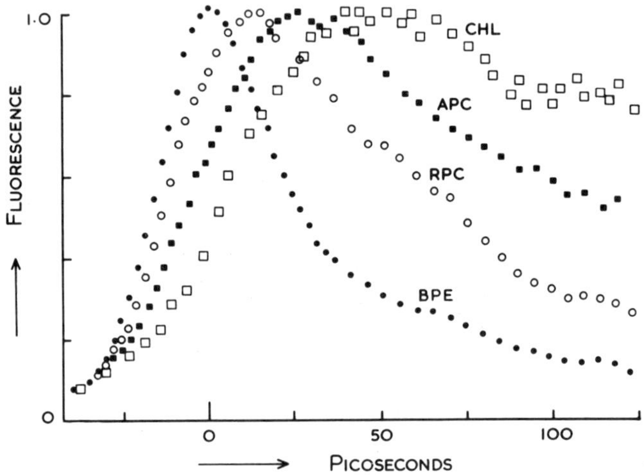

Figure 3. Wavelength resolved pigment fluorescence in P.cruentum. B.P.E. is B-phycoerythrin, RPC is R-phycocyanin, APC is allophycocyanin and CHL is chlorophyll-a.

	Wavelength (nm)	$T_{1/2}$(rise) ps	$T_{1/2}$(decay) ps	
		Algae	Algae	Phycobilisome
PHYCOERYTHRIN	578	0	70 ± 5	70 ± 5
PHYCOCYANIN	640	12	90 ± 10	—
ALLOPHYCOCYANIN	660	22	118 ± 8	4000
CHLOROPHYLL a	685	52	175 ± 10	—

Table 1. Fluorescence of pigments in P.cruentum algae and in phycobilisomes detached from the membrane.

The decay of phycoerythrin fluorescence was not exponential but fitted well the decay law $I(t) = I_0 \exp -2A\, t^{\frac{1}{2}}$. This relation is frequently found to give closer agreement to experimental decays than the exponential law in very fast kinetic studies and is predicted theoretically for the early stages of time-dependent diffusion processes and for Förster type energy transfer to a random array. Its significance in photosynthetic antennae must await more precise kinetic measurements as well as better information on the structure of these systems. At present these experiments confirm very nicely the stepwise energy transfer which occurs in multipigmented algae. Similar results have been obtained for the blue-green alga Anacystis nidulans (18).

Yoshihara et al. (19) have recently repeated our measurements using very low excitation intensities of 10^8 photons/cm^2/pulse from a dye laser with single-photon counting giving time resolution of about 25 ps. The fluorescence was measured at 200 points over the spectrum. Their time-resolved spectra elegantly confirm our conclusions, reached from measurements at only four wavelengths, including the exp $(- 2At^{\frac{1}{2}})$ decay law.

LIGHT HARVESTING IN THE CHLOROPHYLL ANTENNA OF CHLORELLA AND CHLOROPLASTS

Owing to their larger size and slower transfer, these antennae are, as we have seen, more sensitive to bimolecular exciton processes than the smaller antennae of red and blue-green algae. Single pulse measurements at intensities of 3×10^{13} photons/cm^2 do, however, give reproducible lifetimes which are independent of intensity (over a small range) and which agree with the lifetimes expected from fluorescence yield measurements. These are 450 ps and 1800 ps for chlorella respectively dark adapted and preillumined with DCMU to close the reaction centre traps (20). The precision of single pulse measurements at these very low pulse intensities is poor and the intensity is still on the threshold where exciton annihilation processes begin. In our later work we replaced the solid state single pulse laser by a mode-locked synchronously pumped argon/ion dye laser system with single-photon counting. Intensities were in the range 10^9- 10^{11} photons/cm^2/pulse, which are four and two orders of magnitude below the threshold where intensity-dependence begins (21).

The fluorescence of dark adapted chlorella (Figure 4a) was very near to a single exponential function over a thousand-fold intensity, with a lifetime of 492 ps. Chloroplasts isolated from pea (Pisum cativum), both dark adapted and illuminated with DCMU or DCMU + Mg^{2+} ion, to close the reaction centres, were biexponential, (Figure 4b) the shorter lifetime being similar to that of chlorella and the longer lifetime being in the range 1330-1460 ps (see Table 2).

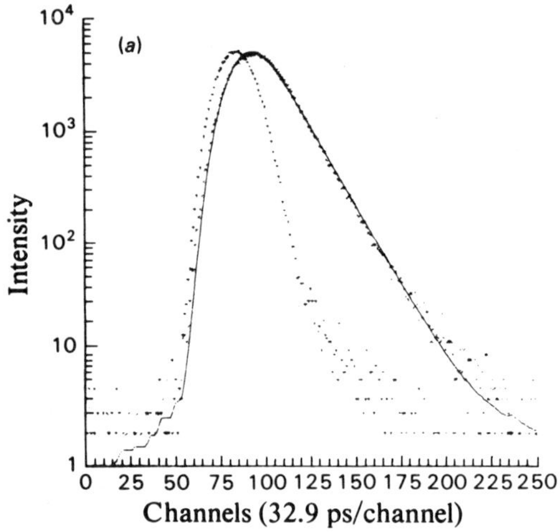

Figure 4a. Time resolved fluorescence emission of dark adapted chlorella

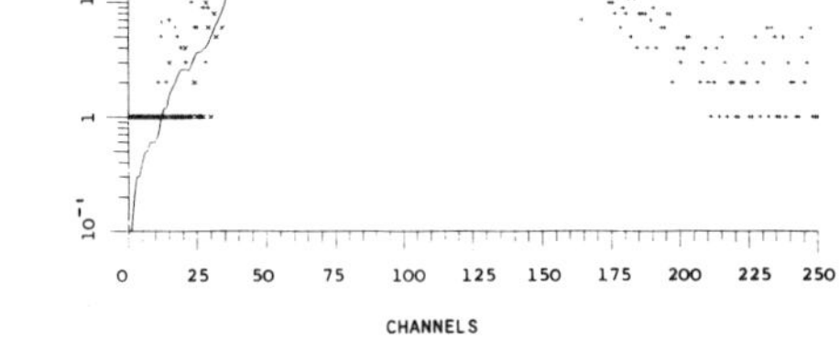

Figure 4b. Time resolved fluorescence emission of dark adapted pea chloroplasts

31.9 ps/channel EXCITATION = 620 nm

τ_1 = 410 ps EMISSION 660 nm

τ_2 = 1460 ps (3.9%)

	τ_1(ps)	τ_2(ps)	% $\frac{\tau_2}{\tau_1}$	ϕ calc
CHLORELLA				
Dark adapted	490	-	-	0.03
CHLOROPLASTS				
Dark adapted	410	1460	3.8	0.02
Light + DCMU	450	1330	9.9	0.03
Light + DCMU + Mg^{2+}	460	1340	36.6	0.04
SLV FRACTION	110	1190	3-9	0.006

Table 2. Lifetimes of short (τ_1) and long (τ_2) fluorescence and calculated quantum yields ϕ in chlorella and chloroplasts. DCMU and Mg^{2+} have the effect of partially closing the traps. The SLV fraction is mainly PSI.

The effect of closing the reaction centre traps was to increase the proportion of the longer decay component without significantly changing the lifetime. We assign the shorter lifetime to P.S.Us in which the trap is open and the longer lifetime to P.S.Us in which the trap is closed. Accepting that there are about 300 antennae chlorophyll molecules to each reaction centre, two models of the P.S.U. have been proposed. In the "puddle" model the P.S.Us are separate entities with no transfer between P.S.Us being possible whilst in the "lake" model many or all of the chlorophyll antennae are interconnected so that, if the nearest trap is closed, the excitation can migrate further until it encounters an open trap. The lake model would result in a single exponential decay whose lifetime increased as the proportion of closed traps increased whilst the puddle model would result in two distinct constant exponential lifetimes, the proportion of the longer lifetime increasing as the proportion of closed traps increased. Our results indicate that the puddle model is the one which is appropriate in the chloroplasts that we have studied.

These measurements all refer to the antennae of Photosystem II. To study the much weaker fluorescence from Photosystem I, it is necessary to isolate chloroplast fragments in which the proportion of PS I/PS II has been increased. In stroma lamellae vesicle fractions we have observed a shorter-lived component with fluorescence lifetime of 113 ± 10 ps, which we attribute to PS I, whose fluorescence intensity is $\frac{1}{4}$ that of PS II.

COMPARISON WITH THEORETICAL ESTIMATES OF EXCITON MIGRATION

The simplest model of a P.S.U. is a regular square or cubic array of 300 chlorophyll molecules with one trap, transfer to which is irreversible. All the indications from spectra and from the lifetimes involved are that coupling between the chlorophyll molecules is very weak and that Förster type energy-transfer will therefore predominate. The single transfer time t_1 between chlorophyll molecules is easily calculated as a function of separation, from the well known expression of Förster (22) and the rate of migration to the trap in a regular array is given by the simple expressions of Montroll (23) which are

$$\frac{\tau_T}{t_1} = \frac{N \ln N}{4\pi} \quad (1) \qquad \text{and} \qquad \frac{T}{t_1} = \frac{1.5164\, N}{6} \quad (2)$$

for two and three dimensional regular arrays respectively, and where N is the number of antenna molecules/trap and τ_T is the average trapping time.

The trapping times as a function of separation of chlorophyll molecules in the array are given in Table 3 from which it is seen that the lifetime of 450 ps, which we have found for P.S.Us with open traps, corresponds to a lattice distance of 17A on a two dimensional lattice and 18A on a cubic lattice.

TRANSFER TIMES FOR P.S.U

300 Chl/trap ; R_o = 70Å ; τ_o = 18 ns

L(Å)	10	12	14	17	20
t_1(p.s.)	0.152	0.46	1.15	3.68	9.79
τ_T(p.s.).D = 2	20	62	157	502	1333
τ_T(p.s.).D = 3	11.5	35	88	286	743

Table 3. Calculated trapping times for a P.S.U with 300 Chl/trap. L is the lattice distance separating chlorophyll molecules, t_1 is the time for a single transfer and τ_T is the trapping time for square (D = 2) and cubic (D = 3) lattices

This is a few A greater than what little is known of the average chlorophyll separation in the smallest chlorophyll-protein complexes but is a reasonable value for the average separation in a larger array composed of an aggregate of many smaller chlorophyll-proteins. It appears therefore that Förster transfer in randomly oriented array adequately accounts for the rate of transfer in a light harvesting antenna of chlorophyll in vivo. We have examined the effects of other distributions, particularly random arrays. The differences are small and do not allow any further conclusions to be made about the structure of the antenna

THE REACTION CENTRE

In the intact cell or chloroplast, the antenna pigments dominate the spectrum and studies of the reaction centre are difficult. For fast time-resolved studies, which generally do not have the sensitivity of steady state measurements, it is therefore necessary to isolate the reaction centres before satisfactory measurements of the transient electron-transfer processes can be made. Up to the present, isolation of the reaction centres from green plants and algae has not been possible but such centres have been prepared from photosynthetic bacteria and kinetic studies of these, using absorption spectroscopy,have been one of the notable achievements of picosecond spectroscopy, although it must be said that the success owes much to the plant biochemists who isolated and characterised the reaction centres.

The characterisation, as well as the picosecond kinetic studies, are far from complete but, at the present time, it appears that the primary receptor or absorber is a bacteriochlorophyll dimer and the electron transfer from this molecule to the adjoining bacteriopheophytin occurs in a time of 4 ps. There is some evidence for an intermediate monomeric chlorophyll to which transfer occurs in about 1 ps but these interpretations are still in question (24). The next electron transfer from pheophytin to a quinone/iron complex occurs in 250 ps, whilst transfer of an electron to the oxidised bacteriochlorophyll dimer from cytochrome-c and subsequent steps occur on increasingly long time scales. (It should be noted however that this is a necessary consequence in the experimental study of sequential reactions; a reaction which is faster than one of the preceding reactions will not be observed). These slower steps have already been studied in depth by flash photolysis in the milli, micro and nanosecond regimes. The picosecond events of the reaction centre have been reviewed (25) and will not be discussed further here.

MODEL SYSTEMS AND FUTURE RESEARCH

Future progress will probably be along two directions which may be called analysis and synthesis; the chemical problems associated with these are a more limiting factor to progress than the instrumental picosecond techniques themselves. The latter have now attained a time resolution of about 30 femtoseconds (C.V. Shank, this Workshop) which corresponds to an uncertainty in energy of 13 kJ/mole and in wavelength, at 500 nm, of 28 nm. Events faster than this must properly be regarded as instantaneous on the energy scale associated with chemical and biological events, although there is still, of course, much scope for improving sensitivity and the range of available wavelengths. But the most important advances which are needed for the elucidation of events in vivo is the isolation of the individual proteins, and the determination of their composition and structure. The possibility of direct picosecond kinetic studies of these particles has provided further incentive to the biochemists and progress in this field is rapid, although at the present time only one such particle has been separated in crystalline form for X-ray structure determination.

More progress has been made with the second approach, that of synthesising model systems in vitro which enable the individual steps proposed for in vivo photosynthesis to be reproduced and studied. There is a very extensive literature in this field and the following examples are quoted merely as illustrations :

(a) Flash photolysis studies of the individual reactions in homogeneous solution, e.g. the chlorophyll/quinone system (26).

(b) Studies of the same, or similar reactions, in model membranes such as monolayers and bilayers (27), vesicles and micelles (28).

(c) The construction and kinetic study of chemically linked aggregates such as chlorophyll dimers and chlorophyll/quinone bimolecules (29).

(d) Attempts to construct a light harvesting antenna (30).

This last problem is one which has received much attention and has turned out to be surprisingly difficult. It would clearly be an important addition to the armoury of the photochemist if the light harvesting principle could be employed as it is in the chloroplast and, from what has been said earlier, it would, at first sight, seem necessary only to prepare an aggregate, or concentrated solution, of a pigment such as chlorophyll, at a concentration comparable with that in the chloroplast, to ensure that efficient transfer of absorbed light energy occurred between these molecules and eventually to a trap at relatively low concentration. Our attempts to prepare such a system in solution, in membranes, in solids and in lipid bilayer vesicles and liposomes have all failed. This failure is due to the efficiency of concentration quenching which occurs when an excited pigment molecule, such as chlorophyll, encounters a second chlorophyll. An example of the effect, studied by Beddard, Carlin and Porter (31) is shown in Figure 5. The chlorophyll molecules are dissolved in the lipid lecithin and the mixture is made into vesicles.

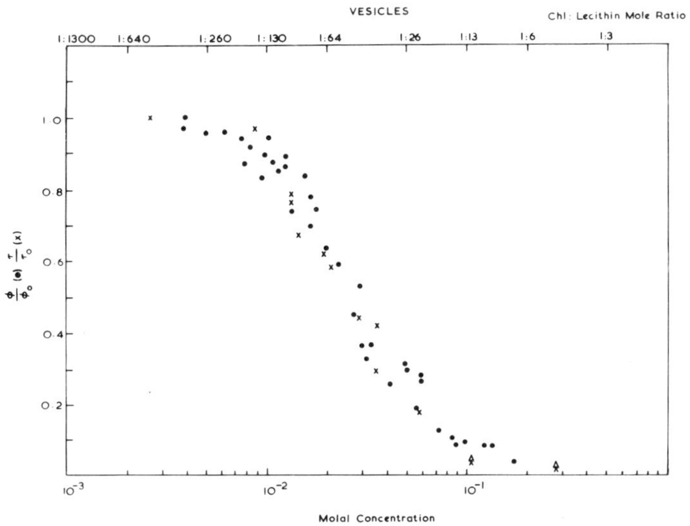

Figure 5. Relative fluorescence yield ϕ/ϕ_o and lifetime τ/τ_o of chlorophyll in lecithin vesicles as a function of concentration (lower scale) or Chl/lecithin molar ratio (upper scale)

Ratios of chlorophyll/lecithin as high as 1/5 can be obtained and no spectral changes are observed. However, at chlorophyll proportions greater than 1% the fluorescence yield and the lifetime decrease, becoming only a few percent of their maximum low concentration value, when the chlorophyll fraction reaches 10%. The lifetime is reduced, over this concentration range, from 5.7 ns to 130 ps. Theoretical studies have shown that the quenching is accounted for by energy transfer between chlorophyll molecules and quenching at a trap; the trap is a pair of chlorophyll molecules which are, for purely statistical reasons, close enough together to form a non-fluorescent quenching excimer and calculation shows that the average distance between centres in the statistical-pair trap is 10A (32).

The question then arises as to how this difficulty is overcome by the photosynthetic antenna. Only in one case, and that not a very typical one, is the structure of a chlorophyll-protein complex known in detail, but this case is very suggestive.

Fenna and Mathews (33) determined the structure of a crystalline bacteriochlorophyll-protein complex composed of a triplet, each containing seven chlorophylls. The average separation of the chlorophylls is 12A and no two chlorophylls are less than 10A apart between centres. It appears likely, therefore, that the photosynthetic unit carefully packages the chlorophylls within the protein so as to avoid the interaction with near neighbours which would lead to quenching and would prevent efficient light harvesting.

The elucidation of the primary events of photosynthesis is still in its early stages; picosecond flash photolysis and fluorescence will play a key role in future progress.

REFERENCES

1. Brody, S.S. and Rabinowitch, E. 1957, Science 125, pp. 555-558.
2. Briantais, J.M., Merkelo, H. and Govinjee 1972, Photosynthetica 6, pp. 133-141.
3. Borisov, A. Yu and Il'Ina, M.D. 1973, Biochim. Biophys. Acta 305, pp. 364-371.
4. Seibert, M. and Alfano, R.R. 1974, Biophys. J. 14, pp. 269-283.
5. Paschenko, V.Z., Protazov, S., Rubin, A.B., Timofeev, K.N., Zamazova, L.M. and Rubin, L.B. 1975, Biochim. Biophys. Acta 408, pp. 143-153.
6. Kollman, V.H., Shapiro, S.L. and Campillo, A.J. 1975, Biochem. Biophys. Res. Comm. 63, pp. 917-923.
7. Beddard, G.S., Porter, G., Tredwell, C.J. and Barber, J. 1975, Nature 258, pp. 166-168.
8. Mauzerall, D., 1976, Biophys. J. 16, pp. 87-91.
9. Campillo, A.J., Shapiro, S.L., Kollman, V.H., Winn, K.R. and Hyer, R.C. 1976, Biophys. J. 16, pp. 93-97.
10. Porter, G. and Wright, M.R. 1958, J. Chim. Phys. 55, pp. 705-712.
11. Breton, J. and Geacintov, N.E. 1976, FEBS Letters 69, pp. 86-89.
12. Porter, G., Synowiec, J.A. and Tredwell, C.J. 1977, Biochim. Biophys. Acta 459, pp. 329-336.
13. Paillotin, G., Swenberg, C.E., Breton, J. and Geacintov, N.E. 1979, Biophys. J. 25, pp. 513-534.
14. Grondell, R.V., Hunter, C.N., Bakker, J. and Kramer, H.J.M. 1983, Biochim. Biophys. Acta 723, pp. 30-36.
15. Porter, G., Tredwell, C.J., Searle, G.W.F. and Barber, J.1978, Biochim. Biophys. Acta 501, pp. 232-235.
16. Searle, G.W.F., Barber, J., Porter, G. and Tredwell, C.J. 1978, Biochim. Biophys. Acta 501, pp. 246-256.
17. Searle, C.W.F. and Tredwell, C.J. 1979, "Chlorophyll Organization and Energy Transfer in Photosynthesis", Excerpta Medica Amsterdam, pp. 257-277.

18. Brody, S.S., Porter, G., Tredwell, C.J. and Barber, J. 1981, Photobiochem. Photobiophys. 2, pp. 11-14.
19. Yamasaki, I.,Mimuro, M., Murao, T., Yamazaki, T., Yoshihara, K. and Fujita, private communication at this meeting to be published in Photochem. Photobiol.
20. Porter, G., Synowiec, J.A. and Tredwell, C.J. 1977, Biochim. Biophys. Acta 459, pp. 329-336.
21. Beddard, G.S., Fleming, G.R., Porter, G., Searle, G.F.W. and Synowiec, J.A. 1979, Biochim. Biophys. Acta 545, pp. 165-174.
22. Förster, T. 1947, Z. Naturfors. 26, pp. 174-182.
23. Montroll, E.W. 1969, J. Math. Phys. 10, p. 753
24. Borisov, A. Yu, Danielus, R.V., Kudzumauskas, S.P., Piskarkas, A.S., Razjivin, A.P., Sirutkaitis, V.A. and Valkunas, L.L. 1983, Photochem. Photophys. 6, pp. 33-38.
25. Windsor, M.W. and Holten, D. 1980, Phil. Trans. R. Soc. Lond. A298, pp. 335-349.
26. Kelly, J.M. and Porter, G. 1970, Proc. Roy. Soc. A319, pp. 319-329.
27. Kelly, A.R. and Porter, G. 1970, Proc. Roy. Soc. A315, pp. 152-161.
28. Kalyanasundaram, K. and Porter, G. 1978, Proc. Roy. Soc. A364, pp. 29-44.
29. Wasielewski, M.R., Studier, M.H. and Katz, J.J. 1976, Proc. Nat. Acad. Sci. USA 73, pp. 4282-4286.
30. Porter, G. 1978, Proc. Roy. Soc. A362, pp. 281-303.
31. Beddard, G.S., Carlin, S.E. and Porter, G. 1976, Chem. Phys. Letters 43, pp. 27-32.
32. Beddard, G.S. and Porter, G. 1976, Nature 260, pp. 366-367.
33. Fenna, R.E. and Mathews, B.W. 1975, Nature 258, pp. 573-577.

PICOSECOND LASER PHOTOLYSIS STUDIES OF PHOTOINDUCED CHARGE TRANSFER AND RELATED PHENOMENA

Noboru Mataga

Department of Chemistry, Faculty of Engineering Science, Osaka University, Toyonaka, Osaka 560, Japan

The fundamental aspect of the solvent effect upon the photoinduced charge transfer phenomena has been discussed on the basis of the ps laser photolysis and transient absorption as well as transient photoconductivity studies of some inter- and intramolecular exciplex systems, and also of some low ionization potential molecules showing monophotonic electron ejection into polar solvent. Moreover, some related processes such as photochemical hydrogen abstraction due to the mechanism of charge transfer followed by proton transfer, charge transfer interaction in the excited hydrogen bonding systems and coupling of the cis trans photoisomerization with the intramolecular charge transfer have been investigated with ps spectroscopy.

Photoinduced charge transfer (CT) or electron transfer phenomena are very important from various viewpoints covering problems ranging from solid state physics to chemistry and biology, and also quite appropriate subject for the study by means of picosecond ∿ femtosecond laser spectroscopy. In the following, several aspects of photochemical CT and related processes studied by means of ps laser photolysis method will be discussed.

SOLVENT EFFECTS UPON PHOTOINDUCED CT, CHARGE SEPARATION AND DEACTIVATION PROCESS FROM THE CT STATE.

In some typical exciplex systems such as pyrene-dimethylaniline and anthracene-dimethylaniline, the photoinduced CT processes in polar solvents are closely coupled with the solvation and the structure of the produced CT state also seem to depend on the solvation (1). The relevant processes may be represented

K. B. Eisenthal (ed.), Applications of Picosecond Spectroscopy to Chemistry, 21–32.

schematically as shown in Fig. 1A. The CT state is stabilized by solvation. However, the ground state which has almost no CT character is destabilized. The important role of solvation for the realization of CT state in the exciplex systems can be demonstrated most clearly by using intramolecular exciplex systems of the type: p-$(CH_3)_2N$-(phenyl)-(CH_2)-(1-pyrenyl) (P_1) or p-$(CH_3)_2N$-(phenyl)-(CH_2)-(9-anthryl) (A_1), because these systems show a typical CT fluorescence band only in polar solvents. For example, in the case of P_1, fluorescence in nonpolar solvent is emitted from the excited state localized in pyrene part (LE state) and we can observe CT fluorescence in only ether as well as more polar solvents (2). In a rough approximation in which the solvent is regarded as a dielectric continuum, the dependence of the CT state energy, E_{CT}, upon the solvent polarity when the solvent reorientation is completed during the excited state lifetime is given by,

$$E_{CT} = \text{Const} - (1/2) \cdot ({\mu_e^{eq}}^2/a^3) \cdot f_\varepsilon \quad (1)$$

where $f_\varepsilon=2(\varepsilon-1)/(2\varepsilon+1)$, μ_e^{eq} is the dipole moment of the equilibrium CT state and a is the cavity radius in the reaction field theory. In the case of A_1 and P_1, the E_{CT} is higher than the energy of the LE state and, with increase of the solvent polarity, level inversion occurs. At the inversion point, there arises the solvent orientational destabilization energy δE_g^{FC} in the Franck-Condon (F.C.) ground state.

$$\delta E_g^{FC} = (1/2) \cdot ({\mu_e^{eq}}^2/a^3) \cdot (f_\varepsilon - f_n) \quad (2)$$

where $f_n=2(n^2-1)/(2n^2+1)$. Therefore, the dependence of the frequency (ν_e) upon the solvent polarity is given by,

$$h\nu_e = \text{Const}' - ({\mu_e^{eq}}^2/a^3) \cdot \{f_\varepsilon - (1/2) \cdot f_n\} \quad (3)$$

and, at the inversion point, the red shift of the CT fluorescence will be especially large owing to the appearance of δE_g^{FC}, which is actually observed in the case of P_1 and A_1 at the value of the solvent polarity parameter $\{f_\varepsilon-(1/2)\cdot f_n\} \sim 0.3$-$0.4$ (2). Moreover, it has been confirmed that eq. (3) is approximately satisfied in these systems (2). Therefore, the existence of the ground state destabilization energy due to the reorientation of polar solvents in the CT state has been directly proved (2). An important implication of such ground state destabilization for the non-radiative deactivation of the CT state in the case of some biologically important exciplex systems such as porphyrin-quinone and flavin-indole in strongly polar solvents will be discussed later. The growing-in of the CT state and the decay of the LE state can be observed directly by means of the ps transient absorption as well as fluorescence measurements for P_1 and A_1 in relatively viscous polar solvents like 2-propanol (3,4). For

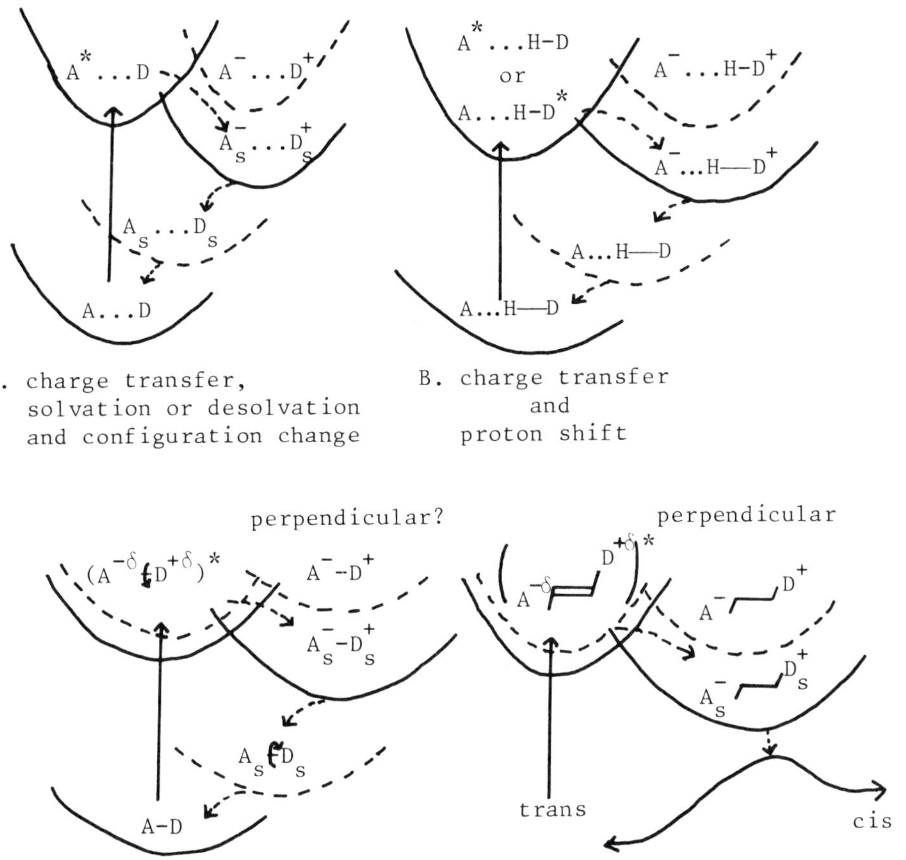

A. charge transfer, solvation or desolvation and configuration change

B. charge transfer and proton shift

C. charge transfer, solvation and conformation change

D. charge transfer, solvation and tr.→cis isomerization

Fig. 1. Photoinduced Charge Transfer, Solvation and Structural Changes

example, the measurement of the LE fluorescence decay and the rise of the CT state absorbance for P_1 have given approximately the same time constant ($\tau_f^{LE} \sim 62 \pm 6$ ps and $\tau_{rise}^{CT} \sim 65$ ps) in 2-propanol solution (3,4). In such less viscous and aprotic polar solvents as acetone and acetonitrile, the CT state formation is realized within the time resolution of the conventional ps laser photolysis apparatus (3). For the direct observation of the CT state rise process, we need subps $\sim$ fs spectroscopy. In contrast to P_1 and A_1 systems which do not have much freedom with respect to the change of mutual distance between two chromophores, extensive conformation change is possible in the case of p-$(CH_3)_2$N-(phenyl)-

$(CH_2)_3$-(1-pyrenyl) (P_3) and p-$(CH_3)_2$N-(phenyl)-$(CH_2)_3$-(9-anthryl) (A_3), which makes possible CT even in nonpolar solvents due to the formation of sandwich configuration. Owing to the extensive conformation change, however, it takes a few ns in hexane solution. The formation of the CT state, however, becomes faster with increase of the solvent polarity since the formation process is assisted by the polar solvent reorientation (3-6). These intramolecular exciplex systems are good examples to show that not only the photoinduced CT process but also the configurations of the formed CT state are affected rather seriously by the interaction with the surrounding solvent molecules. The conclusion derived for the CT state formation process of P_3 in several solvents from the analysis of the time-resolved ps transient absorption spectra as well as fluorescence rise and decay curves measured with ps streak camera may be illustrated as follows (3, 4):

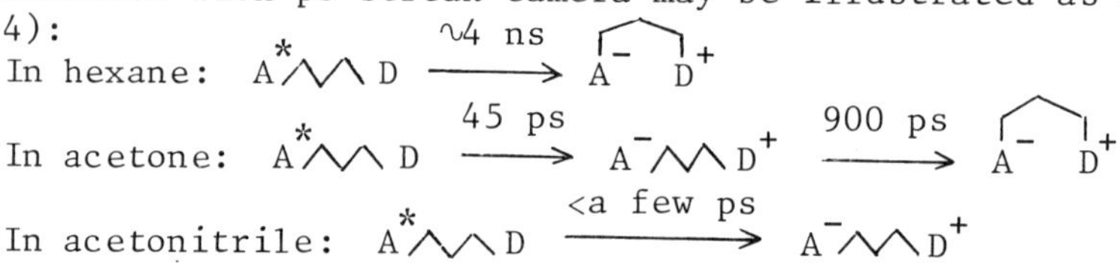

In acetone, not only the solvent reorganization but also slight approach or reorientational motion of two chromophores is necessary. Nevertheless, the CT state formed at first in acetone solution has a much loose structure compared with that in hexane solution, and it changes to a more compact one in the course of time. In acetonitrile solution, the CT state seems to be formed without any conformation change of P_3 and it does not show the change to a more compact structure even at the later stage. Above described solvent-dependent structures of the exciplex or CT state seems to affect considerably the photoinduced CT and ionic dissociation processes in the case of intermolecular exciplex systems in polar solvents. More rapid dissociation and larger ionic dissociation quantum yield are expected in a more polar solvent because of the more loose structure of the CT state. Transient spectral measurement with ps laser photolysis may be one of the most powerful means for the elucidation of the ionic photodissociation or charge separation mechanism in polar solution (7). However, the transient absorption spectra cannot discriminate between exciplexes, geminate ion-pairs and dissociated ions. On the other hand, the ionic dissociation process can be detected by the transient photocurrent measurement. Actually, we have examined the ps laser induced transient photocurrent of pyrene-DMA (N,N-dimethylaniline) as well as pyrene-DCNB (p-dicyanobenzene) systems in various polar solvents and observed directly the dissociation from the geminate ion-pairs of these typical exciplex systems for the first time (8). Some results of photocurrent rise time (τ_r^{ion}) and exciplex fluorescence decay time (τ_f^{ex}) are indicated in Table 1. Although the exciplex fluorescence is largely quenched in solvents of high polarity, we can still observe very weak exciplex fluorescence even in acetone or acetonitrile.

Table 1. Photocurrent Rise Times and Exciplex Fluorescence Decay Times of Pyrene-DMA (A) and Pyrene-DCNB (B) Systems in Several Solvents.

A.	Solvent	ε	[DMA]/M	τ_r^{ion}/ns	τ_f^{ex}/ns
	acetonitrile	37.5(20°C)	0.30	$\lesssim$2	∿2.3
	acetone	20.7(25°C)	0.35	7.0	7.7
	pyridine	12.4(21°C)	0.30	23.8	33.2
B.	Solvent	ε	[DCNB]/M	τ_r^{ion}/ns	τ_f^{ex}/ns
	pyridine	12.4(21°C)	0.058	5.5±0.3	5.7±0.1
	1,2-dichloroethane	10.36(25°C)	0.05	16.4±0.8	19.3±0.2
	dichloromethane	8.93(25°C)	0.05	18.1±1.7	20.9±0.3

The τ_f^{ex} value is rather close to but systematically a little larger than τ_r^{ion} value. This result cannot be explained by assuming "two states" model, which gives for exciplex fluorescence

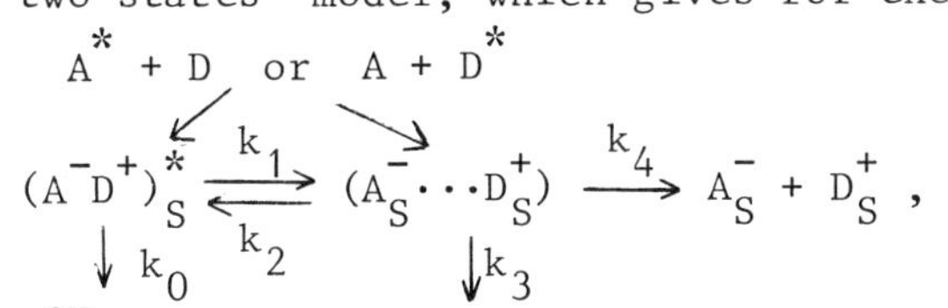

intensity $I_f^{ex}(t)$ and photocurrent rise curve i(t) the following equations.

$$I_f^{ex}(t) = A_1 \exp(-|\lambda_1| t) + A_2 \exp(-|\lambda_2| t)$$
$$i(t) = A_1'\{1-\exp(-|\lambda_1| t)\} + A_2'\{1-\exp(-|\lambda_2| t)\} \quad (5)$$

According to these equations, the exciplex fluorescence decay process must correspond exactly to the photocurrent rise process contrary to the observed results. We propose a tentative model which assumes that there exist multiple exciplexes which differ in energy and structure depending upon the solvation and similarly multiple geminate ion-pairs in accordance with the original idea of the solvent dependent exciplex structure (1) and that the main part of the ion-pairs with the largest statistical weight may not be directly combined with the main part of the exciplexes with the largest statistical weight but there may be intervening exciplex and ion-pair states.

$$A^* + D \quad \text{or} \quad A + D^*$$

$$\underset{\downarrow}{(A^-D^+)^*_S} \rightleftarrows \underset{\downarrow}{(A^-D^+)^*_{S'}} \rightleftarrows \cdots\cdots \rightleftarrows \underset{\downarrow}{(A^-_S\cdots D^+_S)} \rightleftarrows \underset{\downarrow}{(A^-_{S'}\cdots D^+_{S'})} \cdots \rightarrow A^-_S + D^+_S$$

This "multi states model" gives eq. (6) for the exciplex fluorescence decay and photocurrent rise curves.

$$I_f^{ex}(t) = \sum_i A_i \exp(-|\lambda_i| t) \sim A\exp(-|\lambda| t)$$
$$i(t) = \sum_i A'_i\{1-\exp(-|\lambda'_i| t)\} \sim A'\exp\{1-\exp(-|\lambda'| t)\} \qquad (6)$$

These results seem to reflect rather directly the complex and dynamical micro-structures of polar solutions. At any rate, the ionic photodissociation yield of these typical exciplex systems in moderately and strongly polar solvents is rather high. For example, it amounts to almost 80% for pyrene-DMA in acetonitrile. However, it should be pointed out here that, in contrast to these typical exciplex systems, the photoinduced CT state of some biological important systems like porphyrin-quinone (9) and flavin-indole (10) is non-radiatively deactivated very rapidly before dissociation in strongly polar solvents. Fig. 2 shows the result of ps transient absorption measurements for ethyletioporphyrin (EEP)-toluquinone (TQ) system in acetone solution, where the fluorescence of EEP is completely quenched by added high conc. TQ (9). The spectra in Fig. 2 are very similar to the $S_n \leftarrow S_1$ spectra of EEP itself. Therefore, the CT state formed by the interaction between the excited EEP and TQ seems to be deactivated very rapidly without producing dissociated ions. However, in benzene solution with [TQ]∿0.62 M, the excited EEP which is already weakly bound with TQ forms exciplex and shows spectra similar to that of porphyrin cation as shown in Fig. 3 (9). The lifetime

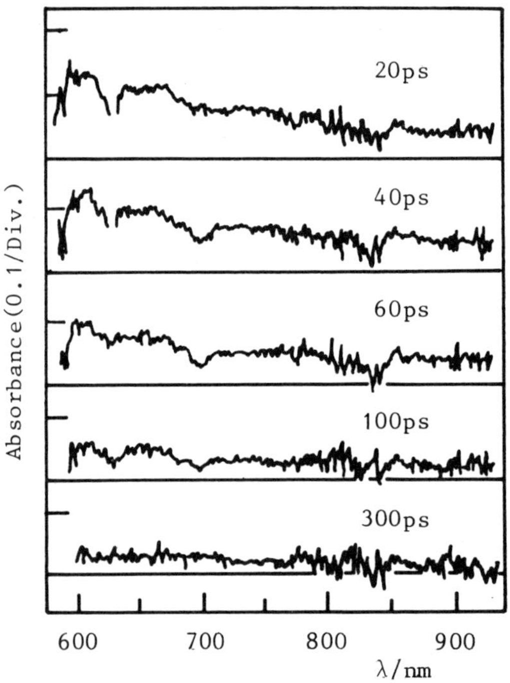

Fig. 2. Transient Absorption Spectra of EEP-TQ System in Acetone

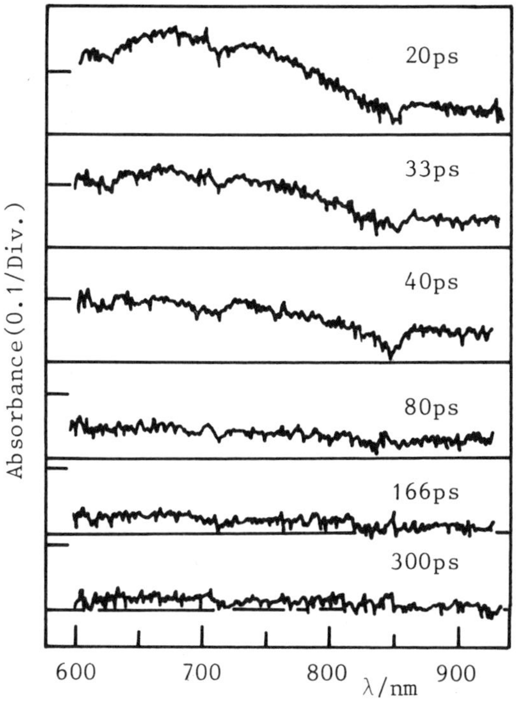

Fig. 3. Transient Absorption Spectra of EEP-TQ System in Benzene

of this EEP-TQ exciplex in benzene is about 40 ps. The CT states of these porphyrin-quinone systems are considerably lower compared with those of the typical exciplex systems, which results essentially shorter exciplex lifetime of the former. The important role of the polar solvent (acetone) in the very rapid non-radiative deactivation of CT state can be understood by the mechanism indicated in Fig. 1A. We have obtained quite similar results also in the case of flavin-indole system (10), which may be explained by the same reasoning. It must be pointed out here that such non-radiative deactivation process becomes much slower in the case of porphyrin-quinone intramolecular exciplex systems of the type: (etioporphyrin)-$(CH_2)_m$-(quinone)-$(CH_2)_n$-(trichloroquinone). For example, in tetrahydrofuran solution, the lifetime of the intramolecular exciplex is ca. 140 ps for m=2, n=4 and ca. 400 ps for m=4, n=4. In addition to these exciplex type interactions and their solvent dependence, we discuss here briefly very important role of solute-solvent interaction in the electron photoejection phenomena. For example, although the photoionization of TMPD (N,N,N',N'-tetramethyl-p-phenylenediamine) in solution was studied quite intensively, the true mechanism of its photoionization in polar solvent does not seem to be made clear by those previous studies (11). The time-resolved transient absorption spectra of TMPD in acetonitrile solution measured with ps laser photolysis method show very clearly that the TMPD cation is produced from the relaxed fluorescent state (12). TMPD $\xrightarrow{h\nu}$ TMPD* (FC state)→TMPD*(fluorescent state) $\xrightarrow{1.2\ ns}$ ionization and fluorescence. The ionization results the formation of cation-acetonitrile dimer anion geminate pair (acetonitrile dimer anion will change to polymer anion in the course of time) which has very long lifetime of μs ∿ 10 μs (12). We have observed similar slow photoionization of TMPD in methanol producing cation-solvated electron ion-pair which dissociates within ca. 15 ns (13). Not only TMPD but also BDATP(bis(dimethylamino)tetrahydropyrene) shows similar slow photoionization in polar solvents (14,15). In acetonitrile (ACN), BDATP cation-acetonitrile dimer anion geminate pair is formed from the fluorescent state ($\tau_f \sim 2$ ns) and this geminate pair dissociates within 9 ns. $BDATP^* \xrightarrow{2\ ns} BDATP^+ \cdots (ACN)_2^- \xrightarrow{9\ ns} BDATP^+ + (ACN)_2^-$. All these results indicate clearly the importance of solute-solvent exciplex type interaction in the so-called electron photoejection in polar solutions.

SOME CHEMICAL REACTION PROCESSES COUPLED WITH PHOTOINDUCED CT.

The aromatic hydrocarbon-secondary amine exciplex is only very weakly fluorescent compared with the exciplex with tertiary amine in nonpolar solvent. The most probable non-radiative path in the former exciplex system was assumed to be the proton transfer in the CT state resulting the ultimate hydrogen transfer (16). We have examined in detail pyrene-N-ethylaniline (NEA) and pyrene-diphenylamine (DPA) systems in nonpolar solvent and detected directly the proton transfer process in the CT state, $(A^- \cdots H\text{-}D^+) \rightarrow (\dot{A}\text{-}H \cdots \dot{D})$, by means of the ps transient absorption measurements (17). Moreover, the effect of the deuteration of the amine hydrogen upon the proton transfer process has been examined for both systems and the kinetic isotope effect has been confirmed (17). The results can be explained by assuming the following reaction scheme and some values of rate constant and exciplex lifetime (τ_{ex}) are given together.

$$^1A^* + H\text{-}D \rightarrow (A^- \cdots H\text{-}D^+) \xrightarrow{k_p} (\dot{A}\text{-}H \cdots \dot{D}) \xrightarrow{k_s} \dot{A}\text{-}H + \dot{D}$$

$$(A^- \cdots H\text{-}D^+) \xrightarrow{k_{isc}} {}^3A^* + H\text{-}D, \quad (A^- \cdots H\text{-}D^+) \xrightarrow{k_f} A + H\text{-}D + h\nu_f, \quad (A^- \cdots H\text{-}D^+) \xrightarrow{k_n} A + H\text{-}D, \quad (\dot{A}\text{-}H \cdots \dot{D}) \xrightarrow{k_n'} A + H\text{-}D$$

$$\tau_{ex}^{-1} = (k_f + k_n + k_p + k_T)$$

pyrene-NEA: $(k_p+k_n)=1.9\times10^8\ s^{-1}$, $k_f=1.5\times10^6\ s^{-1}$, $k_{isc}=1\times10^8\ s^{-1}$, $\tau_{ex}=3.4$ ns

pyrene-NEA-d: $(k_p+k_n)=0.4\times10^8\ s^{-1}$, $k_f=1.5\times10^6\ s^{-1}$, $k_{isc}=1\times10^8\ s^{-1}$, $\tau_{ex}=6.8$ ns

pyrene-DPA: $\tau_{ex} \sim 200$ ps

pyrene-DPA-d: $\tau_{ex} \sim 1.1$ ns

The values of (k_p+k_n) are affected remarkably by deuteration, which is most probably due to the change of k_p by deuteration. Similarly, the change of τ_{ex} by deuteration in the pyrene-DPA system may be also ascribed to the change of k_p. The much smaller τ_{ex} value of pyrene-DPA system compared with the pyrene-NEA system is ascribed to the much faster proton transfer in the CT state of the former, which may be due to the larger acidity of the N-H proton in DPA. It should be also pointed out here that the k_{isc} value of pyrene-NEA system is two orders of magnitude larger than that of the pyrene-DMA exciplex. This result can be explained by assuming non-parallel conformation of pyrene-NEA exciplex, where the N-H or N-D bond is almost perpendicular to the pyrene plane at C_1 position and "hydrogen bonding" interaction between 2pπAO at C_1 of the pyrene anion and N-H of NEA cation is responsible for keeping the oblique configuration of the exciplex (17). This oblique configuration is favorable not only for the proton transfer but also for the fast ISC because of the enhancement of the spin-orbit interaction matrix element between the CT and the pyrene local triplet state in this configuration (7). We have already demonstrated a quite remarkable example of fast ISC from the CT state in the case of the intramolecular exciplex with non-parallel conformation, (1-pyrenyl)-CH_2-$N(CH_3)$-(phenyl), where $k_{isc} \gtrsim 10^{10}$ s^{-1} (18). The hydrogen bonding interaction between the N-H of amine and 2pπ electron of pyrene in the above systems partly leads to the hydrogen abstraction products as discussed above. However, there may arise also some radiationless deactivation to the ground state due to the back CT and this process may be assisted by the N-H bond stretching in the CT state, which destabilizes the ground state making the deactivation easier, as indicated in Fig. 1B. Nevertheless, CT fluorescence can be observed in the case of the above pyrene-secondary amine exciplexes in nonpolar solvents although it is very weak. There is another example of more usual type of hydrogen (H) bonding interaction in the excited state, which causes the quenching of proton donor or acceptor fluorescence. In this case, both proton donor and acceptor are conjugate π-electron systems and they are directly connected by H-bond. As a possible mechanism, we proposed photoinduced CT from proton donor to acceptor resulting formation of a kind of nonfluorescent exciplex (19). Since this nonfluorescent CT state is very short-lived, its detection has become possible only when the method of the ps laser photolysis and transient absorption spectral measurements has become applicable to this problem. We have undertaken a systematic ps laser studies of this problem and actually confirmed the formation of CT state in the excited H-bonded complexes of 2-naphthylamine(NA)-pyridine (20), 13H- and 7H-dibenzocarbazole(DBC)-pyridine (21) and 1-aminopyrene(AP)-pyridine (22). The results for 13HDBC-pyridine will be briefly discussed in the following. The time-resolved ps transient absorption spectra in cyclohexane solution clearly show the change of spectra from those similar to the $S_n \leftarrow S_1$ spectra of 13HDBC H-bonded with TEA (triethylamine) to those quite similar to the spectra of 13HDBC cation with increase of delay time. The time dependence of the absorbance can be reproduced by assuming the reaction scheme,

$$(D^*\text{-}H\cdots A) \rightleftarrows (D^+\text{-}H\cdots A^-) \quad ,$$
$$\qquad\downarrow \qquad\qquad\qquad \downarrow$$

according to which the absorbance at a given wavelength should show a bi-exponential behavior, $A=C_1\exp(-t/\tau_1)+C_2\exp(-t/\tau_2)$. Although no CT fluorescence is observed in this case, weak fluorescence from (D^*-H···A) state can be detected, which should show the same bi-exponential behavior, $[(D^*\text{-H}\cdots A)]=C_1'\exp(-t/\tau_1)+C_2'\exp(-t/\tau_2)$. These predictions have been actually confirmed by means of ps transient absorption and fluorescence measurements and we have obtained τ_1=2.5 ns and τ_2=100 ps for 13HDBC-pyridine in cyclohexane. The nonfluorescent character of the CT state in this type of H-bonding systems compared with the pyrene-secondary amine exciplexes which emit weak CT fluorescence may be ascribed to the much weaker overlap between π-electron systems of donor and acceptor in the former systems leading to very small radiative transition probability which cannot competes with the non-radiative deactivation. The mechanism of the radiationless deactivation from the CT state of this type of H-bonding system can be understood on the basis of Fig. 1B. Namely, the stronger H-bonding interaction in the excited state will lower the energy of the CT state and the CT interaction will make the H-bond proton further shifted toward A, and vice versa. Such a proton shift in the CT state, however, causes a large destabilization of the ground state leading to the enhancement of the nonradiative crossing from the CT state. In the case of 7HDBC-pyridine and AP-pyridine systems, rapid equilibrium between (D^*-H···A) and (D^+-H···A^-) is established and decay is single exponential with τ values of 330 ps and 250 ps, respectively. The double exponential behavior of 13HDBC-pyridine system may be ascribed to the steric hindrance for the H-bonding interaction. Although we have observed clearly the CT state formation in the above H-bond systems, such observation was not possible in some systems like pyrenol-pyridine and azapyrene-aromatic amine (23) where we need subps ∿ fs spectroscopy for more detailed studies. Finally, the effect of the photoinduced intramolecular CT upon the trans→cis photoisomerization will be examined for the compounds of the type: A-CH=CH-D. We have already discussed to some extent the solvation-induced CT coupled with the conformational change in the P_1 and P_3 intramolecular exciplex systems. A special case of such systems is the one where two chromophores are directly combined with a single bond. The famous example is the dimethylaminobenzonitrile (DMABN) and it has been proposed that dimethylamino group is perpendicular to the benzene plane in the strongly polar state produced by solvation with polar solvents as indicated in Fig. 1C (16). However, we can observe more clear-cut structural change coupled with photoinduced intramolecular CT in the case of tr.→cis photoisomerization of the compound: A-CH=CH-D, where A is pyrenyl and D is p-substituited phenyl groups. We have examined compounds substituted with methyl, methoxy, and dimethylamino groups by means of ps and ns spectroscopy as well as reaction quantum yield measurements. These compounds show photoisomerization exclusively from the S_1 state. We have confirmed that the isomerization rate constant increases with increase of the electron donating ability of the substituent and also with increase of the solvent polarity in general (24). For example, the fluorescence decay time τ_f, isomerization quantum yield Φ_{tc} and the isomerization rate constant k_{tc} of the trans methoxy compound is affected by solvent polarity as indicated in the following.

solvent	τ_f	Φ_{tc}	$k_{tc}/10^8\ s^{-1}$
hexane	3.2 ns	0.053	0.17
ethylether	1.3 ns	0.18	1.4
acetonitrile	650 ps	0.27	4.2

These results can be well comprehended on the basis of the mechanism indicated in Fig. 1D, which shows that the lowering of the energy of the intermediate perpendicular configuration with complete electron transfer structure increases with increase of the substituent electron donating ability and also with the increase of the solvent polarity and this energy lowering of the perpendicular configuration will lower the barrier between the tr. fluorescent state and the perpendicular state. The present results indicate clearly the importance of the intermediate CT state with perpendicular configuration in the tr.→cis photoisomerization and the possibility to controle the photoisomerization by changing the degree of the intramolecular CT. Whether the rate of the disappearance of the perpendicular state is dependent upon the substituent electron donating ability as well as the solvent polarity is beyond the scope of this work.

REFERENCES

(1) Mataga, N. Okada, T. and Yamamoto, N. 1967, Chem. Phys. Lett. 1, pp. 119-121.

(2) Masaki, M., Okada, T., Mataga, N., Sakata, Y., and Misumi, S. 1976, Bull. Chem. Soc. Jpn. 49, pp.1277-1283.

(3) Migita, M., Okada, T., Mataga, N., Sakata, Y., Misumi, S., Nakashima, N. and Yoshihara, K. 1981, Bull. Chem. Soc. Jpn. 54, pp. 3304-3311.

(4) Okada, T., Migita, M. Mataga, N., Sakata, Y. and Misumi, S. 1981, J. Am. Chem. Soc. 103, pp.4715-4720.

(5) Mataga, N., Okada, T., Masuhara, H., Nakashima, N., Sakata, Y. and Misumi, S. 1976, J. Luminescence 12/13, pp.159-168.

(6) Okada, T., Saito, T., Mataga, N., Sakata, Y., and Misumi, S. 1977, Bull. Chem. Soc. Jpn. 50, pp. 331-336.

(7) Mataga, N. 1983, Radiat. Phys. Chem. 21, pp. 83-89.

(8) Hirata, Y., Kanda, Y. and Mataga, N., 1983, J. Phys. Chem. 87, pp. 1659-1662.

(9) Karen, A., Okada, T. and Mataga, N., to be published.

(10) Karen, A., Ikeda, N., Mataga, N. and Tanaka, F. 1983, Photochem. Photobiol. 37, pp. 495-502.

(11) Lesclaux, R. and Joussot-Dubien, J. 1973, "Org. Molecular Photophysics" Birks, J. B. ed., Wiley-Interscience, London, 1, pp. 455-487.
(12) Hirata, Y. and Mataga, N. 1983, J. Phys. Chem. 87, pp. 1680 -1682.
(13) Hirata, Y. and Mataga, N., to be published.
(14) Hirata, Y., Mataga, N., Sakata, Y. and Misumi, S. 1982, J. Phys. Chem. 86, pp. 1508-1511.
(15) Hirata, Y., Mataga, N., Sakata, Y. and Misumi, S. 1983, J. Phys. Chem. 87, pp. 1493-1498.
(16) Mataga, N. 1981, "Molecular Interactions" Ratajczak, H. and Orville-Thomas, W. J. ed., John-Wiley & Sons, London, 2, pp. 509-570.
(17) Okada, T., Karaki, I. and Mataga, N. 1982, J. Am. Chem. Soc. 104, pp. 7191-7195.
(18) Okada, T., Karaki, I., Matsuzawa, E., Mataga, N., Sakata, Y. and Misumi, S. 1981, J. Phys. Chem. 85, pp. 3957-3960.
(19) Mataga, N. and Tsuno, S. 1956, Naturwiss. 10, pp. 304-305.
(20) Ikeda, N., Okada, T. and Mataga, N. 1981, Bull. Chem. Soc. Jpn. 54, pp. 1025-1030.
(21) Martin, M. M., Ikeda, N., Okada, T. and Mataga, N. 1982, J. Phys. Chem. 86, 4148-4156.
(22) Ikeda, N., Miyasaka, H., Okada, T. and Mataga, N., 1983, J. Am. Chem. Soc., in press.
(23) Ikeda, N. and Mataga, N., to be published.
(24) Maeda, Y., Okada, T. and Mataga, N., to be published.

APPLICATION OF PICOSECOND SPECTROSCOPY TO ORGANIC REACTION MECHANISMS

Kevin S. Peters

Department of Chemistry, Harvard University

INTODUCTION

Organic reaction mechanisms are normally formulated in terms of bond rearrangements often without reference to the relevant time scale. This is due in part to the lack of information regarding the dynamics of structural rearrangements. For the most part the kinetics of reactions have been studied employing technologies that have inherent time resolutions that preclude direct observation of bond formation and cleavage processes. With the development of picosecond laser techniques one can begin to address in greater depth questions pertaining to the dynamics of chemical transformation. No longer is it necessary to describe the dynamics of structural rearrangements in terms of "greater than diffusion controlled".

In this paper we will detail an experiment directed towards establishing the mechanism and timescale for carbon-oxygen bond formation by examing the photochemical cycloaddition reaction of an olefin and a ketone. Furthermore we present the new experimental technique of photoacoustic calorimetry utilized in ascertaining the energetics of reactive intermediates that are often encountered in organic reactions.

PHOTOCHEMICAL CYCLOADDITION REACTIONS

The mechanism of the photocycloaddition reaction of an olefin to a ketone, forming an oxetane, has been extensively investigated since Buchi's confirmation of the original studies of Paterno and Chieffi[1]. The first evidence for the existence of a metastable intermediate on the reaction coordinate followed from

K. B. Eisenthal (ed.), Applications of Picosecond Spectroscopy to Chemistry, 33–40.

Arnold's studies[2] of the photoreduction of benzophenone with cis or trans 2-butene which yielded the same mixture of cis and trans oxetanes regardless of the initial stereochemistry of the starting olefin. These results suggested the intermediacy of a 1,4 biradical species whose lifetime allowed for the free rotation about one of the terminal bonds.

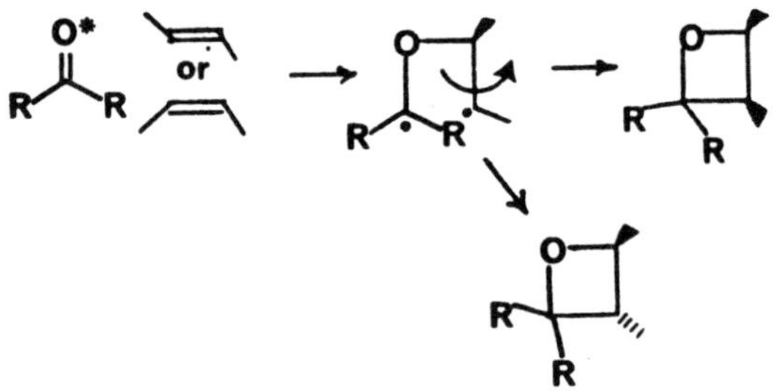

Despite the success of the biradical intermediate hypothesis there are a number of pieces of evidence which intimated a more complicated picture of the reaction profile. Results of isotope studies have been interpreted as evidence against rate limiting formation of a 1,4 biradical[3]. The exceptions to the most stable biradical rule have been troublesome to explain in a consistent manner[4]. Furthermore the high rate of addition and lack of hydrogen abstraction as major side reaction in many cases have prompted the reevaluation of the 1,4 biradical mechanism[5]. The apparent inadequacies of this simple model for photocycloaddition have led investigator to suspect that an intermediate occurs prior to the formation of the 1,4 biradical. The most likely candidate for this transient is believed to be a π-bonded complex involving charge transfer interactions[6].

The nature of the intermediates occuring along the reaction surface have been inferred largely through indirect measurements. However we recently reported the direct observation of a 1,4-biradical in the photocycloaddition reaction benzophenone and dioxene utilizing picosecond laser absorption spectroscopy[7].

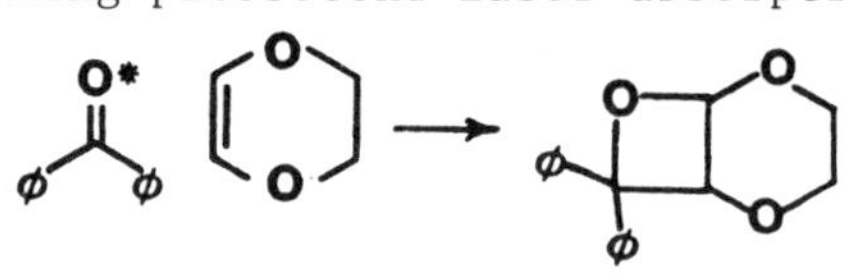

Further direct evidence for the 1,4 biradical was obtained from the photocycloaddition of benzophenone to tetramethylethylene and to ethyl vinyl ether[8]. For the photocycloaddition of benzophenone and dioxene concomitant with the decay of the triplet state of benzophenone, with an absorption maximium at 525 nm, was the appearence of the 1,4 biradical absorbing at 545 nm, Figure 1. The risetime of the biradical was 175 $\pm$ 25 picoseconds. There was no evidencefor the existence of a charge-transfer complex

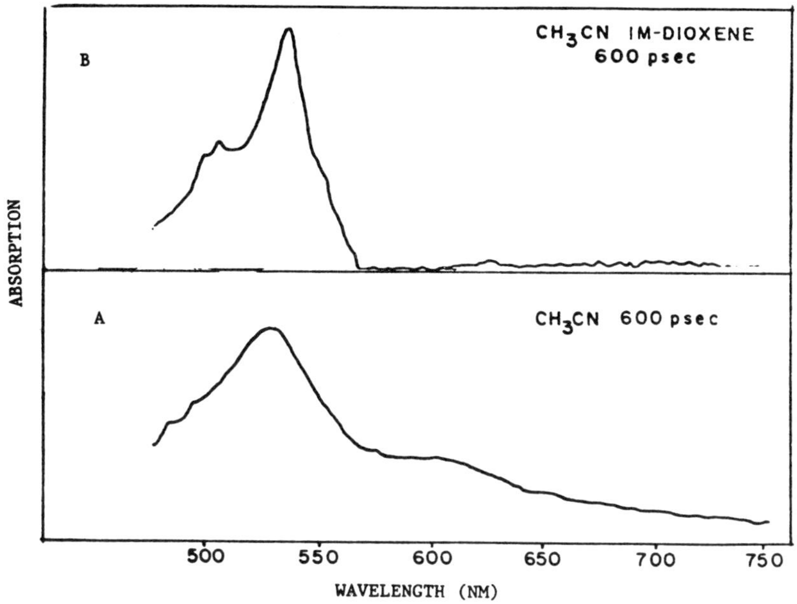

Figure 1. Transient absorption spectrum of (A) 0.2M benzophenone in acetonitrile at 600 ps. OD=0.5 at 525nm. (B) 0.2M benzophenone-1M dioxene in acetonitrile at 600 ps, OD=0.2 at 535 nm.

prior to the biradical formation. Therefore we extended our preliminary picosecond laser studies in an attempt to more fully elucidate the reaction mechanism for the 1,4 biradical formation.

The picosecond absorption spectrometer has previously been described in detail[9]. Briefly, a Quantel YG-400 laser is used to generate 30 picosecond pulses at 1060 nm and 355 nm. The 355 nm light is used for excitation. Transient absorption spectra are recorded using a picosecond continuum generated by focussing the fundamental into a solution of carbon tetrachloride and chloroform. The continuum light is dispersed by a Jarrel Ash monochrometer and imaged onto a vidicon array (PAR-1215-1216-1254) detection system. Transient absorption spectra are recorded on the nanosecond time scale by delaying the continuum light using fiber optics.

The lack of observation of a charge transfer intermediate giving rise to the 1,4 biradical does not preclude its existence as rate limiting electron transfer to the triplet state of benzophenone followed by fast collapse to form the 1,4 biradical would be consistent with the observed kinetics. Therefore if an initial decay of the triplet were to give rise to a charge transfer species, for its observation it would be necessary to stabilize the charge transfer complex relative to the 1,4 biradical. This may be brought about by employing an aromatic ketone which is more easily reduced than benzophenone. For this purpose 4-phenyl-benzophenone has been chosen as its reduction potential of -1.6V is less than benzophenone's -1.8V and therefore should form a more stable charge-transfer complex.

Before investigating the role of charge transfer in the photocycloaddition of 4-phenyl-benzophenone and dioxene it is necessary to establish that benzophenone and 4-phenyl-benzophenone have similiar reactivity towards electron transfer. There has been considerable discussion concerning the relative reactivities of these two molecules towards electron transfer as the electronic configuration of the lowest excited triplet state of these two ketones are different[10]. Benzophenone's lowest triplet state is $n\pi^*$ while that of 4-phenyl-benzophenone is $\pi\pi^*$. The rates of electron transfer determined by picosecond absorption spectroscopy for the amine reduction of 4-phenyl-benzophenone and benzophenone are listed in Table 1. In acetonitrile the rate of electron transfer from the three amines to benzophenone and 4-phenyl-benzophenone varies at most by a factor of two. It is therefore concluded that there is no inherent difference in reactivity for $^3n\pi^*$ and $^3\pi\pi^*$ states towards electron transfer. Consequently if the photocycloaddition reaction were to proceed by initial charge transfer to form the 1,4 biradical, then the triplet states of benzophenone and 4-phenyl-benzophenone should have similiar reactivity towards dioxene in acetonitrile.

Table 1: Electron Transfer Rates for the Amine Reduction of Benzophenone and 4-Phenyl-Benzophenone

Amine	Solvent	$k_{BP}(M^{-1}sec^{-1})$	$k_{4PB}(M^{-1}sec^{-1})$
Diazobicyclooctane	Acetonitrile	1.7×10^{10}	7.3×10^9
	Butyronitrile	1.3×10^{10}	4.9×10^9
	Tetrahydrofuran	1.1×10^{10}	2.1×10^9
Diethylaniline	Acetonitrile	4.2×10^{10}	3.2×10^{10}
Dimethylaniline	Acetonitrile	3.6×10^{10}	7.8×10^9

When 4-phenyl-benzophenone is photolyzed at 355 nm in the presence of 1M dioxene in acetonitrile the triplet state appears in less than 25 picoseconds, Figure 2. However unlike benzophenone the triplet state is not quenched on the picosecond time scale to form the 1,4 biradical but instead persists beyond 250 nanoseconds, the time limit of our interrogation.

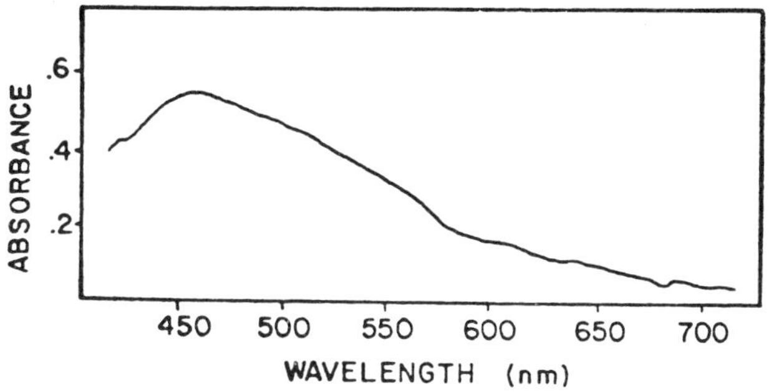

Figure 2. Triplet absorption spectrum of 4-phenyl-benzophenone in presence of 1M dioxene in acetonitrile.

These results when taken in conjunction with our previous studies of benzophenone-dioxene do not support a mechanism whereby a discreet charge transfer complex intervenes between the triplet state of the ketone and the 1,4 biradical. Instead the difference in reactivity of benzophenone and 4-phenyl-benzophenone towards dioxene is indicative of a radical type mechanism. Turro has shown[11] that 4-phenyl-benzophenone is more than a thousand times less reactive than benzophenone towards direct hydrogen abstaction from isopropanol. Therefore we conclude that the mechanism of the addition of triplet benzophenone to dioxene involves direct addition occuring on the time scale of 175 picoseconds at 1M olefin concentration and does not involve an initial one electron transfer.

PHOTOACOUSTIC CALORIMETRY

One of the central problems in the study of the dynamics of organic reactive intermediates has been the inability to determine the change in energy associated with the decay of one intermediate giving rise to another. With the decay of a transient is the generation of heat that is manifested in the form of an acoustic wave. Thus if a relationship between the acoustic wave amplitude and the heat produced by reaction is established, then the energy change associated with the observed kinetics can be deduced. This is the basis of our recently developed photoacoustic calorimeter[12].

The experiment consists of the generation of an intermediate produced by a laser pulse of 10 nanoseconds or shorter in duration and the detection of the resulting acoustic wace amplitude by a 1 MHz piezoelectric transducer where the first excursion of the voltage amplitude, approximately 1 microsecond, is integrated. If a reaction occurs as a result of excition on the timescale of 100 nanoseconds or less, and the resulting reactive intermediate, I, is stable for at least ten microseconds, then the acoustic wave amplitude integrated for one microsecond corresponds to the heat generated from the decay of the excited state to produce I. Thus two conditions must be statisfied: the reactive intermediate of interest must 1) be formed in less than 100 nanoseconds and 2) have a lifetime greater than 10 microseconds.

The photoacoustic calorimeter, Figure 3, employs a Lambda Physic K600 nitrogen laser (5 nanosecond pulse, 337 nm, 0.1 mj); the pulse energy is measured by a Laser Precision Rj-7200 Ratiometer. The photoacoustic wave is detected using a lead-zirconate, lead-titanate transducer, amplified by a PAR 113 Pre-amp, and digitized using a LeCroy WD Waveform Digitizer. The laser power and waveform are inputed into a Digital MINC-23 Lab Computer. The waveform is then normalized to the laser power and averaged for twenty laser shots. The resulting waveform is integrated for 1 microsecond around the first positive excursion.

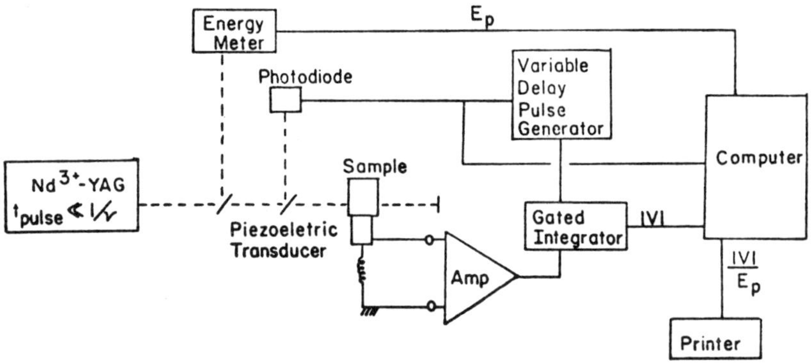

Figure 3. Schematic of the photoacoustic calorimeter.

It remains to be demonstrated the quantitative correspondence between non-radiative energy decay and photoacoustic amplitude. This was examined by using degassed benzene solutions of the compounds listed if Figure 4.[13] The photophysics of these molecules have been well characterized and it thus possible to calculate the quantity of heat released during the first 100 nanoseconds following photolysis. By plotting the observed acoustic wave amplitude with calculated heat released, a linear relationship is established.

One of the first molecules to be examined was diphenyldiazomethane [13]. With the exception of diazomethane producing triplet

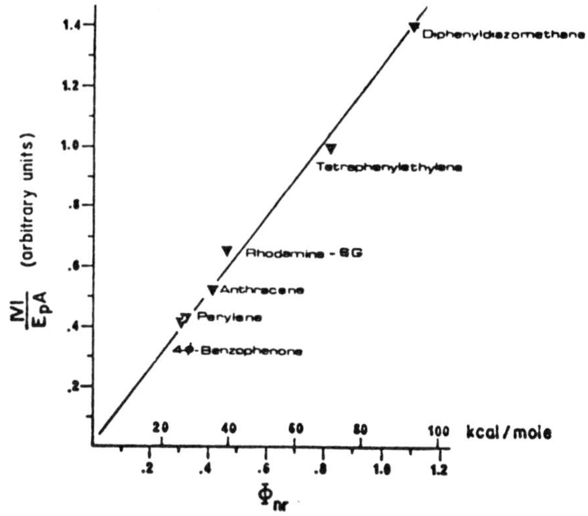

Figure 4. Instrument calibration curve for benzene.

methylene[14], there is nothing known concerning the energy relationship between the azo precusor and the carbene. Photolysis of diphenyldiazomethane in benzene at 337 nm produces singlet diphenylcarbene in less than 10 picoseconds which intersystem crosses producing triplet diphenylcarbene on the timescale of 100 picoseconds[15]. The major decay pathway of the triplet carbene is dimerization and with 1×10^{-6}M solutions employed in these studies the triplet will have a lifetime of approximately 1 millisecond. Thus the conditions for photoacoustic calorimetry are meet.

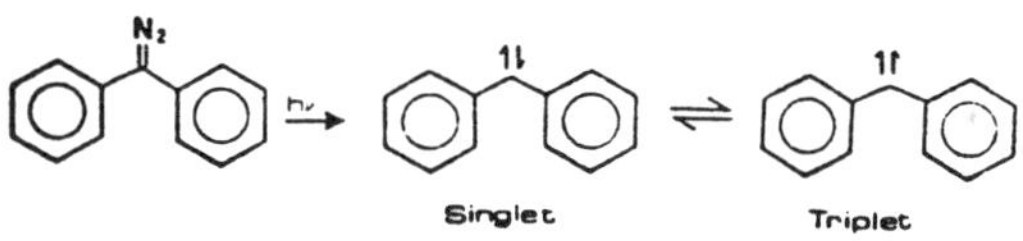

Assuming the quantum yield for photoelimination is .78 then the heat of reaction, Figure 4, for the formation of triplet diphenylcarbene from diphenyldiazomethane in benzene is exothermic by 12 kcal/mole. This is to be contrasted with the endothermic heat of reaction of 30 kcal/mole for the formation of triplet methylene from diazomethane[14].

CONCLUSION

The present report represents only the beginning of the application of picosecond laser spectroscopy to problems in organic reaction mechanisms. There are a great variety of organic reactions that may be activated by light thereby making them amiable to study by picoseconds techniques. Examples include sodium borohydride reduction of ketones and olefin formation from Wittig reagents and aldehydes. When combined with the energetics from photoacoustic calorimetry these studies will lead to a better understanding of the dynamics of bond formation and also will begin to address the important question of the role of solvent participation in these reactions.

ACKNOWLEDGEMENT

This work is supported by a grant from the National Science Foundation, CHE-8117519.

REFERENCES

1. Buchi,G., Inman,C.G.,Lipinsky,E.S. 1954, J. Am. Chem. Soc., 76, pp. 4328.
2. Arnold,D.R., Hinman,R.L. 1962, Tet. Lett., pp. 1425.
3. Caldwell, R.A., James, S.P. 1969, J. Am. Chem. Soc.,91,pp.5184.
4. Caldwell,R.A. 1970, J. Am. Chem. Soc., 92,pp.1439.
5. Jones, G. 1981, Organic Photochemistry, A. Padwa(ed),5, Dekker, New York, p.1.
6. Caldwell, R.A., Sovocool, G.W., Gajewski,R.P. 1973, J. Am. Chem. Soc., 95, p. 2549.
7. Freilich, S.C., Peters,K.S. 1981, J. Am. Chem. Soc.,103,p.6255.
8. Caldwell, R.A., Majima, T., Pac, C. 1982,J. Am. Chem. Soc.,104, p. 629.
9. Simon,J.D., Peters, K.S. 1981, J. Am. Chem. Soc., 103,p.6403.
10. Wagner, P.J. 1976, Topics in Current Chemistry,66,p. 1.
11. Turro, N.J., Lee, C. 1972, Mol. Photochem.,4,p. 427.
12. Rothberg, L.J., Simon, J.D., Bernstein, M., Peters, K.S. 1983, J. Am. Chem. Soc., 105,p.3464.
13. Simon, J.D., Peters, K.S. 1983, J. Am. Chem. Soc.,105,p.5156.
14. Laufer, A.H., Okabe, H.J. 1971, J. Am. Chem. Soc.,93, p.4137.

STUDIES OF THE CHEMICAL INTERMEDIATE DIPHENYLCARBENE: INTERSYSTEM CROSSING, SOLVENT EFFECTS ON DYNAMICS, SPIN STATE SELECTIVE CHEMISTRY, PHOTOCHEMISTRY OF THE EXCITED TRIPLET STATE

E. V. Sitzmann and K. B. Eisenthal

Department of Chemistry
Columbia University
New York, N. Y. 10027

This paper describes the recent results from our laboratory concerning kinetic studies on diphenylcarbene (DPC). A combination of picosecond and nanosecond laser spectroscopy measurements, chemical quenching experiments and triplet sensitization experiments has allowed the determination of the rapid singlet to triplet and slower triplet to singlet intersystem crossing rates for diphenylcarbene in fluid solution at room temperature. It is shown that under the conditions of the kinetic measurements, singlet and triplet diphenylcarbene (^{1}DPC and ^{3}DPC, respectively) are in rapid equilibrium relative to reactions, so that knowledge of the values of k_{ST} and k_{TS} allows determination of the equilibrium constant and change in energy gap for the $^1DPC \rightleftarrows {}^3DPC$ process. It is found that there is a significant solvent effect on the intersystem crossing rates and energy gaps.

The photochemistry arising from the excited triplet state (^{3*}DPC) is examined and compared to the chemistry arising from the ground triplet and thermally accessible singlet states. ^{3*}DPC differs substantially from ^{3}DPC insofar as it is readily attacked by alcohols in an analogous manner to ^{1}DPC. However, like the ground triplet, the excited triplet undergoes facile reactions with olefins, albeit at vastly accelerated rates. Furthermore, novel reaction pathways of the carbene are possible when in its excited triplet state. One example is the charge-transfer reaction of ^{3*}DPC with simple electron donors, a heretofore unknown reaction for the carbene. A model is presented which accounts for the reactivity patterns of ^{3}DPC, ^{1}DPC and ^{3*}DPC.

I. INTRODUCTION

Divalent carbon atom species, commonly referred to as

K. B. Eisenthal (ed.), Applications of Picosecond Spectroscopy to Chemistry, 41–63.

carbenes, have had widespread appeal to the theoretical and experimental chemist [1,2]. The main reason for the attention is that carbenes are proven to be important chemical intermediates which demonstrate a high reactivity toward a host of molecules. A carbene is a divalent carbon atom species that has two nonbonding valence electrons localized on the central methylene carbon atom. Because of the two nonbonding electrons carbenes may exist in two different electronic states, which will differ by spin multiplicity, i.e., singlet and triplet. We have elected to study diphenylcarbene (DPC), which is a prototypical ground triplet aromatic carbene, because of its ease of generation from precursor molecules and the considerable literature of theoretical and

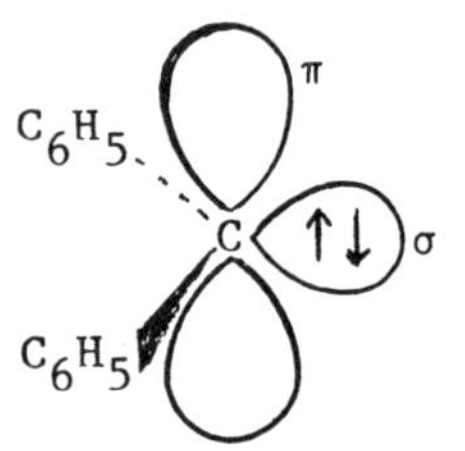

^{1}DPC

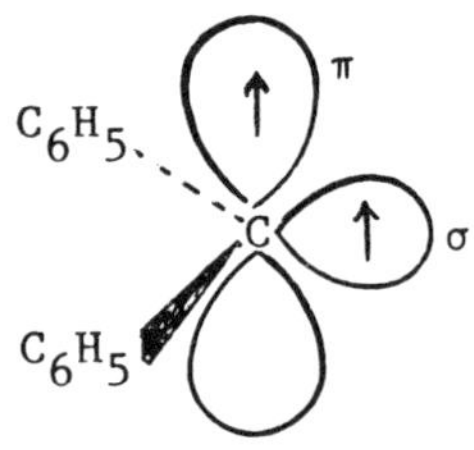

^{3}DPC

spectroscopic studies. The σ orbital, a hybrid orbital having s and p character, is in the plane defined by the carbon-phenyl bonds and the π orbital is perpendicular to the plane of the carbene.

In this paper we would like to center on two main questions that arise in the study of DPC. The first question is the determination of the rates of spin state interconversion. The characterization of the kinetics of the carbene intersystem crossing is important because little is actually known about the process which has direct bearing on the energetics of the carbene, e.g., the singlet-triplet energy gap [3-6], and the distinctive chemistry associated with these states. Coupled to this question is the problem of environmental effects which can serve as perturbations on the intramolecular dynamics of the carbene. This is an important issue and therefore it is desirable to determine the extent to which the medium can influence the properties of the carbene. Finally, in this paper we will examine the nature and reactivity of the photoexcited triplet DPC (^{3*}DPC). Unusual reactivity patterns, significantly different from those found for ^{3}DPC, are observed for ^{3*}DPC. Furthermore we will show that ^{3*}DPC provides new chemical reactions for the carbene, as exemplified by the charge-transfer reactions of ^{3*}DPC with simple electron donors.

II. DYNAMICS AND REACTIVITIES OF DIFFERENT SPIN STATES

2.1 Determination of Singlet to Triplet Intersystem Crossing Rate of DPC

Direct photolysis of diphenyldiazomethane yields singlet DPC upon loss of N_2 from the excited singlet state of the diazo compound. Once formed, ^{1}DPC will undergo energy relaxation via intersystem crossing to produce the ground triplet, ^{3}DPC. From the

$$\mathrm{Ph_2C{=}N_2} \xrightarrow[-N_2]{h\nu} {}^1\mathrm{DPC} \underset{k_{ST}}{\overset{k_{ST}}{\rightleftharpoons}} {}^3\mathrm{DPC} \qquad (1)$$

work of Closs and Rabinow [4] it is known that this exothermic process occurs in the subnanosecond regime and therefore

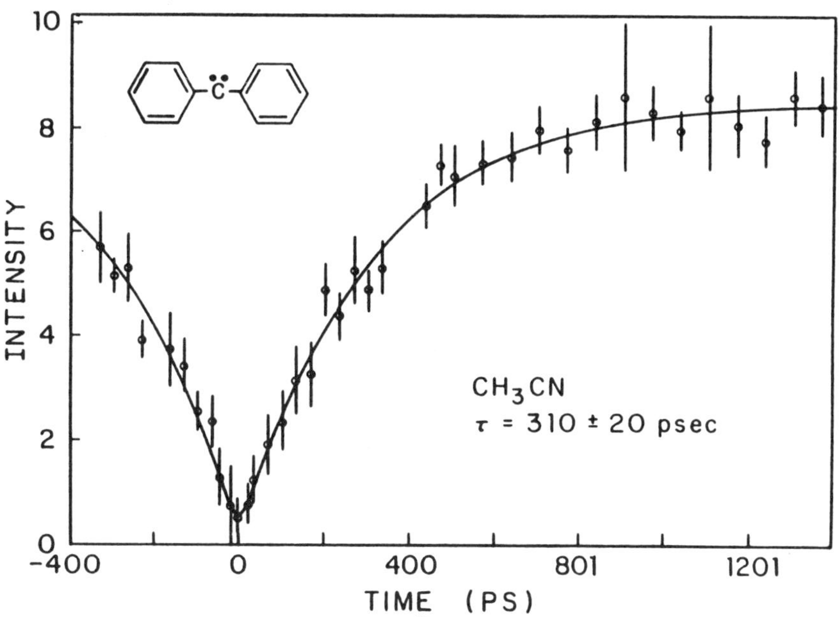

FIGURE 1. Formation Kinetics of ^{3}DPC in Degassed Acetonitrile. The LIF Signal as a Function of Time Delay Between Excitation and Probe Laser Pulses. The Solid Curve is a Calculated Fit.

conventional methods of study are largely inappropriate. However, by using picosecond lasers one can conveniently monitor the kinetic process directly. The experiment makes use of the laser induced fluorescence (LIF) technique for the carbene system [6]. Basically, two laser pulses are used, one for excitation of the diazo compound and the other (which is delayed at variable times) is used for probing the resulting carbene. The experiment consisted of photolyzing a vacuum degassed acetonitrile solution of diphenyldiazomethane (8×10^{-4} M) with a picosecond laser pulse at 266 nm from a passively mode-locked Nd:YAG laser (25-30 ps FWHM, < 0.02 mJ). The formation rate of ^{3}DPC was monitored by laser induced fluorescence using a weak probe pulse at 266 nm at variable delays. The risetime of ^{3}DPC was then obtained by computer simulation of the plot of the ^{3*}DPC -> ^{3}DPC integrated fluorescence intensity [7] versus the time separating the excitation and probe pulses. The results are shown in Fig. 1. The rate constant of the formation of ^{3}DPC obtained in these experiments was $k_{ST} = (3.22 \pm 0.23) \times 10^9\ S^{-1}$.

2.2 Determination of the Triplet-Singlet Interconversion Rate From Competitive Quenching Studies.

As first shown in the elegant study by Closs and Rabinow [4], one can obtain estimates of K_{eq} from a competitive quenching experiment [4,8,9]. The essence of the experiment is to make use of two quenchers, where one is selective for ^{1}DPC and the other is selective for ^{3}DPC. Reaction products will then be generated that arise from the competition between the two spin selective scavengers for the carbene. Clearly, the relative yield of singlet products to triplet products will depend on the relative population of ^{1}DPC and ^{3}DPC, since we are dealing with spin selective traps. Under equilibrium conditions, the relative singlet and triplet spin state populations of DPC will be a function of k_{ST} and k_{TS}. Thus, the relative yields of the products generated by competitive quenching will also be a function of k_{ST} and k_{TS} which can therefore, in principle, provide a means to experimentally determine the relative values of the spin interconversion rate constants k_{ST} and k_{TS}.

In collaboration with Prof. N. J. Turro we have investigated the system described in Scheme 1 by using methanol and isoprene as selective ^{1}DPC and ^{3}DPC traps, respectively. Alcohols are generally accepted to react with ^{1}DPC to yield ethers by O-H insertion [1a,4], and 1,3-dienes have been shown to be efficient quenchers of triplet carbenes [10,11]. We have found that the ratio of the two isomeric cyclopropanes formed upon reaction of DPC with isoprene is independent of isoprene concentration in the range 0.1 to 10 M. Furthermore, the same ratio of cyclopropanes is produced by direct excitation or triplet sensitization of diphenyldiazomethane in the presence of a high concentration of methanol and a low concentration of isoprene. These results define isoprene as a selective trap of ^{3}DPC.

The occurrence of a rapid equilibrium and the spin state selective reactions leads to the following relation for the quantum yield of triplet products ($^3\phi$) to the quantum yield of singlet products ($^1\phi$) as given by Eq. 2.

$$^3\phi/^1\phi = \frac{k_{ST}k_3[IP]}{k_1[CH_3OH](k_{TS} + k_3[IP])} \qquad (2)$$

It is seen that the kinetics of the system are defined by four rate constants (k_{TS}, k_{ST}, k_1 and k_3). From Eq. 2 it is also seen that the ratio $^3\phi/^1\phi$ will have a nonlinear dependence on [IP] for a fixed concentration of methanol. Experimental confirmation of

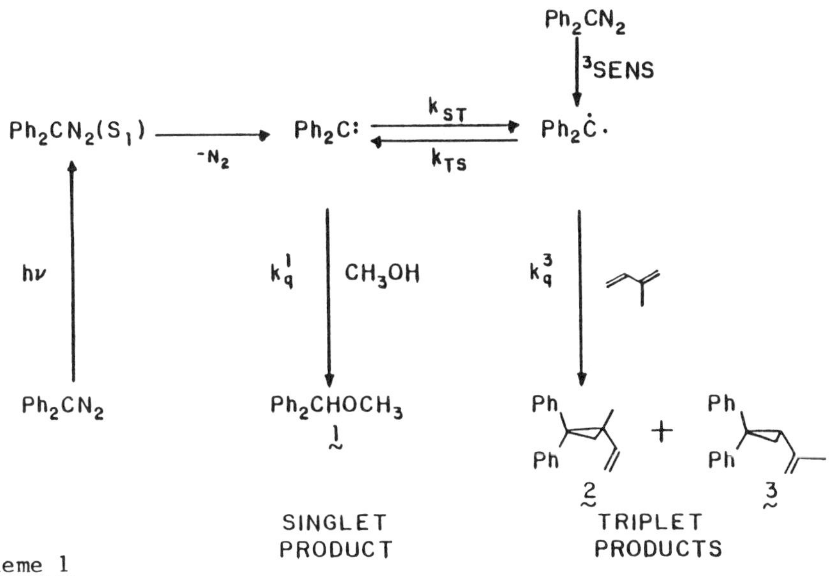

Scheme 1

the predicted behavior is shown in Fig. 2, which shows the ratio of cyclopropanes to ether ($^3\phi/^1\phi$) of DPC in acetonitrile solvent, at fixed (0.05 M) concentration of methanol and variable concentrations of isoprene. Confidence in the kinetic scheme which is used to obtain the values of k_1 and k_{TS} is strengthened by the fact that the $^3\phi/^1\phi$ ratios are independent of whether the experiment is performed under either direct or photosenstized conditions.

One can extract the unknown values of k_1 and k_{TS} from these results by fitting Eq. 2 to the data. The problem is greatly simplified because k_{ST} is known from the picosecond work and k_3 can be measured from nanosecond flash experiments. Analysis of the decay kinetics of the triplet-triplet absorption of ^{3}DPC as a function of isoprene concentration yielded k_3. The value of k_3 was measured as $1.36 \times 10^6\ M^{-1}S^{-1}$ in acetonitrile. Thus, Eq. 2 is fitted to the data in Fig. 2, using $k_{ST} = 3.25 \times 10^9 S^{-1}$ and k_3 =

$1.4 \times 10^6 M^{-1}S^{-1}$ and leaving k_1 and k_{TS} as the adjustable parameters. We find, after optimization by the method of least squares [12], that $k_1 = 7.77 \times 10^9 M^{-1}S^{-1}$ and $k_{TS} = 6.29 \times 10^7 S^{-1}$. It

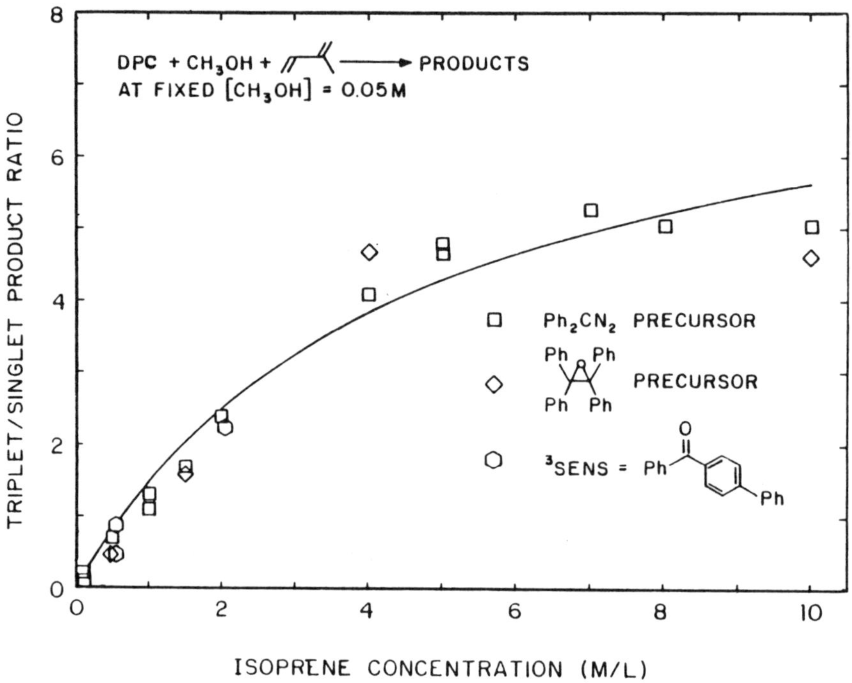

Figure 2. Experimental Behavior of $^3\phi/^1\phi$ as a Function of Isoprene Concentration at a Fixed Methanol Concentration (0.05 M), at 273°K. The Solid Curve is Calculated from Eq. 2 Using $k_{ST} = 3.23 \times 10^9 S^{-1}$, $k_3 = 1.36 \times 10^6 M^{-1}S^{-1}$, $k_1 = 7.77 \times 10^9 M^{-1}S^{-1}$ and $k_{TS} = 6.29 \times 10^6 S^{-1}$.

is interesting to note that Scheme 1 predicts that the ratio of ether to cyclopropane products ($^1\phi/^3\phi$) should be a linear function of methanol concentration, as given by Eq. 3.

$$^1\phi/^3\phi = \frac{k_1[CH_3OH](k_{TS} + k_3[IP])}{k_{ST}k_3[IP]} \qquad (3)$$

This behavior was confirmed (Fig. 3) at a constant isoprene concentration of 1 M, while varying the concentration of methanol from 0.01 M to 0.5 M, at 273°K. Fitting Eq. 3 to the data gave $k_1 = 6.9 \times 10^9 M^{-1}S^{-1}$ and $k_{TS} = 1.0 \times 10^7 S^{-1}$. The k_1 and k_{TS} values are in reasonable agreement to those values obtained from the competition experiment in which isoprene was varied at fixed methanol concentration.

One of the underlying assumptions in the competitive quenching experiment is that the singlet and triplet carbene states have attained equilibrium before their decay via spin selective scavenging. This means that under appropriate conditions of quencher concentration that identical results will be observed in the products arising from the reactions with methanol and isoprene regardless of whether the carbene is initially formed in the singlet or triplet state. To test the extent to which this assumption is valid the carbene was generated by triplet

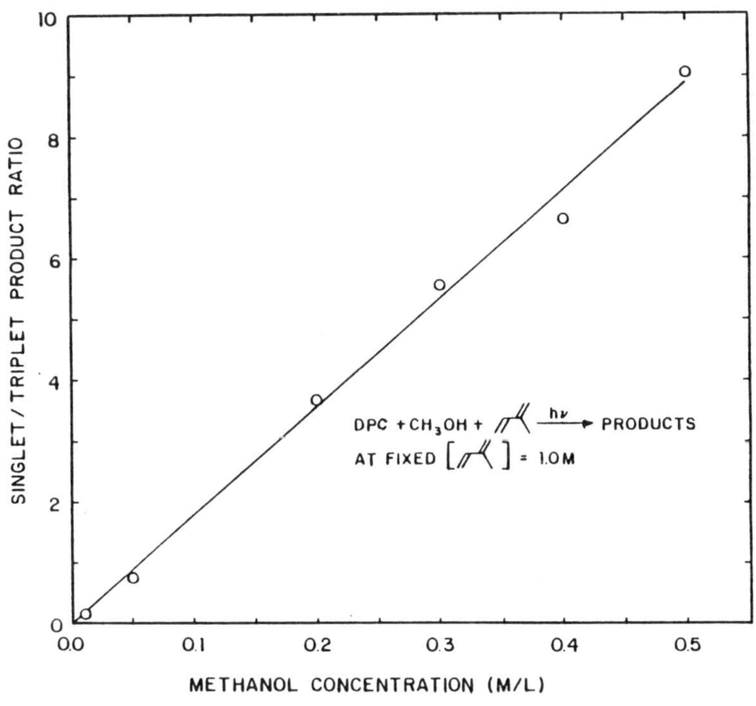

Figure 3. Experimental Behavior of $^{1}\phi/^{3}\phi$ as a Function of Methanol Concentration at a Fixed Isoprene Concentration (1.0 M). The Solid Curve is Calculated from Eq. 3 Using k_{ST} = 3.23 X 10^{9} S^{-1}, k_3 = 1.36 X 10^{6} $M^{-1}S^{-1}$, k_1 = 6.9 X 10^{9} $M^{-1}S^{-1}$ and k_{TS} = 1.0 X 10^{7} S^{-1}.

sensitization of diphenyldiazomethane, using 4-phenylbenophenone as sensitizer.

Using the assumptions of the selective reaction of methanol (singlet scavenger) and isoprene (triplet scavenger), and equilibrium between the spin states, and that photosensitization leads initially to ³DPC, then under steady state conditions the ratio $^1\phi/^3\phi$ will be given by Eq. 4.

$$^1\phi/^3\phi = \frac{k_{TS}k_1[CH_3OH]}{k_3[IP](k_{ST} + k_1[CH_3OH])} \qquad (4)$$

Fig. 4 shows the observed dependence of $^1\phi/^3\phi$ as a function of variable methanol concentration (0.01 - 10.0 M) with fixed isoprene concentration (0.1 M) for triplet sensitized decomposition of diphenyldiaomethane. Fitting Eq. 4 to the data gives $k_{TS} = 1.97 \times 10^7\ S^{-1}$ and $k_1 = 7.4 \times 10^9\ M^{-1}S^{-1}$. These results are in good agreement with the corresponding rate constants evaluated from the direct irradiation experiments, and provide further support for the kinetic model which assumes a rapid rate of equilibration between ¹DPC and ³DPC relative to reaction.

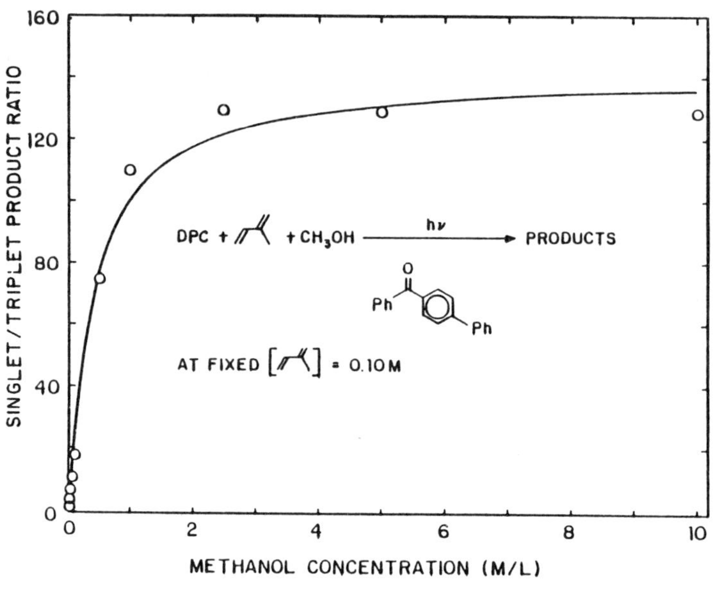

Figure 4. Experimental Behavior of $^1\phi/^3\phi$ as a Function of Methanol Concentration at a Fixed Isoprene Concentration (0.10 M) Obtained From the Photosensitized Decomposition of Diphenyldiazomethane. The Solid Curve is Calculated from Eq. 4 Using $k_{ST} = 3.23 \times 10^9 S^{-1}$, $k_3 = 1.36 \times 10^6 M^{-1} S^{-1}$, $k_{TS} = 1.97 \times 10^7 S^{-1}$ and $k_1 = 7.4 \times 10^9 M^{-1} S^{-1}$.

2.3 The Effect of Solvent.

The results of the two experiments in which either the methanol or the isoprene concentration were fixed are completely consistent with the concept of a rapid equilibrium between the two spin states of the carbene. In view of the large concentrations of isoprene (>4 M) that are required in the experiment with fixed methanol we have examined the effect of solvent on the equilibrium constant. We have chosen to compare the kinetics obtained in the nonpolar solvent isooctane with that in the polar solvent acetonitrile. The value of $^3\phi/^1\phi$ for experiments in which a methanol concentration of 0.05 M was fixed and the isoprene concentration was varied, and the values of $^1\phi/^3\phi$ for a fixed isoprene concentration and various methanol concentrations were measured and are given in Table 1. Additionally the values of $k_{ST} = (1.05 \times 10^{10} S^{-1})$ and $k_3 = (1.52 \times 10^6 M^{-1} S^{-1})$ were determined in isooctane in a manner identical to that previously described for acetonitrile. Analysis of the quantum yield data as before gives values of $k_1 = 1.23 \times 10^{10} M^{-1} S^{-1}$ and $k_{TS} = 3.47 \times 10^6 S^{-1}$ for the fixed methanol experiment and $k_1 = 1.62 \times 10^{10} M^{-1} S^{-1}$ and $k_{TS} = 5.7 \times 10^6 S^{-1}$ for the fixed isoprene experiment. Although these results demonstrate a small but very significant solvent effect [5b,13] on the kinetics of the carbene system, we note that for the solvents used the kinetic model given by Scheme 1 yields good agreement with experiment.

2.4 Determination of the Singlet-Triplet Energy Splitting

Using the values of k_{ST} and k_{TS} appropriate for acetonitrile and isooctane, one can calculate the equilibrium constant, Keq, between the ^{1}DPC and ^{3}DPC states from Eq. 5. The free energy difference between these two spin states follows directly from K_{eq} by use of Eq. 6. The derived value of $-\Delta G$ is 3.2 kcal/mole for acetonitrile and 4.6 kcal/mole for isooctane.

$$K_{eq} = \frac{k_{ST}}{k_{TS}} \qquad (5)$$

$$\Delta G = -RT \ln(K_{eq}) \qquad (6)$$

From the value obtained for ΔG we can now estimate the energy gap between the singlet and triplet states from the thermodynamic relation $\Delta G = \Delta H - T\Delta S$. Since the entropy difference, ΔS, is expected to be primarily due to the multiplicity difference between the singlet and triplet states, and the enthalpy difference, ΔH, and the energy gap, ΔE, are equal to a good approximation in liquids, we obtain,

$$\Delta G = \Delta H - T\Delta S = \Delta E - RT\ln 3 \qquad (7)$$

Consequently, the energy gap is 2.5 kcal/mole (870 cm^{-1}) for

Table 1. Kinetic and Thermodynamic Parameters Obtained from Competitive Quenching Experiments Analyzed According to the Kinetics Given in Scheme 1.

Solvent: ACETONITRILE $k_{ST} = 3.23 \times 10^9 S^{-1}$ $k_3 = 1.36 \times 10^6 M^{-1}S^{-1}$

METHOD	QUENCHER CONDITIONS	$10^{-9}(k_1)$, $M^{-1}S^{-1}$	$10^{-6}(k_{TS})$, S^{-1}	K_{eq}	$-\Delta G$, kcal/mole	$-\Delta H$ kcal/mole	T, (^oK)
Direct Irradiation	vary [IP] fix [CH_3OH]	5.56	11.1	291	3.36	2.7	298
Direct Irradiation	vary [IP] fix [CH_3OH]	7.77	6.29	512	3.38	2.8	273
Direct Irradiation	fix [IP] vary [CH_3OH]	6.9	10.0	321	3.13	2.5	273
Sensitized Irradiation	fix [IP] vary [CH_3OH]	7.43	19.7	164	2.77	2.2	273

Solvent: ISOOCTANE $k_{ST} = 1.05 \times 10^{10} S^{-1}$ $k_3 = 1.52 \times 10^6 M^{-1}S^{-1}$

METHOD	QUENCHER CONDITIONS	$10^{-9}(k_1)$, $M^{-1}S^{-1}$	$10^{-6}(k_{TS})$, S^{-1}	K_{eq}	$-\Delta G$, kcal/mole	$-\Delta H$, kcal/mole	T (^oK)
Direct Irradiation	vary [IP] fix [CH_3OH]	12.3	3.5	3031	4.75	4.1	298
Direct Irradiation	fix [IP] vary [CH_3OH]	16.2	5.7	1845	4.45	3.8	298

acetonitrile and 4.0 kcal/mole (1400 cm^{-1}) for isooctane. Thus the singlet-triplet energy gap is smaller in acetontrile than in isooctane. In other words, the singlet is stabilized relative to the triplet by 1.5 kcal/mole (530 cm^{-1}) in going from the nonpolar isooctane to the polar acetonitrile solvent. This interesting result may be interpreted as arising from the selective stabilization of the polar singlet state relative to the nonpolar triplet by polar enviornments, and hence reduce the relative singlet-triplet energy splitting.

In conclusion, we have combined the use of competitive quenching experiments and picosecond and nanosecond spectroscopy to obtain important information about the physical and chemical properties of DPC in solution. The experimental behavior of the reaction of DPC conforms to the derived kinetics of Scheme 1, which supports the proposal that an equilibrium between carbene spin states is established in a time that is short compared to most reactions. Triplet sensitized experiments have confirmed that under proper conditions the ratio $^3\phi/^1\phi$ is independent of the initially formed spin state because of the presence of this rapid equilibrium. Additionally, the results in isooctane confirm that establishment of the equilibrium is independent of the solvent mixture. Analysis of the kinetic data allows the determination of the free energy difference ($-\Delta G = 3.2 \pm 0.3$ kcal/mole) between the singlet and triplet states of DPC in acetonitrile. Similarly, a value of $-\Delta G = 4.6 \pm 0.3$ kcal/mole is determined in isooctane.

III. PHOTOCHEMISTRY OF CARBENES: THE EXCITED TRIPLET DPC MODEL FOR REACTIONS OF ^{3}DPC, ^{1}DPC AND ^{3*}DPC.

It is well recognized that carbenes are often classified according to their reactivity, and that the reactivity of a given carbene is often spin state selective [1]. In the example of DPC, it was shown that there is a significant chemical difference between the singlet and triplet states, even though they have comparable energies. A natural question arises as to why these states should behave so differently, and whether the spin selective nature of a given carbene scavenger for the low-lying carbene states would be preserved for the higher energy carbene electronic states. A simple model which may be used to account for the different reactivity patterns of the ground triplet, lowest

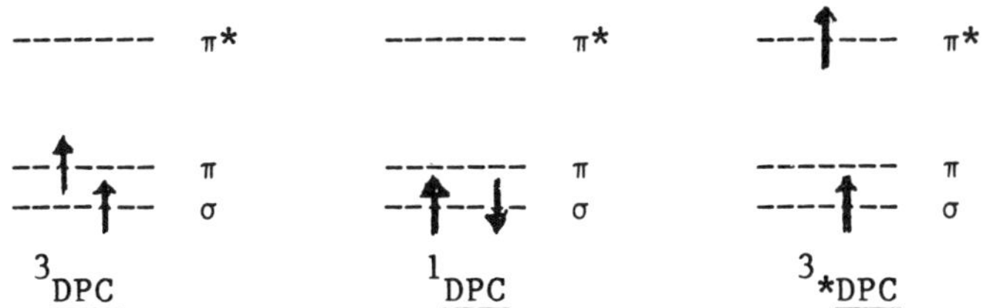

Scheme 2

singlet and excited triplet is given by the orbital filling scheme involving the highest nonbonding molecular orbitals σ and π and the antibonding π^* orbital of the carbene (Scheme 2). The key point is the presence of an empty low-lying π orbital for both ^{1}DPC and ^{3*}DPC electronic states. It should be noted that ^{3}DPC does not have an empty low-lying orbital, whereas the excited triplet does, due to promotion of an electron into an upper π^* orbital. Since the ground triplet has both σ and π orbitals singly occupied, it is expected that it should behave like a free radical in its reactions. This expectation is borne out experimentally by the fact that ^{3}DPC undergoes hydrogen abstraction reactions and stepwise addition to olefins. Because the singlet has a $\sigma^2\pi^0$ electron configuration, it will be able to accept a pair of electrons into its low-lying empty π orbital. Chemically this means that the singlet will be able to undergo O-H insertion reactions with alcohols to form ethers via the donation of a pair of electrons from oxygen of the alcohol into the vacant π orbital

$$^1Ph_2C \quad + \quad R_2CHOH \quad \longrightarrow \quad Ph_2CHOCR_2$$

"Singlet Reaction"

$$^3Ph_2C \quad + \quad R_2CHOH \quad \longrightarrow \quad ^3[Ph_2CH \quad + \quad R_2COH]$$

"Triplet Reaction"

of the carbene. The ground triplet carbene does not have an empty low-lying orbital and thus would not react with alcohols by an insertion reaction but rather by a hydrogen atom abstraction reaction mentioned above. Therefore, Scheme 2 offers justification for the observed difference between the reactions of the ground triplet and lowest singlet state of the carbene.

Although the chemistry of the low-lying electronic states ^{3}DPC and ^{1}DPC have been extensively studied, the chemistry of the higher states remains unknown in large part due to the reactivity of the thermally accessible singlet and ground triplet states. In order to delineate the properties of the excited triplet state and, equally important, to establish its chemistry we have used picosecond methods to study the kinetic behavior of 3*DPC in the presence of known spin selective carbene scavengers [14-17].We note that Scheme 2 predicts that the excited triplet should undergo certain reactions that the ground triplet would not. For example, ^{3*}DPC having an empty low-lying π orbital, as does ^{1}DPC, should resemble ^{1}DPC in certain of its reactions, e.g., with alcohols.

DPC was generated by UV laser irradiation of the precursor molecule diphenyldiazomethane (10^{-3} M) in a degassed acetonitrile solution. Excitation of the precursor molecule with a single laser pulse from a frequency quadrupled Nd:YAG laser (266 nm) produces two fragments, diphenylcarbene and N_2. A second UV laser pulse also at the same wavelength was time delayed by 8 ns after the first laser pulse and was used to excite the carbene fragment from its ground state, ^{3}DPC, to its excited triplet state, ^{3*}DPC. The reactions of ^{3*}DPC were monitored by measurement of the ^{3*}DPC fluorescence lifetime, Fig. 5, using a streak camera optical multichannel analyzer system. The lifetime of ^{3*}DPC in acetonitrile with no quencher present was found to be 3.8 $\pm$ 0.2 ns

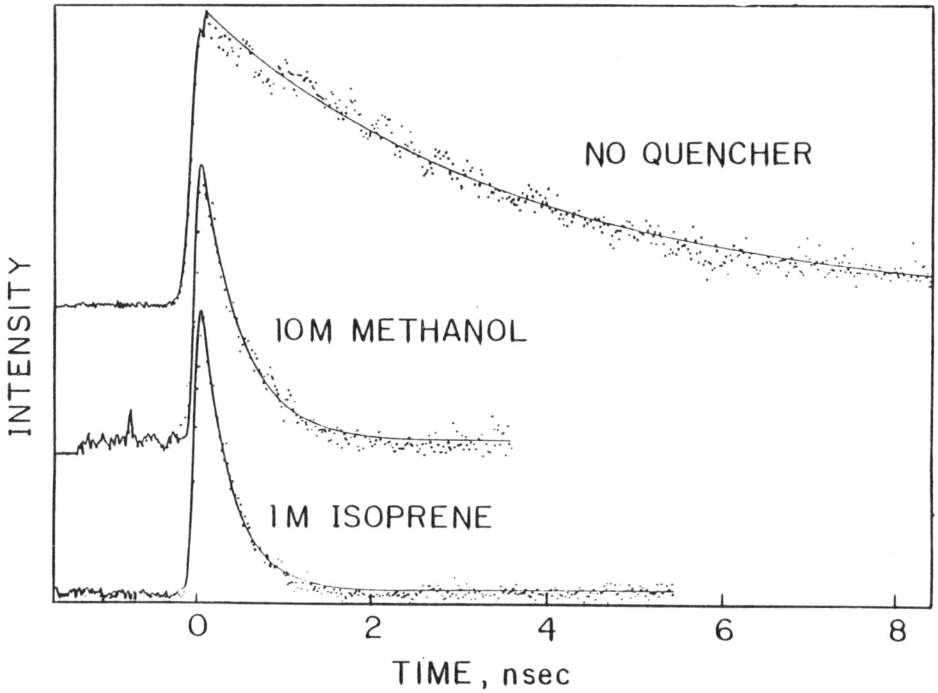

Figure 5. The Decay Curves of Excited Triplet Diphenylcarbene with (1) no Quencher, (2) 10 M Methanol, and (3) 1 M Isoprene in Degassed Acetonitrile at Room Temperature. The Points are Experimental and the Solid Curves are Theoretical.

Its decrease as a function of the added quencher concentration yielded the bimolecular reaction rate constant, ^{3*}k.

$$^{3*}\text{DPC} + \text{Quencher} \xrightarrow{^{3*}k} \text{product} \qquad (8)$$

3.2 Excited Triplet Carbene Reactions with Alcohols

By determining the kinetic behavior of ^{3*}DPC in the presence of a series of alcohols we can characterize whether the mode of reaction for the excited triplet follows the singlet mechanism (attack of the O-H bond), the triplet mechanism (attack of the C-H bond), or some other mechanism.

Table 2 gives the reaction rate constants as determined from an analysis of the fluorescence lifetime changes of ^{3*}DPC induced by the alcohols at several different concentrations (Fig. 6) [14-16]. A comparison of the observed rate constants for methanol, isopropanol, and t-butanol in their reaction with ^{3*}DPC shows that there is an ordering of reactivity. It is seen that methanol is the most reactive, while t-butanol is the least. The observed ordering of the alcohol reactivities with ^{3*}DPC closely parallels the relative lability/acidity of the alcohols [8,18-21]. This observation supports the mechanism that ^{3*}DPC reacts with the O-H bond of the alcohol. It is noted that if the C-H bond of the alcohol were attacked by ^{3*}DPC, as in the ground triplet state reaction [8], then one would not expect to see the observed trend in the rates. For example, since isopropanol has a more labile C-H bond than does methanol, it would be anticipated, if hydrogen abstraction were taking place, that isopropanol would react faster than methanol with ^{3*}DPC; however such is not the case. The interpretation that ^{3*}DPC attacks the O-H bond rather than the C-H bond of the alcohols is further supported by two additional experimental results. First, diethylether or tetrahydrofuran was substituted for the alcohols to see if they could affect the fluorescence of ^{3*}DPC. It is noted that the ethers should be similar to the alcohols insofar as their ability for the oxygen to serve as an electron donor for ^{3*}DPC and the presence of adjacent

<u>Table 2.</u> Reaction Rate Constants and Relative Reactivities of Alcohols with ^{3*}DPC and ^{1}DPC in CH_3CN at 20°C.

	^{3*}DPC [16]		^{1}DPC [5b]	
alcohol	10^{-8} (^{3*}k), $M^{-1}S^{-1}$	rel reac-tivity	10^{-9} (^{1}k), $M^{-1}S^{-1}$	rel reac-tivity
CH_3OH	3.1 ± 0.4	1.0	5.6	1.0
CH_3OD	2.02 ± 0.08	0.7		
i-C_3H_7OH	1.46 ± 0.04	0.5	3.5	0.6
i-C_3H_7OD	0.92 ± 0.05	0.3		
t-C_4H_9OH	0.83 ± 0.08	0.3	1.5	0.3

C-H groups. It was found that neither diethyl ether nor tetrahydrofuran (THF) influences the normal fluorescence lifetime of ^{3}*DPC. The lack of any discernable quenching of ^{3}*DPC by the two ethers, in contrast to the alcohols, indicates that simple C-H abstraction α to the oxygen is not important. Second, kinetic isotope effects are observed for methanol and isopropanol [16]. Substitution of the hydroxyl proton with deuterium on the alcohol leads to a reduction in the rate of reaction pointing to involvement of the O-H bond in the excited state reaction [16,22]. Finally, comparing the results of the relative rates of reaction of ^{3}*DPC vs. ^{1}DPC with methanol, isopropanol, and t-butanol [5b] is revealing. Although ^{1}DPC is more reactive than ^{3}*DPC by approximately an order of magnitude, the relative reactivities of the alcohols are strikingly similar. The conclusion one may draw from this correlation is that ^{3}*DPC probably reacts with alcohols in a manner analogous to the ^{1}DPC reaction with alcohols, i.e., the carbene attacks the O-H bond of the alcohol.

At this point it is interesting to consider that if ^{3}*DPC follows a concerted mechanism in its reaction with alcohols, as it has been considered for ^{1}DPC, then it is expected that the resulting ether will be formed in an excited triplet state [16]. From an estimate of the energetics one would predict that the concerted reaction of ^{3}*DPC with methanol leading to the triplet benzhydryl methyl ether (^{3}*Ph_2CHOCH_3) is exothermic [16]. Using

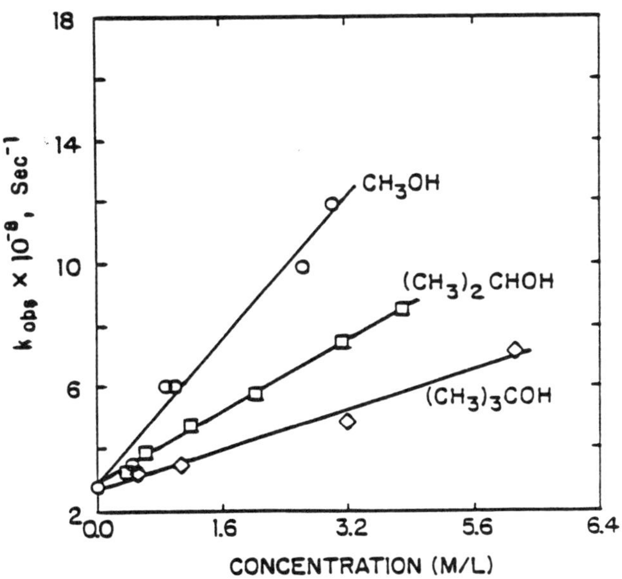

Figure 6. ^{3}*DPC Decay Rate as a Function of Alcohol Concentration in Degassed Acetonitrile.

gas phase thermochemical data and Benson's group additivity method we find that the heat of reaction ΔH_{rx} for $^{3*}DPC + CH_3OH \rightarrow$ $^{3*}Ph_2CHOCH_3$ is given by:

$$\Delta H_{rx} = \Delta H_f(Ph_2CHOCH_3) + E_T(^{3*}Ph_2CHOCH_3) - \Delta H_f(^{3}DPC) - E_T(^{3*}DPC) - \Delta H_f(CH_3OH)$$

$$\sim [(-6) + (80) - (108) - (59) - (-48)] \text{ kcal/mole}$$

$$= -45 \pm 15 \text{ kcal/mole} \qquad (9)$$

Although these estimates neglect solvation effects we can use them to compare the energetics of the ground triplet, singlet and excited triplet state reactions, since the primary differences are due to the energies of ^{3}DPC, ^{1}DPC and $^{3*}DPC$. It is seen that the ground triplet would not be as energetically favored as the excited triplet by the magnitude of the energy $E_T = E(^{3*}DPC) - E(^{3}DPC)$, whereas the singlet would be more favored since a ground singlet rather than an excited triplet ether product would be generated.

In summary, our results show that $^{3*}DPC$ is readily quenched by alcohols and that unlike the ground triplet, the preferred route of reaction involves attack on the O-H bond, rather than the C-H bond of the alcohols. Furthermore, it has been demonstrated that the pattern of the relative rates of reaction of $^{3*}DPC$ and ^{1}DPC are the same and follows the relative acidity of the alcohols (CH_3OH > iPrOH > tBuOH). Scheme 2, shows that ^{1}DPC and $^{3*}DPC$ have in common an empty low-lying orbital and thereby provides a simple explanation for the observed behavior.

3.3 Excited Triplet Reactions with Amines

Primary amines, like alcohols, have been shown to selectively scavenge ^{1}DPC to form mainly N-H insertion type products [19,20,23]. In a study by Bethell et al. [19,20,23], a reaction mechanism involving an ylide intermediate was proposed for ^{1}DPC reactions with primary amines (Eq. 9). As given by Eq. 10 ylide formation occurs by bond formation between the carbene and an amine that donates a pair of electrons to produce a zwitterionic

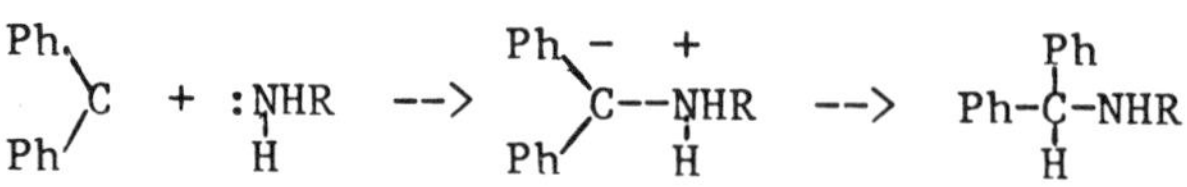

(10)

species [24-20]. The ylide intermediate, however, should be differentiated from species produced by single electron transfer processes, such as a charge-transfer complex or an ion pair. Since $^{3*}DPC$ appears to mimic qualitatively the behavior of ^{1}DPC with alcohols in attacking the O-H bond [5b,15,16], the question is whether this similarity will also be present in the reactions of ^{3*}DPC with amines. The issue addressed here is important, not only in establishing the chemistry of ^{3*}DPC with amines, but also to determine whether a triplet state of a carbene can undergo a reaction with electron donors, and if so, establish if it follows the proposed singlet mechanism (ylide mechanism) or some other mechanism. In order to answer this question the kinetic behavior of ^{3*}DPC with a series of amines was investigated [17]. The results showed that the excited triplet undergoes a charge-transfer process with electron donors, a heretofore unknown pathway in carbene chemistry.

Table 3 gives the quenching rate constants of ^{3*}DPC as determined from a Stern-Volmer analysis of the changes in fluorescence lifetime induced by the addition of the amines at several different concentrations. Inspection of the data reveals a significant variation in the rate of quenching in the series of amines studied. For example, tertiary amines, such as DABCO and

Table 3. Quenching of ^{3*}DPC by Electron Donors.

Donor	$^{3*}k$ ($M^{-1}S^{-1}$) [17]	IP (eV) [30]
1,4-Diaza[2,2,2]bicyclooctane, DABCO	$(2.6 \pm 0.2) \times 10^{10}$	7.20
Triethylamine	$(5.4 \pm 0.4) \times 10^{9}$	7.5
1-Aza[2,2,2]bicyclooctane, ABCO	$(2.7 \pm 0.3) \times 10^{9}$	7.7
Diethylamine	$(9.8 \pm 1.2) \times 10^{8}$	8.01
Dimethylamine	$(9.2 \pm 0.6) \times 10^{8}$	8.24
di-n-Butylsulfide	$(3.2 \pm 0.6) \times 10^{8}$	8.30
n-Butylamine	$(5.3 \pm 0.6) \times 10^{7}$	8.71
t-Butylamine	$(2.7 \pm 0.4) \times 10^{7}$	8.64

triethylamine quench ^{3*}DPC at or near the diffusion controlled limit, while the primary amines quench the excited triplet state at a rate which is three orders of magnitude slower. To determine whether a charge-transfer interaction is responsible for the ordering of reactivity a plot was made of log(^{3*}k) versus the ionization potential of the quenchers given in Table 3. As shown

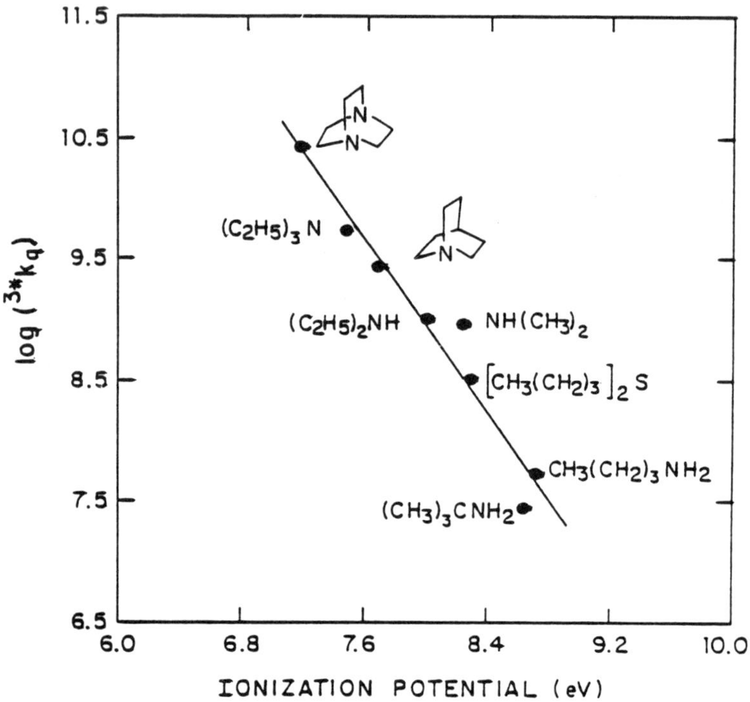

Figure 7. Quenching of Excited Triplet Diphenylcarbene as a Function of Donor Ionization Potential in Acetonitrile, at 20°C.

in Fig. 7, the logarithm of the quenching rate constants is a linear function of the quencher ionization potential with a slope = $-(0.079 \pm 0.006)$ mole/kcal. Such behavior can be accounted for by a charge-transfer mechanism which predicts that log($^{3*}k$) will be proportional to the ionization potential of an electron donor quencher with negative slope [31]. Indeed, a result similar to the one observed here is also found for the quenching of triplet benzophenone with amines [32,33], which is known to proceed by full single electron transfer [33].

On the basis of the dependence of the observed $^{3*}k$ on ionization potential we therefore favor the mechanism in which ^{3*}DPC undergoes photoreduction by amines, A_M, thereby forming a triplet charge-transfer intermediate as the primary step in quenching (Eq. 11).

$$^{3*}DPC + A_M \longrightarrow {}^{3}(\overset{-}{DPC}\ \overset{+}{A_M}) \qquad (11)$$

Concerning the fate of the triplet charge-transfer (CT)

intermediate, it is reasonable to assume that it will decay by one of several pathways, as proposed in Scheme 3. The triplet CT intermediate can suffer annihilation in the triplet manifold either by back electron transfer to produce ^{3}DPC + amine or by triplet state product production. Alternatively, the triplet CT intermediate can decay by intersystem crossing into the singlet

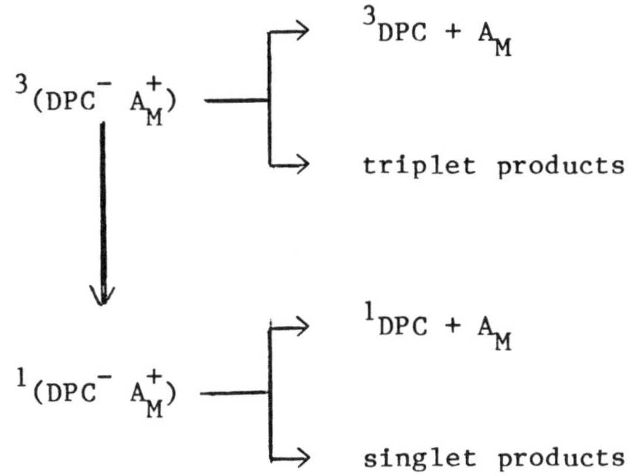

Scheme 3. Proposed Decay Pathway of the Triplet Charge-Transfer Intermediate Involving Carbene and Amine.

state manifold producing either a singlet charge-transfer complex, an ion pair or ylide. They could then collapse into either ^{1}DPC + amine or singlet reaction products. The production of either triplet or singlet products following the initial photoreduction of ^{3*}DPC will of course depend on the type of amine. It is expected, for example, that N-H insertion type products could be formed for the case of primary and secondary amines. In this way, ^{3*}DPC would appear to follow the mechanism proposed by Bethell [19,20,23] for the ^{1}DPC reaction with primary amines. It is emphasized, however, that the ability of an arbitrary electron donor to quench ^{3*}DPC will basically depend on two factors. At low ionization potential it is expected that a charge-transfer mechanism will dominate. This is the situation for the series of amines studied here as shown by the linearity of the log(^{3*}k) vs. IP plot as seen in Fig. 7. However, at high ionization potentials, the charge-transfer mechanism will become energetically disfavored and the rate of quenching is expected to be controlled by other mechanisms, such as abstraction and/or one step insertion processes as proposed for the quenching of ^{3*}DPC by alcohols, having IP > 9 eV (e.g., alkyl alcohols). In the latter case, it was found that quenching occured in alcohols since they have a labile O-H bond, thus permitting O-H insertion reactions, but not for ethers which have only strong C-H bonds [16,17].

3.4 Excited Triplet Carbene Reactions with Olefins

For isoprene and tetramethylethylene, which are two well known triplet scavengers, we find that ^{3*}DPC is quenched at a rate of 2.1 X 10^{9} $M^{-1}S^{-1}$ and 1.5 X 10^{8} $M^{-1}S^{-1}$, respectively. Since ^{3}DPC reacts with isoprene with a rate constant of 1.5 X 10^{6} $M^{-1}S^{-1}$ [5b], we find that the reactivity of the carbene has been enhanced by roughly a factor of 10^{3} when in its photoexcited triplet state. These results are significant because they show that the excited triplet state reacts efficiently with both "triplet" carbene scavengers (i.e., isoprene) and "singlet" carbene scavengers (i.e., methanol). ^{3*}DPC thus represents a carbene state that challenges the classification of the spin selectivity nature of carbene quenchers that have been used effectively to discriminate between the lower electronic spin states of the carbene.

In conclusion, we have found that the excited triplet diphenylcarbene has important new features compared with the ground triplet and lowest singlet state of diphenylcarbene. They include significant excess energy (E_T = 58 kcal/mole) and what is proposed to be central to its chemistry, the presence of an empty low-lying orbital which can accept electrons. ^{3*}DPC is found to be readily quenched by olefins, alcohols, and amines. A pronounced increase in reactivity is found for the excited triplet state relative to the ground triplet state in its reactions with olefins. However, unlike the ground triplet the excited triplet can be quenched by typical singlet carbene scavengers, e.g., amines and alcohols. The ordering of reactivity of ^{3*}DPC with amines is given by tertiary > secondary > primary, which follows inversely the ionization potential ordering. It is pointed out that the mechanism of reaction with amines is not the same as for its reaction with alcohols. Alcohol reactivities are not ordered by their ionization potentials, but rather by their relative acidities.

IV. CONCLUSION AND OUTLOOK.

Because of their importance as reactive chemical intermediates, the novelty of their reactions and their unusual spectroscopy, carbenes have been and will continue to be the focus of intense study. The study of intramolecular relaxation processes of carbenes by time resolved experiments provides important insight into the properties inherent to the carbene system. Employing picosecond spectroscopy the time evolution of the rapidly interconverting singlet and triplet carbene states can be mapped out, thus allowing the determination of the spin equilibration rate between singlet and triplet states which is of key importance in determining the nature and distribution of chemical products and providing an estimate of the energy separation between these states. Furthermore, environmental effects on the energetics and dynamics of the carbene system have been found. We expect that these effects will be also seen for other carbene derivatives, in addition to DPC. From the standpoint of characterizing carbenes, the extent to which solvent effects on carbene dynamics are expressed should thus be important. Because the medium affects the energy gap and spin equilibration rates of the ground triplet state and of the thermally accessible singlet state, the chemistry of the carbene system will in turn be affected. In this way one can effectively control the fate of the singlet and triplet states merely by choice of solvent. The study of optically excited states of carbenes is still in its infancy. However, unusual properties of the excited triplet carbene state are evident, and thus its examination offers an unique opportunity to test current chemical and physical theories of divalent carbon atom species.

ACKNOWLEDGEMENTS

We thank the National Science Foundation, the Air Force Office of Scientific Research and the Joint Services Electronics Program 29-82-K-0080 for their support of this work. We also express our thanks to Prof. N.J. Turro and to Dr. I.R. Gould for stimulating discussions and collaboration on the competitive quenching measurements. We also thank J. Langan and Dr. Y. Wang for their assistance in the picosecond measurements.

REFERENCES

1) a) W. Kirmse, Carbene Chemistry, 2nd ed.(Academic Press, New York,1971); b) R.A. Moss and M. Jones Jr., Eds., Carbenes, Vols. 1,2 (Wiley-Interscience, New York, 1975); c) H. Durr, Topics Curr. Chem., 87 (1975) 255; d) D. Bethell, in: Organic Reactive Intermediates, ed. S.P. McManus (Academic Press, New York, 1973); e) H.D. Roth, Acc. Chem. Res., 10 (1977) 85

2) a) L. Salem and C. Rowland, Angew. Chem. Int. ed. 11 (1972) 92; b) J. Metcalfe and E.A. Halevi, Perkin Trans. II, (1977) 634; c) R. Hoffman, G.D. Zeiss and G.W. Van Dine, J. Am. Chem. Soc., 90 (1968) 1485; d)J.F. Harrison, R.C. Liedtke and J.F. Liebman, J. Am. Chem. Soc., 101 (1979) 7162; e) J.F. Harrison, Acc. Chem. Res., 7 (1974) 378; f) W.T. Borden and E.R. Davidson, Ann. Rev. Chem., 30 (1979) 125; g) R.K. Lengel and R.N. Zare, J. Am. Chem. Soc., 100 (1978) 7495.

3) a) R.W. Murray, A.M. Trozzolo, E. Wasserman, and W.A. Yager, J. Am. Chem. Soc., 84 (1962) 3213; b) R.W. Brandon, G.L. Closs, and C.A. Hutchinson, J. Chem. Phys., 37 (1962) 1878; c) R.W.R. Humphreys and D.R. Arnold, Can. J. Chem., 55 (1977) 2286.

4) G.L. Closs and B.E. Rabinow, J. Am. Chem. Soc., 98 (1976) 8190

5) a) K.B. Eisenthal, N.J. Turro, M. Aikawa, J.A. Butcher Jr., C. Dupuy, G. Hefferon, W. Hetherington, G.M. Korenowski, M.J.

McAuliffe, J. Am. Chem. Soc., 102 (1980) 6563; b) K.B. Eisenthal, N.J. Turro, E.V. Sitzmann, I.R. Gould, G. Hefferon, J. Langan, and Y. Cha, submitted for publication.
6) C. Dupuy, G.M. Korenowski, M.J. McAuliffe, W. Hetherington and K.B. Eisenthal, Chem. Phys. Lett., 77 (1981) 272
7) a) W.A. Gibbons and A.M. Trozzolo, J. Am. Chem. Soc., 88 (1966) 172; b) G.L. Closs, C.A. Hutchinson and B. Kohler, J. Chem. Phys., 44 (1966) 413; c) W.R. Ware and P.J. Sullivan, J. Chem. Phys. 49 (1968) 1445
8) D. Bethell, G. Stevens and P. Tickle, J. Chem. Soc. D. (1970) 792
9) P.P. Gasper, B.L. Whitsel, M. Jones Jr. and J.B. Lambert, J. Am. Chem. Soc., 102 (1980) 6108
10) P.P. Gasper, G.S. Hammond, in: Carbenes, eds. R.A. Moss and M. Jones Jr. Vol. 2, p. 320 (Wiley-Interscience, New York 1975)
11) M. Jones, Jr., W. Ando, M.E. Hendrisck, A. Kulczycki, Jr., P.M. Howley, K.I. Hummel and D.S. Malament, J. Am. Chem. Soc., 94 (1972) 7469
12) The data analysis was done using the subroutine STEPIT, written by C. Chandler. STEPIT is available from Quantum Chemistry Program Exchange (QCPE Program No. 307).
13) E.V. Sitzmann, J. Langan and K.B. Eisenthal, submitted for publication.
14) Y. Wang, E.V. Sitzmann, F. Novak, C. Dupuy and K.B. Eisenthal, J. Am. Chem. Soc., 104 (1982) 3238
15) E.V. Sitzmann, C. Dupuy, Y. Wang and K.B. Eisenthal, Picosecond Phenomena, eds. K.B. Eisenthal, R.M. Hochstrasser, W. Kaiser and A. Laubereau (Springer-Verlag: Berlin, 1982, Vol III).
16) E.V. Sitzmann, Y. Wang and K.B. Eisenthal, J. Phys. Chem., 87 (1983) 2283
17) E.V. Sitzmann, J. Langan and K.B. Eisenthal, Chem. Phys. Lett., 102 (1983) 446.
18) W. Kirmse, Annalen, 666 (1963) 9
19) D. Bethell A. Newall, G. Stevens and D. Whittaker, J. Chem. Soc. B., (1969) 749
20) D. Bethell, A. Newall and D. Whittaker, J. Chem. Soc. B., (1971) 23
21) D. Bethell and R. Howard, J. Chem. Soc. B. (1969) 745
22) a) K. Wiberg, Chem. Rev., 55 (1955) 713; b) E. Lewis and R. Grinstein, J. Am. Chem. Soc., 84 (1962) 1158; c) F. Hawthorne and E. Lewis, J. Am. Chem. Soc., 80 (1958) 4296
23) D. Bethell, J. Hayes, A.R. Newall, J. Chem. Soc. Perkin II (1974) 1037.
24) P.C. Wong, D. Griller and J.C. Scaiano, Chem. Phys. Lett. 83 (1981) 69
25) P.C. Wong, D. Griller and J.C. Scaiano, J. Am. Chem. Soc. 103 (1981) 5034
26) P.C. Wong, D. Griller and J.C. Scaiano, J. Am. Chem. Soc., 104

(1982) 6631

27) D. Griller, C.R. Montgomery, J.C. Scaiano, M. S. Platz and L.M. Hadel, J. Am .Chem. Soc., 104 (1982) 6814

28) L. M. Hadel, M.S. Platz and J.C. Scaiano, Chem. Phys. Lett., 97 (1983) 446

29) B. E. Brauder, P.R. Grasse, K.J. Kaufmann and G.B. Schuster, J. Am. Chem. Soc., 104 (1982)6814

30) J.I. Franklin, J.C. Pillard, H.M. Rosenstock, J.T. Herron, K. Droxl and F.H. Field, Ionization Potentials, Appearance Potentials and Heats of Formation of Gaseous Positive Ions, NSRDS-NBS No. 26 (Nat.Bur.Std. Washington, 1964)

31) a) D. Rehm and A. Weller, Ber. Bunsenges Physik. Chem., 73 (1969) 834; b) F. Scandola, V. Balzani and G.B. Schuster, J. Am. Chem. Soc., 103 (1981) 2519; c) F.M. Martins, J.W. Verhoeven, R.A. Gase, U.K. Pandit and Th. J. de Boer, Tetrahedron, 34 (1978) 443

32) a) S.G. Cohen and A.D. Litt., Tetrahedron Lett. (1970) 837; b) S.G. Cohen and J.B. Guttenplan, Tetrahedron Lett. (1968) 5353; c) J.B. Guttenplan and S.G. Cohen, J. Am. Chem. Soc., 94 (1972) 4040; d) H. Parola, A.W. Rose and S.G. Cohen, J. Am. Chem. Soc., 97 (1975) 6202; d) P.J. Wagner and R.A. Leavitt, J. Am. Chem. Soc., 95 (1973) 3669; e) S. Inbar, H. Linschitz and S.G. Cohen, J. Am. Chem. Soc., 102 (1980) 1419; f) J.B. Guttenplan and S.G. Cohen, Tetrahedron Lett. (1972) 2163

33) a) J.D. Simon and K.S. Peters, J. Am. Chem. Soc., 103 (1981) 6403; b) K.S. Peters, S.C. Frielich and C. G. Schaeffer, J. Am. Chem. Soc., 102 (1980) 5701; c) C.G. Schaeffer and K.S. Peters, J. Am. Chem. Soc., 102 (1980) 7566.

PART II ENERGY RELAXATION AND MOLECULAR MOTIONS IN CONDENSED MEDIA

PHOTOCHEMICAL ISOMERIZATION VIEWED AS A MODEL FOR ACTIVATED BARRIER CROSSING IN SOLUTION

G.R. Fleming, D.H. Waldeck, K.M. Keery and S.P. Velsko

Department of Chemistry and James Franck Institute,
The University of Chicago, 5735 S. Ellis Avenue,
Chicago, Illinois 60637, U.S.A.

ABSTRACT. Measurements of the influence of solvent on the rate of photochemical isomerization are described. The results are discussed in terms of models of one-dimensional activated barrier crossing. The possible influence of a frequency dependent friction is suggested for isomerizations with sharp barriers.

BARRIER CROSSING IN SOLUTION

Introduction

The idea that chemical reactions can be modeled as an activated crossing over a potential barrier is, of course, an old one. Recently there has been considerable interest in the description of the influence of solvent on such barrier crossing processes both from the theoretical [1-5] and experimental points of view [6-12]. The solvent has two antagonistic roles; it promotes the reaction by providing the solute with sufficient energy to overcome the barrier, but if the reaction involves large amplitude motion the frictional forces exerted by the solvent will also impede the reaction. The balance of these two forces in determining the rate of the barrier crossing depends on the value of the friction coefficient (or alternatively on the collision rate) and on the curvature of the potential surface. At low values of friction the rate should increase with increasing friction until a maximum value (which is expected to be less than the transition state theory value of the rate) is reached. The rate then turns over and begins to decrease with increasing friction.

K. B. Eisenthal (ed.), Applications of Picosecond Spectroscopy to Chemistry, 67–77.

In our work we have studied the viscosity, temperature and pressure dependence of photochemical isomerization in solution. Our hope was that these readily measurable processes would provide model systems with which to test the concepts and theories mentioned above. We studied the isomerization of diphenyl butadiene (DPB) on the excited singlet surface in two different classes of solvents; nonpolar [9] and polar [11], and the isomerization of DODCI on the excited state surface [8] and the return isomerization on the ground state surface [10]. The large time scale separation (ns to ms) makes it possible to measure both processes in the same molecule. In carrying out this program we had in mind the following questions: To what extent is a one-dimensional barrier crossing picture applicable to the photoisomerization of large molecules in solution? What are the relative time scales of the observed isomerization and the correlations among forces exerted by the solvent? How accurate are simple hydrodynamic models for the microscopic motions involved in isomerization?

Experimental

The fluorescence lifetimes of DPB and DODCI were measured by the time correlated single photon counting technique [9]. The decays fit well to single exponentials in all cases. Lifetimes were measured in a range of solvents and over a range of temperatures. In the case of DPB experiments were also carried out at pressures up to 3 kbar in octane solution. Quantum yields were obtained from integrated fluorescence spectra. The return isomerization of DODCI and the initial photoisomer yield were measured via conventional flash photolysis using a 20 μs flash from a flash lamp pumped dye laser.

The Internal Barrier

If we assume that the isomerization rate takes the form

$$k_{iso} = F(\eta)\exp(-E_0/RT) \qquad (1)$$

where $F(\eta)$ is a universal function of viscosity and E_0 is an intrinsic molecular barrier height, then by plotting $\ell n(k_{iso})$ vs 1/T at constant viscosity one obtains a value for E_0. In a given solvent type the implicit assumption that E_0 is not a strong function of solvent appears to be a good one. For example, for DPB in alkanes we find that the reduced isomerization rate k^*_{iso} (= $k_{iso}\exp(E_0/RT)$) is a universal function of viscosity regardless of whether viscosity is changed by varying solvent, temperature, or pressure [9]. The values for the internal barrier height for the systems studied are given in Table I. Note the much larger barrier for ground state return isomerization than for excited state isomerization in DODCI, and the sharp decrease in barrier for DPB (excited state) in alcohol vs alkane solvent.

The twisted configuration of the diphenyl polyenes is believed to have considerable charge transfer character [13] and should be stabilized by polar and polarizable solvents. The absorption and emission maxima are unaltered by the solvent type consistent with the ideas that (1) the energy of the B_u state is little influenced by solvent and (2) that both absorption and emission take place from this state.

Given that k^*_{iso} (or $F(\eta)$) seems to be a universal function of viscosity it seems reasonable to confront the various theories of barrier crossing with our data, making the hydrodynamic approximation for the friction. By the hydrodynamic approximation we simply mean that the correlation time for the relevant motion is assumed inversely proportional to the solvent shear viscosity.

Brownian Motion Theories of Barrier Crossing

At the present state of knowledge about liquid state dynamics we have decided to begin the interpretation of our data through statistical mechanical models of barrier crossing. These models idealize the reaction as motion on a one dimensional harmonic potential surface, and describe solvent friction by means of a single (constant) parameter. Clearly this approach ignores a great deal of the personality of the solvent and solute. In addition the relationship of the parameters derived by fitting the theoretical expressions to experimental data to molecular properties - such as vibrational frequencies - is not obvious since the coordinate along which the reactive motion occurs is not clearly defined. These comments are not intended to imply that the use of simple Brownian motion pictures for interpretation of barrier crossing processes is without merit. On the contrary, they provide important information on the relative time scales of 'free' motion on the potential surface and of the solvent frictional forces. The concept of reaction coordinate and of the transition state,so valuable in the interpretation of gas phase reactions, should also be applicable to solution phase reactions, and it seems possible that even a full multidimensional theory would still have a single 'effective' coordinate along which the reaction proceeds, although coupling to other modes would be an important feature of such a theory.

The transition state value for the rate of barrier crossing is given by

$$k_{TST} = \frac{\omega_0}{2\pi} \exp(-E_0/RT) \qquad (2)$$

where ω_0 is the frequency of the reactant well minimum, i.e. the 'attempt frequency' and E_0 the barrier height. Kramers expression for the barrier crossing rate [14] can be written as

$$k_{Kr} = k_{TST} \left[\frac{1}{2\omega'\tau_v} \left[(1 + (2\omega'\tau_v)^2)^{1/2} - 1 \right] \right] \quad (3)$$

where ω' is the curvature at the barrier maximum of τ_v is the velocity relaxation time of the relevant coordinate. The term in square brackets is always less than unity and approaches unity for large $\tau_v\omega'$ and zero for small $\tau_v\omega'$. In Kramers theory the factor in square brackets accounts for the decrease in overall rate caused by repeated crossing and recrossing of the barrier i.e. it corrects the transition state theory rate for the influence of collisions in the region of the barrier. τ_v is related to the friction coefficient ζ by

$$\tau_v = \mu/\zeta \quad (4)$$

with μ being the effective mass of the coordinate. If a hydrodynamic model is assumed for ζ, then

$$\zeta \propto \eta \quad (5)$$

and τ_v becomes inversely proportional to viscosity. Then, for fixed ω', at low viscosity equation (3) reduces to equation (2) and at high viscosity the Kramers rate approaches zero. In fact for $\omega'\tau_v \ll 1$ (3) simplifies to

$$k_{SM} = k_{TST}\, \omega'\tau_v \propto \frac{1}{\eta} \exp(-E_0/RT) \quad (6)$$

Expression (6) is called the Smoluchowski limit of Kramers equation and might be expected to apply at high viscosity or for a potential surface which is rather flat (small ω').

Expression (3) has at least one serious shortcoming, in that it predicts $k = k_{TST}$ at $\eta = 0$. In fact the rate is expected to approach zero as η approaches zero. The reason for this is the collision rate approaches zero as η approaches zero and so at low enough viscosity there will be insufficient collisions to provide the necessary energy to overcome the barrier. Kramers himself realized this problem and derived a low friction expression in which the rate increases linearly with viscosity [14]. The position of the turnover is of considerable interest, recent stochastic trajectory calculations by Singer and Bagchi [15] place the turnover in the region of 0.05 cp for DPB. Jonas and coworkers have (using high pressure nmr techniques) observed the turnover in the ring isomerization of cyclohexane [12] and Millar and Eisenthal have some evidence for a turnover in binaphthyl [16]. We have constructed a cell which will enable us to reach viscosities of 10^{-2} cp and hope to build up a detailed picture of the dynamics in this inertial or energy controlled regime [17,18].

When the curvature of the barrier top is very large the reactive motion may be sufficiently rapid that the solvent is unable to respond in its 'normal' low frequency fashion. In other words, for very sharp barriers, the friction may no longer be Markovian. Grote and Hynes have provided a theoretical discussion of this problem [1]. These authors suggest that when the forces on the isomerization coordinate relax on a time scale comparable to the velocity relaxation time (τ_v), the frequency dependent (or time dependent) friction should be evaluated at the reactive frequency $\lambda_r = 2\pi k^*_{iso}$. Grote and Hynes find a self consistency relationship between k^*_{iso} and $\zeta(\omega)$ which for unequal ω_0 and ω' reads:

$$k^*_{iso} = \frac{1}{2\pi}\,\frac{\omega_0^2}{2\pi k^*_{iso} + (\omega_0/\omega')[\hat{\zeta}(2\pi k^*_{iso})/\mu]} \tag{7}$$

It should be noted that this expression cannot be valid at low friction since (7) approaches the transition state value at zero friction, i.e. $2\pi k^*_{iso} \sim \omega_0$. Nitzan and Carmelli [19] have recently provided a discussion of non-Markovian effects.

The influence of barrier height with respect to k_BT is of interest. Kramers theory is generally presented as only being valid for $E_0 \gg k_BT$. Theoretical studies of double minimum potentials show considerable time dependence in the rate for $E_0 \sim k_BT$ [2]. However, in the case of isomerization the second minimum contains a sink. This alters the behavior considerably since the time dependence referred to above arises essentially from rebounding. Bagchi, Fleming and Oxtoby [20], and more recently Bagchi, Singer and Oxtoby [15] have considered the situation where there is no barrier at all and the initial state is not in thermal equilibrium. In general, as expected, the decay is nonexponential and for a position dependent sink (e.g. Gaussian [20]) a fractional quantum yield dependence on viscosity is found [20]. At low viscosities the position dependent sinks (as opposed to pinhole or delta function-like sinks) exhibit exponential decay [20]. Stochastic trajectory calculations find a nonmonotonic viscosity dependence in the low viscosity region [21]. More recently Bagchi and Singer have carried out very detailed stochastic trajectory calculations for a range of barriers and potentials [15]. These calculations should provide the definitive picture of barrier crossing models derived from the Fokker-Planck equation.

Experimental Results

We analyze our data in two ways, first by comparing the activation energy of the nonradiative process, E_{nr}, with E_η the solvent viscosity activation energy, and secondly by fitting plots of k^*_{iso} vs η to the various theoretical expressions and to one empirical relation originally suggested by Gegiou et al. [22]

$$k^*_{iso} = \frac{A}{\eta^a} \tag{8}$$

Expression (8) fits our data well under all circumstances, however the value of a varies from 0.26 (DODCI ground state) to 0.92 ± 0.05 for DPB (alcohol solvents) (see Table I).

Table I

Summary of Barrier Crossing Data

System	Barrier height (kcal)	Effective frequency at 1 cp (s^{-1})	$\omega'\tau_v$ at 1 cp	a^b
DODCI				
GS	13.7	4×10^{12}	7.5	0.26
ES	2.7	8×10^{11}	0.8	0.43
[Rotation]	0	8×10^{9}	...	0.99
DPB				
ES alkanes	4.7	1.5×10^{12}	0.71	0.69
ES alcohols	∿0.5	1.5×10^{10}	...[a]	0.92
Fav 2R				
ES (alcohols) (Ref. 7)	∿0.2	9×10^{10}	...[a]	0.98

[a]In the Smoluchowski limit only the product $\omega_o\omega'\tau_v$ is significant (Ref. 11).
[b]Obtained from $k\propto(1/\eta^a)$.

As far as fitting to expression (3) or (6) our data fall into two groups. The DPB in alcohols data fit (3), (6) or the expression of Skinner and Wolynes [3] very well [11]. The best fit lines for the three expressions are superimposable and the only region where the three expressions should be identical is in the high friction or Smoluchowski limit [11]. The very small barrier in DPB (alcohols) from isoviscosity plots is confirmed by the finding that the nonradiative rates in particular alcohols measured as a function of temperature lie on the same smooth curve (vs viscosity) as the nonradiative rates measured in the same series of alcohols at room temperature. In this system we find $E_{nr} = E_\eta$ within experimental error [11] (as predicted by (6) if $E_o \ll E_\eta$).

The two DODCI data sets and the DPB (alkanes), however, fit poorly to expression (3) [9,10]. If the low viscosity rates are fit to the Kramers expression, then the rates at high viscosities are consistently larger than theory predicts. On the other hand, if high viscosity points are fit to the Kramers expression then theory predicts too low values for rates at low viscosities. In other words, Kramers theory does not have the correct curvature necessary to describe the experimental results at low and high viscosities. In these systems E_{nr} is always greater than E_{η} (but less than the sum of E_{η} and E_{0}). A typical example is the DODCI excited state data shown in Figure 1. In the case of DODCI we

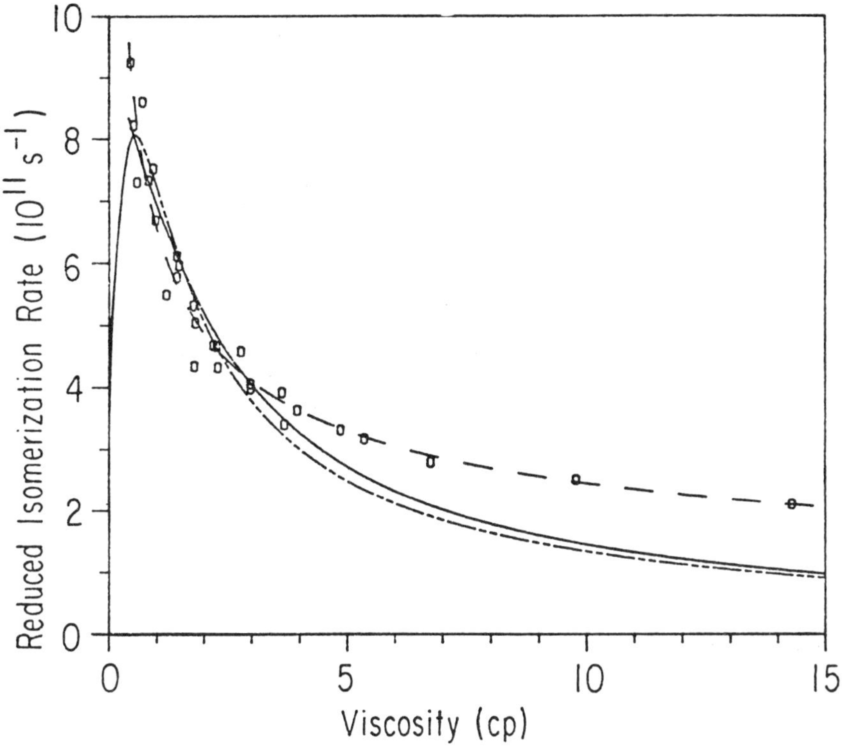

Fig. 1. Reduced isomerization rate vs viscosity for the excited state of DODCI. The solid line is a fit of Kramers equation to all the points. The dotted line of the Skinner and Wolynes function [3] to all the points. The dashed line is a fit to expression (7). The value of a is 0.43. An excellent fit is also obtained to the ground state data with a = 0.26.

attempted to improve the fit by replacing viscosity in (3) by directly measured rotational reorientation times [23] (a full description of the procedure is given in [10]). On fitting the reduced reorientation times (again with the intrinsic temperature dependence removed) we find that the data fit well to equation (8) with a = 0.99, i.e. Stokes law is obtained. In view of this it is hardly surprising that using the reduced reorientation times as a measure of the friction does not improve the fit.

Before attempting an explanation of the apparent failure of Kramers expression to describe these three isomerizing systems, a few general comments on the data in Table I are in order. First, within a single molecular species, a smaller barrier implies a smaller reduced rate or effective frequency k^*_{iso}. Second, a larger barrier implies a less strong dependence on viscosity as measured by a or by the product $\tau_v\omega'$. In barrierless (or low barrier) cases the Smoluchowski limit of a $1/\eta$ dependence is approached. (We also find this for the Fast Acid Violet (FAV) data of Tredwell and Osborne [7].) It seems likely that as the barrier gets smaller ω' will decrease and our data are consistent with the notion that the high barrier cases are more potential controlled or inertial, and the low barrier cases are more diffusive or viscosity controlled.

Bagchi and Oxtoby [4] and more recently Rothenberger, Negus and Hochstrasser [24] have calculated the isomerization rate via the Grote and Hynes expression (7). A general feature of the Bagchi and Oxtoby calculation, using pure slip or pure stick boundary conditions is that the viscosity dependence is well described by k^*_{iso}/η^a with $0.1 < a < 1.0$. The result of a particular calculation, compared with Kramers equation for the same barrier frequency are shown in Figure 2. As with the experimental results the Grote-Hynes theory shows a much weaker dependence on viscosity at higher viscosity values; eventually the rate reaches a constant value [4]. The physical reason for this effect is that for sharp barriers, that part of the friction which comes from low frequency motion does not effect the reactive motion across the barrier.

Although, in the authors' opinion, it is premature to attempt quantitative fits of our data using the Grote-Hynes theory (because of lack of data on viscoelastic response at high frequencies and because of lack of good experimental values for the barrier curvature), a frequency dependent friction based on the sharpness of the intramolecular potential barrier does seem to provide the most consistent explanation of our results.

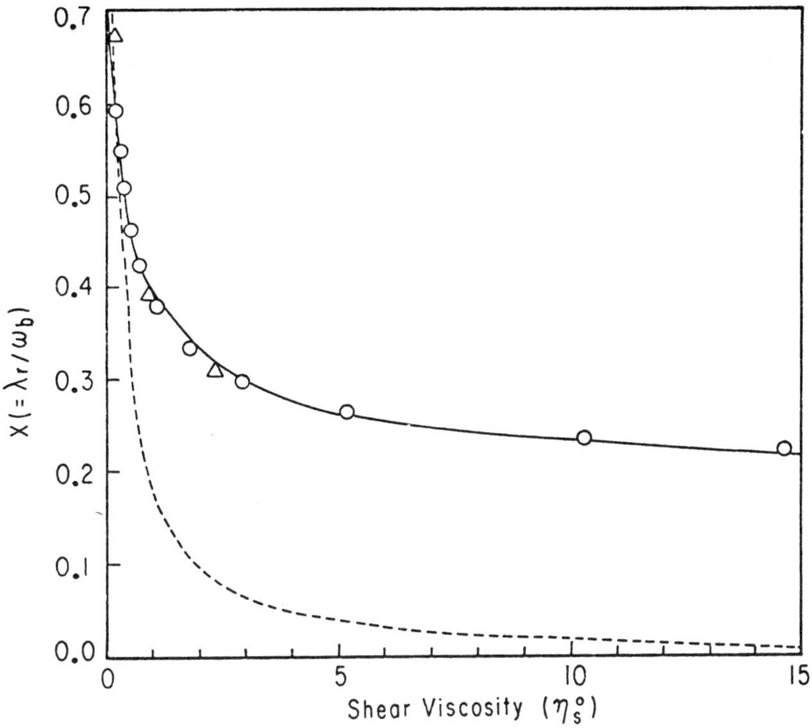

Fig. 2. The rate parameter $x(= \lambda_r/\omega')$ vs shear viscosity η_s^0 for DPB in octane (Δ) and in butanol (O). $\omega' = 6.0 \times 10^{12}\ s^{-1}$. The solid line is a fit to equation (7) with a = 0.25. The dashed line is the Kramers theory result with the same value of the barrier frequency. From reference (4) with permission.

Summary

We have observed rates of photochemical isomerization both in accord with simple theoretical predictions and at variance to these theories. We suggest that the shape of the intramolecular potential surface is responsible for the two types of behavior. We feel that our data favors an explanation involving frequency dependent friction, but further work is clearly necessary before the breakdown of the Kramers equation can be definitely stated.

As a final point we note that the process that has the slowest observed rate; DODCI (ground state) (∿ 3 ms, ethanol, 25°C) apparently requires a frequency dependent friction to explain the observed behavior, whereas the fastest rate we have studied, DPB excited state isomerization (∿ 60 ps, ethanol, 25°C) only requires that we use the zero frequency limit of the friction. This arises because the effective time scale for friction is determined by k_{iso}^* not k_{iso} itself. These results imply that high frequency motions and the frequency response of the medium could have important implications on chemical reactions which occur at modest rates.

ACKNOWLEDGEMENT

This work was supported by a grant from the National Science Foundation. We would like to thank David Oxtoby and Biman Bagchi for many hours of discussion, and for permission to reproduce Figure 2.

REFERENCES

1. Grote, R.F. and Hynes, J.T. 1979, J. Chem. Phys. 69, p. 2143.
2. Montgomery, J.A., Chandler, D. and Berne, B.J. 1979, J. Chem. Phys. 70, p. 4056.
3. Skinner, J.L. and Wolynes, P.G. 1978, J. Chem. Phys. 69, p. 2143.
4. Bagchi, B. and Oxtoby, D.W. 1983, J. Chem. Phys. 78, p. 2735.
5. Carmeli, B. and Nitzan, A. 1983, Phys. Rev. Lett., 51, p. 233.
6. Hochstrasser, R.M. 1980, in Picosecond Phenomena II, p. 259.
7. Tredwell, C.J. and Osborne, A.D. 1980, J. Chem. Soc. Faraday Trans. II, 76, pp. 1627-1638.
8. Velsko, S.P. and Fleming, G.R. 1982, Chem. Phys. 65, p. 59.
9. Velsko, S.P. and Fleming, G.R. 1982, J. Chem. Phys. 76, p. 3553.
10. Velsko, S.P., Waldeck, D.H. and Fleming, G.R. 1983, J. Chem. Phys. 78, p. 249.
11. Keery, K.M. and Fleming, G.R. 1983, Chem. Phys. Lett., 93, p. 322.
12. Hasha, D.L., Eguchi, T. and Jonas, J. 1982, J. Amer. Chem. Soc., 104, p. 2290.
13. Baraldi, I., Bruin, M.C. Momicchioli, F. and Ponternini, G. 1980, Chem. Phys. 52, p. 415.
14. Kramers, H.A. 1940, Physica, 7, p. 284.
15. Singer, S. and Bagchi, B. unpublished results.
16. Eisenthal, K.B. and Millar, D., private communication.
17. Grote, R.F. and Hynes, J.T. 1982, J. Chem. Phys. 77, p. 3739.
18. Carmeli, B. and Nitzan, A. 1982, J. Chem. Phys. 76, p. 5321.
19. Carmeli, B. and Nitzan, A. 1983, J. Chem. Phys. 79, p. 393.
20. Bagchi, B., Fleming, G.R. and Oxtoby, D.W. 1983, J. Chem. Phys. 78, p. 7375.
21. Bagchi, B., Singer, S. and Oxtoby, D.W. 1983, J. Chem. Phys., in press.
22. Gegiou, G., Muzkat, K.A. and Fisher, E. 1968, J. Amer. Chem. Soc. 90, p. 12.

23. Waldeck, D.H. and Fleming, G.R. 1981, J. Phys. Chem. 85, p. 2614.
24. Rothenberger, G., Negus, D.K. and Hochstrasser, R.M. 1983, J. Chem. Phys., in press.

DYNAMICS OF THE RADIATIONLESS RELAXATIONS IN TRIPHENYLMETHANE MOLECULES AND THE INFLUENCE OF SOLVENT-SOLUTE INTERACTIONS

Villy Sundström and Tomas Gillbro

Department of Physical Chemistry
University of Umeå
S-901 87 Umeå, Sweden

ABSTRACT

We have used picosecond transient absorption techniques to study the radiationless relaxation processes of several triphenylmethane molecules in n-alcohol solution. A kinetic model for the relaxations is presented and the dynamics of a process identified as a large amplitude motion of the phenyl rings is further studied. The influence of temperature, solvent viscosity and solvent-solute interactions on this process is investigated. It is concluded that the large amplitude motion can be described as a rotational motion with very low internal energy barrier, largely controlled by diffusion with slip or sub-slip boundary conditions.

1. INTRODUCTION

The triphenylmethane (TPM) molecules are a class of dye molecules that has been extensively studied through the years (1-9). They display highly efficient radiationless relaxations which quenches the fluorescence emission to such an extent that even under laser excitation some dyes are apparently non-fluorescent. The quantum yield of fluorescence and excited state lifetimes are strongly viscosity and temperature dependent. Excitation of a TPM-molecule to its S_1 electronic state is believed to induce a synchronous rotation of the phenyl rings to a new equilibrium position (1) which has a high rate of radiationless relaxation. The viscous drag of the solvent hinders the rotation of the phenyl rings. High solvent viscosity therefore results in a high fluorescence quantum yield. Earlier studies were performed in solvents or solvent mixtures of high viscosity.

K. B. Eisenthal (ed.), Applications of Picosecond Spectroscopy to Chemistry, 79–98.

We have undertaken a thorough study of several TPM molecules (see Figure 1) in n-alcohol solutions in order to obtain a comprehensive picture of the relaxation mechanism and the effects of the solvent on the relaxation dynamics.

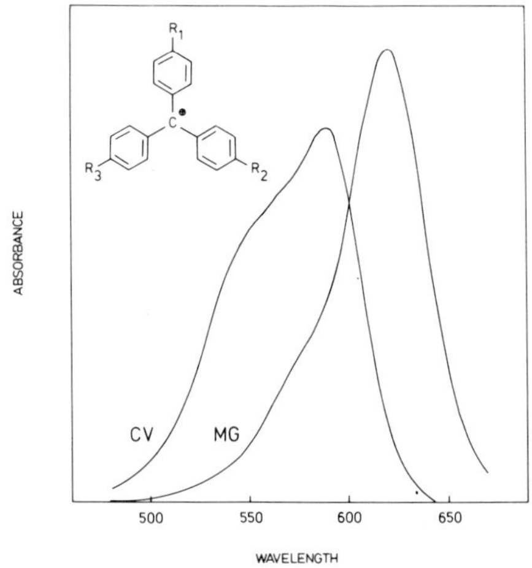

Figure 1. Structure and typical absorption spectra of the studied molecules.

Crystal violet CV, $R_1 = R_2 = R_3 = N(CH_3)_2$;
ethylviolet EV, $R_1 = R_2 = R_3 = N(C_2H_5)_2$;
parafuchsin PF, $R_1 = R_2 = R_3 = NH_2$;
malachite green MG, $R_1 = R_2 = N(CH_3)_2$, $R_3 = H$;
brilliant green BG, $R_1 = R_2 = N(C_2H_5)_2$, $R_3 = H$.

Our picosecond time-resolved absorption recovery data allow us to suggest a kinetic model (11) for the relaxation processes. From this model we identify a radiationless process which corresponds to a large amplitude rotational motion of the phenyl rings. The major goal of this work is to study the effect of solvent viscosity and solvent-solute interactions on this process. We also compare our experimental results with theoretical predictions obtained by using some current theories for isomerization dynamics in solution (12-14).

2. PHOTOPHYSICS OF TPM MOLECULES IN N-ALCOHOL SOLUTION

2.1 Experimental techniques

Two slightly different transient-absorption techniques were used to study the relaxation dynamics of the TPM molecules shown in Figure 1. The principle of these two techniques is that of flash-photolysis; the sample molecules are excited with a picosecond light-pulse and the resulting absorbance changes are detected with a much weaker probe-pulse or continuous light. In the technique schematically illustrated in Figure 2 A, the traditional pump and probe technique (15), both excitation and probe light are derived from the same pulse-source, a synchronously pumped, mode-locked and cavity-dumped dye laser.

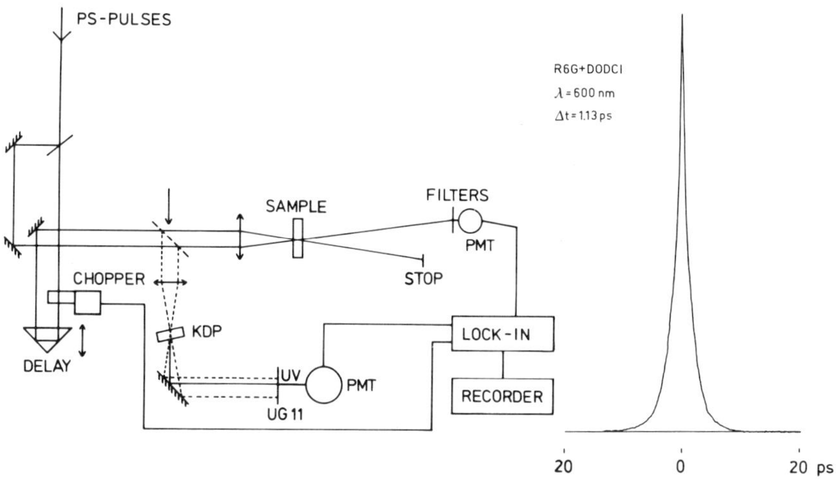

Figure 2 A. Experimental arrangement for transient absorption measurements with λ_{exc} and λ_{probe} identical. To the right: typical pulse used.

This implies that the wavelength of the probe-pulses is bound to be the same as that of the excitation-pulses. Light in the interval 550-675 nm can easily be generated by using the three dyes R110, R6G and RB. Excitation and probe-pulse trains are formed by dividing the dye laser beam into two parts with an intensity ratio of ca 95/5. The polarizations of both beams are controlled by prism polarizers and a Soleil-Babinet compensator. Time-resolution is accomplished by delaying the analysing pulses, relative to the excitation pulses in a step-motor controlled optical delay line. The energy content of the excitation-pulse at the sample, ca 5 nJ, results in an excitation degree of ca 1 %. This necessitates modulation of the excitation beam at a low rate, ca 1000 Hz, and phase-sensitive detection of the analyzing light. A signal-to-noise ratio better than 200:1 is obtained with this method. It is therefore especially well suited for precise kinetic measurements. To measure transient absorption spectra with this technique is difficult since both excitation and probe wavelengths change when the dye laser wavelengths is tuned. In order to overcome this problem the technique illustrated in Figure 2 B has been developed (16-18).

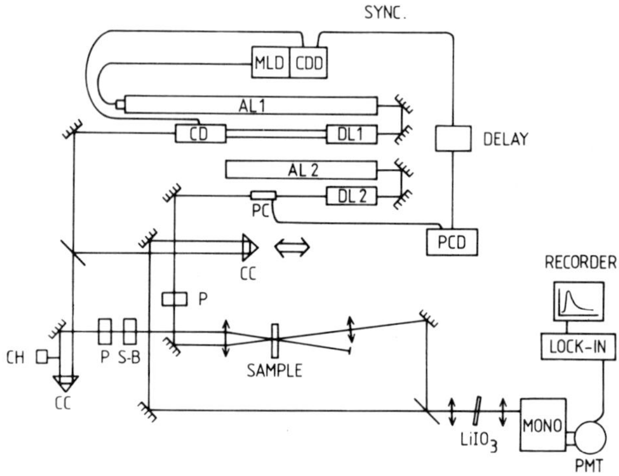

Figure 2 B. Experimental arrangement for transient absorption measurements with independent λ_{exc} and λ_{probe}.

Here, the probe light is generated in an independent dye laser (DL2) pumped by a small CW-Ar^+-laser (AL2). Normally, an average power of 50-100 mW from DL2 is used. To reduce the power incident on the sample the probe-beam is modulated by a Pockels cell (PC) into a train of ca 20 ns long pulses synchronized at the sample to the picosecond pulses by means of an electric delay line of proper length. Polarizing optics is used in both excitation and probing beams to gain full control over the directions of polarization. The modulation of sample absorbance caused by the excitation pulses is viewed by the probe-ligth and results in a modulation of its intensity,but the picosecond temporal information is not directly available since it is stored on a high back-ground. This back-ground is removed by sum-frequency generation with a small part of the ps-pulses in a non-linear $LiIO_3$ or KDP crystal, followed by phase sensitive detection of the resulting UV-signal. Sensitivity and signal-to-noise ratio of this technique are comparable to those of the first method. This technique is particularly useful for resolving multiple exponential decays and to obtain absorption spectra of new transients (18).

2.2 Photophysics of TPM molecules.

The measured absorption recovery kinetics of the studied TPM molecules at a particular wavelength can in most cases be fitted to a double exponential decay with a major fast component of lifetime τ_1

and a slower component of lower intensity having the lifetime τ_2 (11). The molecules fall into two chategories (A and B) of which CV and MG respectively, are typical members. The members of the first group, (A) consisting of CV, EV and PF, are characterized by an absorption spectrum having a pronounced short-wavelength shoulder (see Figure 1) and wavelength dependent lifetimes τ_1 and τ_2; exciting into the short-wavelength shoulder results in increased relaxation lifetimes by about a factor of two as compared to excitation of the long-wavelength band. This wavelength dependence of τ_i is attributed to the existense of two ground-state conformers having different relaxation rates. MG and BG, lacking the shoulder in the absorption spectrum and displaying wavelength independent lifetimes, constitute the second group (B). It should be noted that the dyes belonging to the first group have identical substituents in all three phenyl rings (see Figure 1) while the dyes of the second group have one ring different from the other two. Thus, it seems as if this structural difference somehow causes the observed differences in the number of stable molecular conformations.

The τ_1 absorption recovery lifetime measured in the red part of the absorption spectrum of the molecules belonging to group A agrees well with fluorescence lifetimes obtained in the same solvents (9-11), and τ_2 is not observed in the fluorescence measurements (10-11). For the molecules belonging to group B there are no time-resolved fluorescence measurements available for n-alcohol solvents. In glycerol, however, there is good agreement between lifetimes obtained with absorption recovery and time-resolved fluorescence measurements, respectively (9,11).

The magnitude of the measured lifetime τ_1 is strongly correlated to the size of the aryl group (see Table I); lifetimes are increasing in the order PF < CV < EV and MG < BG respectively

Table I The influence of substituent group size and moment of inertia on the k_1 relaxation rate for TPM molecules in hexanol at 296 ± 1 K.

Dye	$V_p/Å^3$	$I_p \times 10^{45}$/kg m^2	τ/ps
PF	82	1.49	14 ± 1
CV	123	2.18	19 ± 1
EV	156	2.94	28 ± 1
MG	75, 123	1.46, 2.18	10 ± 1
BG	75, 156	1.46, 2.94	16 ± 1

Thus, by combining our transient absorption results with those obtained from time-resolved fluorescence measurements we arrive at the kinetic model shown in Figure 3. The measured short lifetime τ_1,

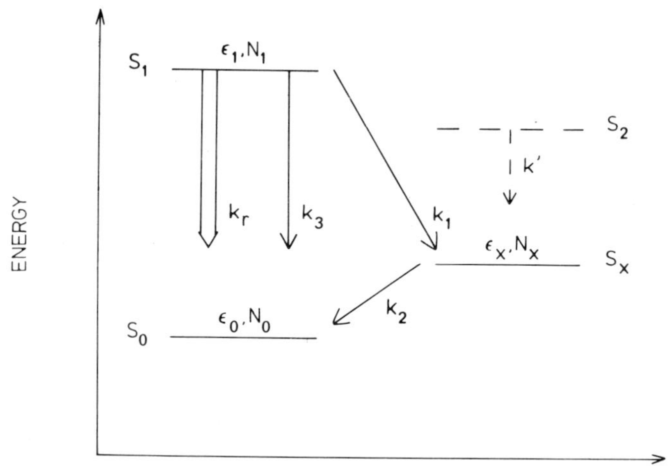

Figure 3. Kinetic model for the TPM relaxations.

is seen to represent a large amplitude rotational motion to a twisted eletronic state, whereas τ_2 is the following relaxation back to the initial ground state S_0. Details of this scheme have been published elsewhere (11).

3. EFFECTS OF TEMPERATURE, VISCOSITY AND SOLVENT ON THE τ_1 LIFETIME

Presently there is a vivid experimental (9-24) and theoretical (13, 14, 25-28) interest in the reaction dynamics of molecules in solution, thought to undergo radiationless relaxation through internal large amplitude motions. Experimentally the interest is focused on molecules, such as cyanines (19-20), stilbenes (23), and diphenylbutadiene (22), that undergo photochemical isomerization or conformational changes in the excited state. To understand the mechanism of the large amplitude motions that are responsible for these relaxations several aspects of the problem have to be studied. Information concerning the importance of intramolecular potential surface topology can be gained by studies of different types of molecules.

The rate of large amplitude motion depends critically on the kind of frictional forces excerted by the solvent molecules on the rotating group. Size and shape of the relaxing group, as well as nature of the solvent, are determining the friction. Thus, by studying the radiationless relaxations of molecules having various twisting groups in various solvents information about the frictional forces can be obtained. Moreover, the large amplitude motion can be imagined as a translational or rotational motion, or a mixture of both. Which one of these processes that is active in the relaxation process is probably important for the interaction between the moving group and the solvent molecules.

In the light of these considerations it is interesting to investigate the influence of various solvent parameters on the lifetime τ_1 of the TPM molecules, which according to the kinetic model represents a large amplitude twisting motion of the phenyl rings. We have done that in n-alcohol solutions where the viscosity was varied in two different ways. a) by varying the temperature of a single solvent, and b) by using the solvent series methanol-decanol at constant temperature. The lifetime data obtained in these studies were also used to perform Arrhenius plots and isoviscosity plots (Arrhenius plots at constant viscosity) of τ_1 in order to obtain information about the intramolecular activation energy for the large amplitude motion, E_0.

3.1 Activation energy and the potential barrier, E_0

In Figure 4 Arrhenius plots for EV in methanol, ethanol and hexanol are shown. The plots are seen to be straight lines within the

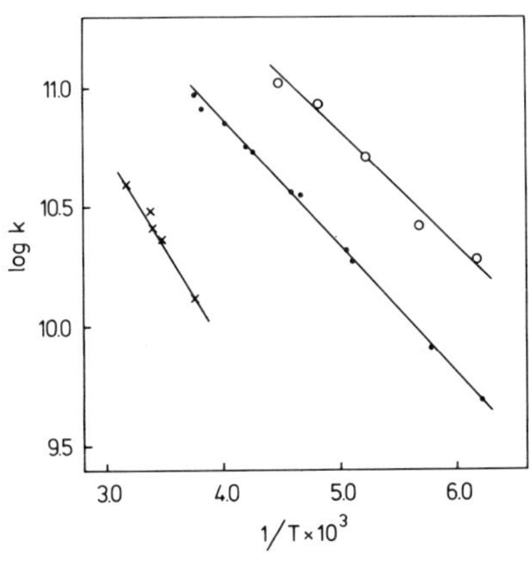

Figure 4. Arrhenius plots of EV in
o methanol; • ethanol; x hexanol.

measured temperature interval. This indicates that only one relaxation process is responsible for the measured rate constant. The activation energy for relaxation, E_a, is obtained from the slope of these plots. E_a and the activation energy of viscosity, E_η, for the n-alcohols used are collected in Table II. From these results it is evident that $E_a \leqslant E_\eta$. For the lower alcohols (MeOH, EtOH) E_a is very close to E_η, wheras for the higher n-alcohols E_a is significantly lower than E_η, by ca 5-8 kJ $mole^{-1}$. E_a increases slowly through the series of solvents MeOH - OctOH. The crucial observation here is that $E_a \leqslant E_\eta$, which indicates that there must be a low internal barrier for relaxation.

Table II Activation energy, E_a, for EV in ROH and activation energy of viscosity, E_η, for the same solvents.

Solvent	E_a/kJ $mole^{-1}$	E_η/kJ $mole^{-1}$	$\Delta E(E_a - E_\eta)$
MeOH	11.3 ± 0.8	10.6	+ 0.7
EtOH	10.0 ± 0.5	13.0	- 3.0
BuOH	12.6 ± 0.5	19.1	- 6.5
HexOH	15.1 ± 1.0	23.5	- 8.4
OctOH	20.1 ± 1.0	26.9	- 6.8

If the relaxation process is assumed to be a large amplitude motion as we have described it above it is reasonable to assume that it is controlled by a viscosity term and a temperature term. This is also the general form of the theoretical expressions used to calculate the rate constants (see below for details). Thus the measured rate constants can generally be expressed as

$$k = f(\eta)\cdot\exp(-E_0/RT) \qquad (1)$$

where $f(\eta)$ is a function of viscosity and E_0 is the internal barrier height. From this expression it is evident that E_0 should be obtained from an Arrhenius plot at constant viscosity. Two such plots are shown in Figure 5 for EV/ROH at 5 and 10 cp, respectively.

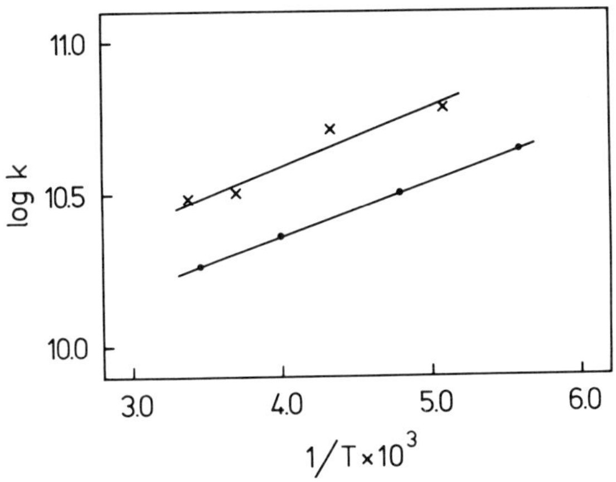

Figure 5. Isoviscosity plots of EV in n-alcohols at x 5 cp; o 10 cp.

A negative value, $E_0 = -3.5 \pm 0.2$ kJ mole^{-1} is obtained for both viscosities. This is quite an interesting result, it means that the rate of relaxation process increases with decreasing temperature, when the viscosity is kept constant. The following question now arises: Is this feature characteristic of the relaxation process itself or is it merely an apparent effect induced by some other factor? In an effort to clarify this point isoviscosity measurements were also performed in glycerol-water solutions. Here a "normal" positive activation energy $E_0 \approx 1 \pm 0.5$ kJ mole^{-1} was obtained. These results show that the internal energy barrier to relaxation in the excited state is very low, probably on the order of 0 - 1 kJ mole^{-1} (ca 0.2 kcal mole^{-1}).

At this point it is appropriate to discuss the implications of the very low E_0 value. The obvious question to answer is: is there at all a barrier on the potential surface separating the initial S_1 state and final twisted S_2 state? The alternative would be that the relaxation proceeds "downhill" throughout its path, which could occur if the S_0 and S_1 surfaces are strongly displaced. This is the model assumed by Förster and Hoffmann (1). The more or less arbitrary assumption that the relaxation rate from S_1 is proportional to the square of the displacement of the twisting coordinate from the initial position leads to the unvariable $\eta^{2/3}$ viscosity dependence of the relaxation lifetime and the never-observed $\exp(at^3)$ timedependence of the excited state relaxation rate. A different description of the relaxation process might lead to a more realistic result for this kind of potential surface situation.

In the case of strongly displaced S_0 and S_1 energy minima one would expect a time varying fluorescence spectrum, shifting towards the red as time proceeds. A time-integrated spectrum is expected to be broad, structurless and display a large Stokes shift. Nothing of this kind is observed experimentally (9-11); a fluorescence spectrum typical of many dye molecules is rather observed. Thus, this observations argues against the model with displaced energy minima and all-downhill relaxation, and favours the model where the initial state is a local shallow energy minimum, separate from the final state by very low barrier.

The magnitude of the potential barrier is an important parameter to consider when relating the experimental results to theory. The theoretical expressions of the rate constant for barrier passage which we will use (see below) are all derived under the assumption that the barrier is high relative to kT. For temperatures around room temperature this assumption implies that $E_0 >> 2.5$ kJ mole^{-1}. If this condition is not met the relaxation rate might not be characterizable by a well-defined rate constant, but rather behave as a time-varying rate. In our case E_0 is obviously of the same order of magnitude as kT, thus great caution is motivated. The observed kinetics are, however, in all cases well described by a double exponential decay and have several features of kinetics originating from two separate energy levels (11). Therefore we believe, as mentioned above, that the measured rate constant, labeled k_1, is characteristic of the large amplitude relaxation in the excited state. Consequently we will use the theoretical expressions derived with the above assumptions to calculate rate constants which will be compared to those experimentally obtained. This is done in the following paragraphs.

Here we have only used existing theories and fitted our experimental data to the theoretical expressions available in the literature with the assumptions mentioned above. Obviously it would be desirable to modify the various theories to include the low barrier case ($E_0 \approx 0$) as well.

3.2 Kramers' equation

In the intermediate and high friction region, Kramers' equation for activated barrier crossing takes the form (12)

$$k = \frac{\omega}{4\pi\omega'\tau_v}\left[(1 + (2\omega'\tau_v)^2)^{\frac{1}{2}} - 1\right] \exp(-E_0/RT) \qquad (2)$$

where ω and ω' are the frequencies of the potential near the initial state minimum and the top of the barrier, respectively; E_0 is the barrier height and τ_v is the momentum relaxation time for the large amplitude motion (system memory), which is related to the friction ξ and moment of inertia μ by eq (3),

$$\tau_v = \mu/\xi \tag{3}$$

if the test particle is treated as a Brownian particle. If in addition, the friction ξ between the test particle and the bath can be described by hydrodynamic theory we have

$$\xi \propto \eta \tag{4}$$

Specifically, if stick boundary conditions (BC) are prevailing

$$\xi = 8\pi r^3 \eta \tag{5}$$

for the rotational motion of a spherical particle of radius r. In the limit of perfect slip BC, $\xi = 0$ for a spherical particle. For particles of other shapes ξ is determined by the amount of solvent displaced during the reorientation. Hu and Zwanzig (28) have calculated values of ξ for ellipsoids of various axial ratio. Using the hydrodynamic expression for the friction, Kramers' equation can be written (22)

$$k = \frac{A}{B}\eta \left[\left(1 + \left(\frac{B}{\eta}\right)^2\right)^{\frac{1}{2}} - 1 \right] \exp(-E_0/RT) \tag{6}$$

where $A = \omega/2\pi$ and $B = 2\omega'\eta\tau_v$.

We have performed fits of eq (6) to our experimental data of k_1 for TPM dyes in n-alcohols. In these fits we assume that the exponential term, $\exp(-E_0/RT)$, equals unity on account of what was said about the barrier height above. The results of these fits to the data of EV/EtOH and CV/EtOH at $\lambda = 560$ nm (temperature varied) are shown in Figure 6. The data and fits are represented as τ versus η plots. When fitted to all data points eq (6) agrees reasonable well with experiment, except for the highest viscosities ($\eta \gtrsim 50$ cp) where longer lifetimes than measured are predicted. At low viscosity there are only minor deviations. The deviation at high viscosity (low T) is not a consequence of a failure of the assumption $E_0 = 0$, since that would result in slower theoretical relaxation rates and consequently an even larger deviation. Despite an almost linear viscosity dependence in a considerable viscosity interval the fitted values of the B-parameter indicate that the relaxation is not in the Smoluchowski limit (12), since $\omega'\tau_v$ (=B/2 at η = 1 cp) is not much less than unity. A fit to the low viscosity data points only, is steeper and has a lower value of the B-parameter, indicating closer correspondence to the diffusion limit. See section 3.4 below for a further discussion of this point. This failure of the Smoluchowski equation to describe these results, although the viscosity dependence of τ is nearly linear, can be traced back

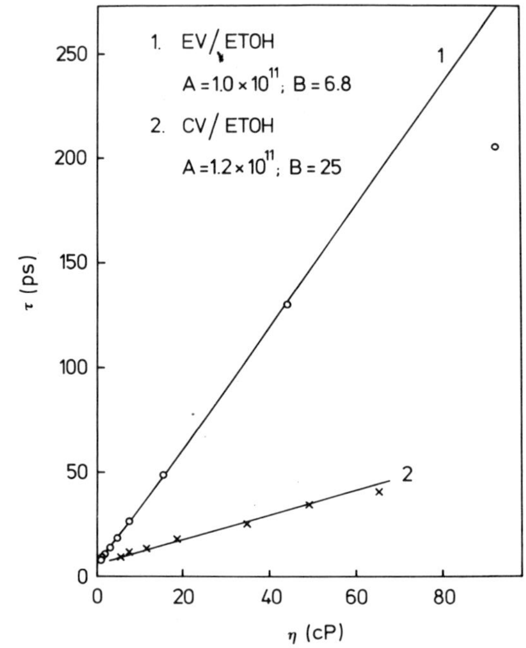

Figure 6. Viscosity dependence of the relaxation lifetime of 1. EV/EtOH; 2. CV/EtOH. The solid lines are fits to Kramers' equation.

to the non-zero intercept of the τ vs. η plots (see Figure 6). Extrapolating the experimental data to $\eta = 0$ yields a zero-viscosity intercept $\tau_0 \approx 3$-7 ps depending on the TPM molecule and the solvent. If one assumes that the measured, nearly linear viscosity dependence of τ is indicative of Smoluchowski behaviour (implying high friction) it is expected that extrapolating this dependence into the low viscosity (and friction) region would give a zero intercept. Obviously, this is not the situation at hand. On the contrary, the observed behaviour is very similar to that expected in the intermediate friction region, as judged by the relatively good fits to eq (6). If the intercept τ_0 is interpreted as the free-rotor correlation time, a property of the rotating group, this variation in τ_0 with solvent could be rationalized as a consequence of solvent association; association of, for example, one solvent molecule would result in a varying effective moment of inertia through the n-alcohol series. As discussed later on, see section 3.4 below, solvent association could also affect the viscosity dependence of the relaxation rate by changing the volume and shape of the relaxing group and thus affect the boundary conditions.

Similar fits of eq (6) to the experimental data of EV and CV in the n-alcohol series (at constant temperature) were also performed. The results for EV are shown in Figure 7 A. It is quite evident

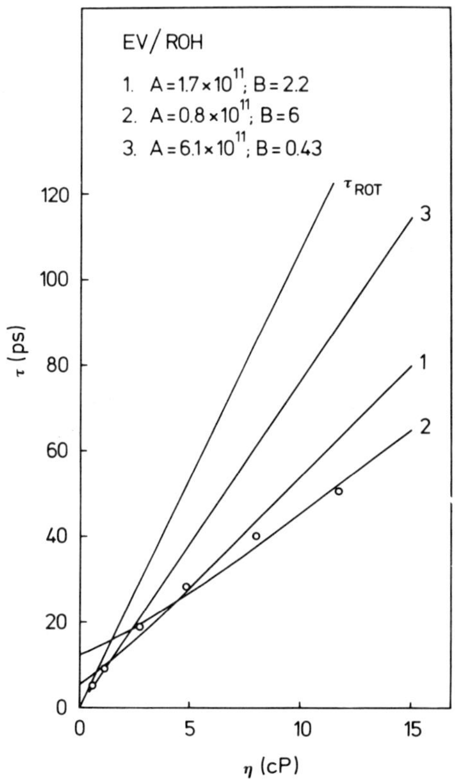

Figure 7 A. Viscosity dependende of the relaxation lifetimes of EV in the ROH series, at constant temperature. The solid lines 1-3 are fits to Kramers' equation yielding the various indicated values of the A and B parameters. τ_{ROT} is the theoretical viscosity dependence for a free rotational reorientation of the aryl group of EV, assuming slip BC.

that the fits can not reproduce the curvature of the measured data points. A fit over the full viscosity range (0.58 - 11.7 cp) used, shows that the experimental τ's are lower than those calculated in the low and high viscosity regions (Figure 7 A, fits 1 and 2). A fit to only the low viscosity data points yields a much steeper fit with a B-value indicating approach to Smoluchowski behaviour.

These results show that there is a distinct difference between the viscosity dependence of the measured relaxation rates, produced by varying the temperature of a single solvent and by using a solvent series, respectively. In the first case a linear dependence over a wide viscosity interval and good fits to the hydrodynamic Kramers equation are obtained, whereas in the latter case a non-linear viscosity dependence and pronounced deviations between experiment and theory are observed. Possible reasons for these deviations are discussed in section 3.4 below.

3.3 The equation of Skinner and Wolynes

Skinner and Wolynes have derived an expression (13) which produces the expected viscosity dependence of the rate constant in both the low and intermediate to high viscosity regions. By inserting the hydrodynamic assumption, $\xi \propto \eta$, into this expression it can be written on a form similar to that of the hydrodynamic Kramers equation

$$k = \frac{4\pi A}{B}\eta\left[1 + \frac{2\pi}{B}\eta + \frac{8\pi}{B^2}\eta^2\right]^{-1} \exp(-E_0/RT) \qquad (7)$$

We have performed fits of eq (7) to our experimental data of CV and EV in n-alcohol solutions. The exponential factor in eq (7) is set equal to unity, for the same reasons as in the fits to Kramers' equation. An example of the fits obtained is shown in Figure 7 B. A fit to Kramers' equation, with the same values of

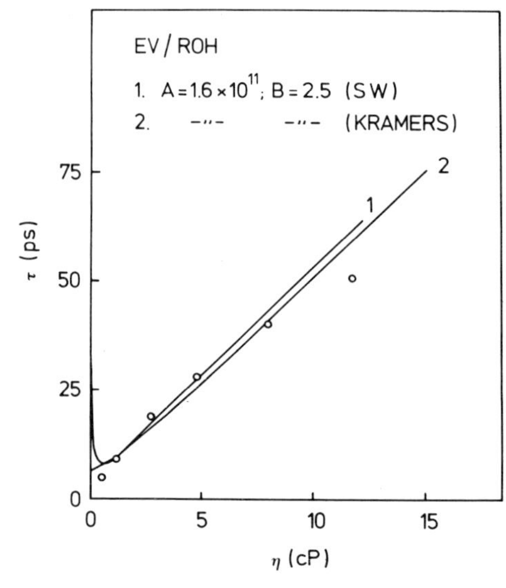

Figure 7 B. Comparison between Kramers and SW fits to the EV/ROH data.

the corresponding parameters, is also included in Figure 7 B for comparison. It is seen that if the parameters are chosen to give a reasonable fit to the low viscosity data points, there are rather severe deviations in the high viscosity region. There is a small but clearly distinguishable difference between the τ's obtained from the Kramers and the SW equations, respectively. For viscosities above ca 1 cp Kramers' equation yields a somewhat lower value (0 - 5 %) of τ than does the SW equation. At very low viscosities, ($\eta \leq 0.5$ cp), the low friction region of the SW equation is entered, and the calculated values of τ turns over and starts to increase with decreasing η. By choosing somewhat different parameters in the two types of fits it is possible to obtain equal good fits in both cases. The Kramers and SW theories are expected to yield somewhat different results in the intermediate friction region (13). Consequently, the large amplitude motion of the TPM molecules in n-alcohols belong to this type of friction. In the next section we will discuss possible reasons for the deviations of the theoretical fits from the experimental data.

3.4 Boundary conditions and the hydrodynamic assumption

To obtain the hydrodynamic Kramers and SW equations, we made the hydrodynamic assumption. This assumption implies that the volume of the solute molecule (V_p) is much larger than the volume of the solvent molecules (V_s), and the friction felt by the test molecule is proportional to solvent viscosity (eq.4).

Measurements of rotational relaxation times of large dye molecules (30 - 32) with $V_p \gg V_s$, have demonstrated that the hydrodynamic stick BC correctly describes the reorientational motions of such molecules in solution in many cases. For several small molecules, in the absence of strong solvent-solute interactions, the measured rotational relaxation times have been shown to have a much lower viscosity dependende than that, calculated with stick BC (i.e. Stokes-Einstein equation)(33-35). Assuming hydrodynamic slip BC and using the friction coefficients calculated by Hu and Zwanzig (29), for various axial ratios of ellipsoids, Bauer et al, (33) have demonstrated good agreement between calculated and experimentally obtained rotational relaxation times for several small molecules. Small, nearly spherical, molecules are expected to deviate especially strong from the Stokes-Einstein equation (29); in many cases the measured relaxation times and those calculated with slip BC are more then a factor of ten lower than the values obtained using stick BC (33).

These considerations motivate an investigation concerning the importance of boundary conditions in the large amplitude relaxation of TPM molecules. We have collected molecular volumes of the relaxing groups and of solvent molecules in Table III. It is quite evident that V_p/V_s ratio changes considerably through the n-alcohol

Table III Molecular volumes

Solvent or group	$V_m/Å^3$ a)	$V_w/Å^3$ b)	$\Delta V/Å^3$ c)
MeOH	67.2	40	27.2
EtOH	96.8	57	39.8
BuOH	152	91	61
HexOH	209	125	84
OctOH	262	159	103
DecOH	317	193	124
glycerol	88.5	121.3	32.8
H_2O	20.6	29.6	9.0
φ-NH_2 (PF)		82	
φ-$N(CH_3)_2$ (CV)		123	
φ-$N(C_2H_5)_2$ (EV)		156	

a) V_m is the molecular volume obtained from molecular weights and density data.
b) V_w is the molecular volume obtained from van der Waals increments.
c) The difference $V_m - V_w = \Delta V$ is approximately equal to the free volume V_f.

series and that $V_p \lesssim V_s$ for all the investigated molecules in the higher alcohols. For PF, the TPM molecule with the smallest substituents, $V_p \lesssim V_s$ for all solvents. A large amplitude motion of a φ-NR_2 fragment of a TPM molecule, is very likely to be constituted by a rotational motion. Consequently, what was said above concerning BC for free rotational relaxation should also apply to the large amplitude motion. When the V_p/V_s ratio decreases through the n-alcohol series the BC will vary and thus the friction felt by the relaxing molecule will vary in a non-hydrodynamic fashion. The deviation of the friction from its hydrodynamic value is thus expected to increase with decreasing V_p/V_s ratio. This will then produce the kind of deviations from the hydrodynamic Kramers and SW equation seen in Figure 7 , i.e. the measured relaxation times increase slower with viscosity then expected for hydrodynamic interactions. When $V_p/V_s < 1$, and the solvent in addition can form three-dimensional structures through hydrogen bonds the probe molecule can reorient inside cavities formed by the structured solvent. If V_p approaches the free-volume (V_f) of the solvent, cavities large enough that such reorientation might occur are formed (36). A probe molecule inside such a cavity has very little contact with the solvent molecules, and the viscosity dependence of the relaxation rate will thus be very low. From Table III we see that such effects might be active in the higher n-alcohols. In this context it is interesting to note that the relaxation times of PF, which has the smallest substituents, exhibit the lowest viscosity dependence of all the investigated molecules; changing the viscosity from 1 to 60 cp, by lowering the temperature of an ethanol solution, causes an increase in relaxation time by only a factor of four. The measured relaxation rates in the glycerol-water solutions also have a very weak and non-linear viscosity dependence. This can be explained with arguments similar to those used to explain the behaviour in the ROH series. In the glycerol/water solutions there are many possibilities for hydrogen bonding between the solvent molecules, and thus formation of a net-work liquid structure. Reorientation of the relaxing group inside cavities formed by these structures are, according to the discussion above, expected to have a very low η-dependence.

In the Kramers fit to the low-viscosity region for EV/ROH we obtain parameters ($\omega'\tau_v = 0.21$, $A = 6.1\cdot10^{11}$) which indicate that the relaxation process is approaching the diffusion limit in this region. We have shown above that a likely explanation to the discrepancies between experimental and calculated relaxation times is changing boundary conditions. The obvious question now is: what are these BC? Stick, slip or something else. One simple way to obtain an answer to this question should be the following. From the Arrhenius and isoviscosity plots we know that $E_0 \approx 0$. This implies that the large amplitude motion could intuitively be considered as a rotational relaxation (around one axis). Such a relaxation is described by

$$\tau_{rot} = \frac{8\pi r^3\eta}{6kT} f_p f_{slip} = C\cdot\eta\cdot f_{slip} = \tau_{stick}\cdot f_{slip} \qquad (8)$$

in the rotational diffusion limit. The factor f_{slip} expresses the ratio between the actual rotational relaxation time for a particular BC and the rotational relaxation time for stick BC, and f_p is the Perrin correction factor that accounts for the non-spherical shape of the molecule. In case of perfect slip, f_{slip} is a function of the axial ratio of the molecule modelled as an ellipsoid (29). For a spherical molecule $f_{slip} = 0$.

We have calculated τ_{rot} for the various φ-NR_2 fragments of the studied TPM molecules for stick and slip BC, respectively. Stick BC yields much too high values of τ_{rot} and a much too strong η-dependence (C) compared with the experimental results. Slip BC, with f_{slip} taken from ref. 29 , yields a much lower C-value and τ_{rot} which are in fair agreement with measured values for MeOH and EtOH. This is illustrated in Figure 7A for EV. It is also seen that the fit of Kramers' equation to the low viscosity data and the calculated τ_{rot} are rather close to each other. It is also worthwhile to note that the Smoluchowski equation, which is the diffusion limit of Kramers' equation,

$$k = \frac{\omega\omega'\tau_v}{2\pi} \exp(-E_0/RT) \qquad (9)$$

yields an expression of the same form as eq (8) if $\omega = \omega'$, eq.(4) (the hydrodynamic assumption), and $E_0 \approx 0$ are substituted into eq.(9). This indicates that our simplified description above, of the large amplitude motion as a rotational relaxation process is reasonable.

The most direct way to obtain information concerning the BC of the relaxation process would of course be to compare calculated values of the B-parameter of the hydrodynamic Kramers (or SW) equation, for various types of friction , with fitted values. Unfortunately, the parameter B contains ω', the frequency at the top of the barrier, which is difficult to obtain.

However, by assuming $\omega' = \omega$ we should obtain a reasonable value of ω' from our fits to Kramers' and SW's equations. From the fitted A and B parameters for EV we obtain $\omega' \approx 1\cdot10^{12}$ s^{-1} which yields $\tau_v \approx 10^{-13}$ s. This value agrees well with that calculated from eq. (3) defining τ_v ($\tau_v = 0.5 - 1\cdot10^{-13}$ s), using slip BC (an axial ratio $\varphi = 0.55$ for the relaxing group (33), yielding f_{slip} = 0.20 (29) was used in this calculation).

Thus in summary, there are several arguments strongly suggesting that the large amplitude motion in these TPM molecules can be described as a rotational motion with $E_0 \approx 0$, largely controlled by diffusion and slip BC, for the lower n-alcohols. For the higher n-alcohols the observed f_{slip} decreases, and is lower than that calculated for perfect slip, sub-slip. As argued above this can be explained with a changing BC due to decreasing V_p/V_S ratio and solvent network structures. The f_{slip} factor in eq.(8) is in addition very sensitive to variations in molecular shape(29). An ellipsoid axial ratio approaching unity results in a low value (< 0.1) of f_{slip}, which yields a very low η-dependence. A changing molecular shape, and thus f_{slip}, due to varying degree of solvent attachment could thus also be a source of changing viscosity dependence. Indications of solvent attachment were also obtained from the solvent dependence of the zero-viscosity intercept (see above). This discussion also shows that solvent series have to be used with great care when evaluating the influence of viscosity on relaxation processes induced by large amplitude motions.

4. ACKNOWLEDGEMENTS

Financial support from the Swedish Natural Science Research Council is greatfully acknowledged.

REFERENCES

1 Förster, T. and Hoffman, G. 1971, Z. Physik Chem. NF 75, p. 63.
2 Brey, L.A., Schuster, G.B. and Brickamer, H.G. 1977, J. Chem. Phys. 67, p. 2648.
3 Mastrangelo, C.J. and Offen, H.W. 1977, Chem. Phys. Letters 46, p. 588.
4 Magde, D. and Windsor, M.W. 1974, Chem. Phys. Letters 24, p. 144.
5 Ippen, E.P., Shank, C.V. and Bergman, A. 1976, Chem. Phys. Letters 38, p. 611.
6 Cremers, D.A. and Windsor, M.W. 1980, Chem. Phys. Letters 71, p. 27.
7 Yu, W., Pellegrino, F., Grant, M, and Alfano, R.R. 1977, J. Chem. Phys. 67, p. 1766.
8 Hirsch, M.D. and Mahr, H. 1979, Chem. Phys. Letters. 60, p. 299.
9 Beddard, G.S., Doust, T. and Windsor, M.W., in Picosecond Phenomena II, Springer Series in Chemical Physics, Vol 14. eds. Hochstrasser, R.M., Kaiser, W. and Shank, C.V. (Springer Berlin 1980) p. 167.
10 Doust, T. 1983, Chem. Phys. Letters 96, p. 522.
11 Sundström, V., Gillbro, T. and Bergström, H. 1982, Chem. Phys. 73, p. 439.
12 Kramers, H.A. 1940, Physica 7, p. 284.
13 Skinner, J.L. and Wolynes, P.G. 1978, J. Chem. Phys. 69, p. 2143.
14 Skinner, J.L. and Wolynes, P.G. 1980, J. Chem. Phys. 72, p. 4913.
15 Ippen, E.P. and Shank, C.V. 1977, Ultrashort Light Pulses, ed. Shapiro, S.L. (Springer) p. 83.
16 Wiesenfeld, J.M. and Ippen, E.P. 1979, Chem. Phys. Letters 67, p. 213.
17 Gillbro, T. and Sundström, V. 1980, Chem. Phys. Letters 74, p. 188.
18 Sundström, V. and Gillbro, T. 1983, Chem. Phys. Letters 94, p. 580.
19 Sundström, V. and Gillbro, T. 1981, Chem. Phys. 61, p. 257.
20 Sundström, V. and Gillbro, T. 1982, J. Phys. Chem. 86, p. 1788.
21 Velsko, S.P. and Fleming, G.R. 1982, Chem. Phys. 65, p. 59.
22 Velsko, S.P. and Fleming, R.R. 1981, J. Chem. Phys. 76, p. 3553.
23 Hochstrasser, R.M. 1980, Pure Appl. Chem. 52, p. 2683.
24. Tredwell, C.J. and Osborne, A.D. 1980, J. Chem. Soc. Faraday Trans. 2 76, p. 1627.
25 Grote, R.F. and Hynes, J.T. 1980, J. Chem. Phys. 73, p. 2715.
26 Grote, R.F. and Hynes, J.T. 1981, J. Chem. Phys. 74, p. 4465.
27. McCaskill, J.S. and Gilbert, G. 1979, J. Chem. Phys. 44, p.389.

28 Wilhelmi, B. 1982, Chem. Phys. 66, p. 351.
29 Hu, C.-M. and Zwanzig, R. 1974, J. Chem. Phys. 60, p. 4354.
30 Von Jena, A. and Lessing, H.E. 1979, Chem. Phys. 40, p. 245.
31 Porter, G., Sadkowski, P.J. and Tredwell, C.J. 1977, Chem. Phys. Letters 49, p. 416.
32 Waldeck, D.H. and Fleming, G.R. 1981, J. Phys. Chem. 85, p. 2614.
33 Bauer, D.R., Brauman, J.I. and Pecora, R. 1974, J. Am. Chem. Soc. 96, p. 6840.
34 Fury, M. and Jonas, J. 1976, J. Chem. Phys. 65, p. 2206.
35 Pajak, Z., Latanowicz, L. and Jurga, K. 1983, Ber. Bunsenges. Phys. Chem. 87, p. 143.
36 Dole, J.L., Kivelson, D. and Schwartz, R.N. 1981, J. Phys. Chem. 85, p. 2169.

VERIFICATION OF OPTICAL REDFIELD THEORY IN DOPED MOLECULAR SOLIDS

Laurens W. Molenkamp and Douwe A. Wiersma
Picosecond Laser and Spectroscopy Laboratory
of the Department of Physical Chemistry
University of Groningen, Nijenborgh 16
9747 AG GRONINGEN, The Netherlands

ABSTRACT

We report results of an optical and picosecond photon echo study on the zero-phonon line of photosite I of pentacene in benzoic acid. The results show that optical dephasing in this system proceeds via uncorrelated phonon scattering processes from the ground and optically excited state to singly-excited librational levels. It is also shown that the data can not be consistently interpreted using coherence exchange theory. The librational frequencies and lifetimes at low temperature are determined to be 12 cm^{-1} and 18 psec in the ground state and 16.7 cm^{-1} and 2.5 psec in the singlet excited state. It is argued that population decay of librational overtones proceeds in single quantum steps down the librational ladder which excludes these modes from playing a role in optical dephasing.

From a temperature-dependent lineshape study of the libronic transition in absorption it is concluded that the low-temperature librational lifetime is primarily due to resonant decay into acoustic phonons. At elevated temperature cubic libration-phonon anharmonicity dominates the librational lifetime. These lifetime shortening processes are reflected in the temperature dependent dephasing of the origin. With the dephasing model presented in this paper for photosite I, existing controversies concerning optical dephasing in the unburned origin of pentacene in benzoic acid can be resolved.

Finally lineshape studies performed by other groups are discussed in light of the results reported in this paper.

K. B. Eisenthal (ed.), Applications of Picosecond Spectroscopy to Chemistry, 99–125.

I. INTRODUCTION

In the past decade the development of tunable dye-lasers has greatly stimulated the field of coherent optical spectroscopy. Frequency and time-domain experiments in the form of (photochemical) hole burning, photon-echo and coherent Raman scattering have yielded in this period a wealth of information on the optical dynamics in molecular and ionic solids. In this presentation we will limit our discussion to one specific system, pentacene in benzoic acid; but the findings and the ideas presented should have general application in the field of doped molecular crystalline solids. It is specifically the intention of this paper to examine the role of molecular librations in the optical dephasing of the zero-phonon line in such systems.

McClure (1) was the first one to recognize that many of the sharp peaks in the phonon sideband structure of electronic and vibronic spectra in mixed crystals are transitions that involve librations of the guest molecule in a cage of host molecules. In the specific case of the naphthalene-in-durene mixed crystal he suggested that a 18 cm^{-1} phonon peak observed in the lowest excited singlet state is due to a naphthalene libration around the out of plane axis. This suggestion was recently confirmed in a Monte Carlo type calculation of the librational potential for this naphthalene motion by Koehler and Schmidt (2). McClure further noted that this libration in the naphthalene spectrum also acted as a "false" origin which indicated that this libration is highly localized at the guest site.

It was more than a decade later that, through spin echo (3), photon echo (4) and photochemical hole burning (5) experiments, it became clear that these guest librations play a crucial role in the spin-, Raman-, and optical dephasing processes that determine the homogeneous lineshape. Following van't Hof and Schmidt (6), several theories have been advanced (7-11) to explain the observed quasi-exponential activation of the low-temperature homogeneous linewidth of these transitions. Völker et al. (5) showed that in the case of free base porphin doped into Shpol'skii host crystals, *exchange theory* quite nicely explains the observed optical linewidth and shift with increasing temperature. Wiersma and coworkers claimed that in many chemically mixed crystals (12) *uncorrelated phonon scattering* in the ground and excited state determines optical dephasing. De Bree and Wiersma (dBW, 10) recently presented a detailed microscopic theory of optical dephasing using a four-level system coupled to an anharmonic phonon bath where next to the ground and electronically excited state the lowest frequency libration in each state was "projected out" of the phonon bath.

From this theory it became clear that coherence exchange and uncorrelated phonon scattering (UPS) are limits of a more general relaxation theory, coined by de Bree and Wiersma as optical Redfield theory. They further emphasized that exchange of coherence between the cold and librationally-hot transition can only occur when these transitions spectrally overlap. On optical (electronic) excitation the librational potential may undergo substantial changes as noted from a change in molecular geometry. It is therefore expected that these "exchange conditions" are more likely to be met for spin- and Raman- than for optical transitions. A striking illustration of this point was recently presented by Duppen et al. (13) in a comparative study of vibrational and vibronic dephasing in pentacene doped into naphthalene. Raman dephasing within the electronic ground state manifold was shown to be ineffective due to *coherence exchange* between the cold and librationally-hot vibrational transition. In contrast, vibronic dephasing, involving the *same* libration, increased rapidly with temperature through UPS processes in the initial and final state. It therefore seems that the dBW theory of optical dephasing in a crystalline environment gives a satisfactory description of the results obtained. However, what presently is lacking is a *clear* demonstration of the bi-exponential activation of the homogeneous linewidth as required by the UPS mechanism and direct measurement of the (temperature-dependent) lifetimes of the librational modes which are, in addition to the librational frequencies, the essential parameters in both exchange and uncorrelated phonon scattering (dBW) dephasing theory. Moreover, the dBW theory is basically a four-level theory and therefore does not include the effects of overtones of the active libration, nor does it incorporate contributions of the other guest librations to the homogeneous lineshape. As complete verification of the Redfield-UPS limit necessitates dephasing measurements up to temperatures where population of these other modes no longer can be neglected (ca. 20 K), a further study of optical dephasing in mixed molecular crystals was deemed necessary.

In a recent paper by Duppen et al. (14), results of psec photon echo experiments on pentacene in benzoic acid were reported. These results seemed to indicate that next to librational dephasing, involving a libration of 16.5 cm^{-1}, adiabatic coupling to acoustic host phonons (15) was also of importance in this system. This conclusion was drawn on basis of comparison of the phonon sideband in absorption and emission, which both seem to have a progression in a 16.5 cm^{-1} mode. In a later paper by Olson et al. (16) results of low-temperature photochemical holeburning experiments indicated that, up to 5 K, the homogeneous linewidth increased exponentially with temperature with an activation of 11 cm^{-1} and pre-exponential constant of 0.3 cm^{-1}. Andrews and Hochstrasser (17) concluded from a coherent Stokes Raman scattering study that the homogeneous lineshape was activated with an energy of 13.8 cm^{-1}. In view of these controversies it seemed worthwhile to reinvestigate this system.

Here the photon echo experiments are extended to a photosite of pentacene in benzoic acid. When pentacene in benzoic acid is excited to its electronically excited state, proton rearrangements in the host occur that lead to the appearance of several metastable pentacene sites (16) in the crystal.

We have performed echo experiments on the most red shifted photo-induced pentacene site (I) for the following reasons. First, in contrast to the unburned origin, this site shows a distinct progression in a 16.7 cm^{-1} libration. Furthermore, the intensity of the fundamental libronic transition is such that photon echo experiments on this band seemed feasible, which is important in a direct determination of its lifetime. Finally, the fluorescence spectrum of this site shows more pronounced structure than that of the unburned origin which is essential in assigning the librational mode in the ground state.

The results presented in this paper show that the UPS-limit of the dBW theory correctly describes the homogeneous lineshape of this site up to 20 K, where the homogeneous linewidth exceeds the low-temperature limit by a factor of 10^4. The librational frequencies in the ground and excited state are determined to be 12 and 16.7 cm^{-1} with low-temperature lifetimes of 18 and 2.5 psec respectively. It is further shown that the low-temperature lifetime of the libration in the excited state is mainly due to direct decay into acoustic phonons at the same frequency. At higher temperature, phonon up- and down-conversion processes reduce the librational lifetime.

Finally, it is concluded from the great similarity between the dephasing of photo-induced site I and the unburned origin that the same dephasing mechanism holds for both transitions, which resolves existing controversies.

II. THEORETICAL BACKGROUND

In order to understand the significance of the results presented in section IV, it is necessary to briefly discuss the de Bree-Wiersma dephasing model (10) which schematically is shown in Fig. 1. In this figure $|1>$ is the ground state and $|2>$ the electronically excited state of the isolated guest impurity molecule in the host lattice. Levels $|3>$ and $|4>$ are associated with local-

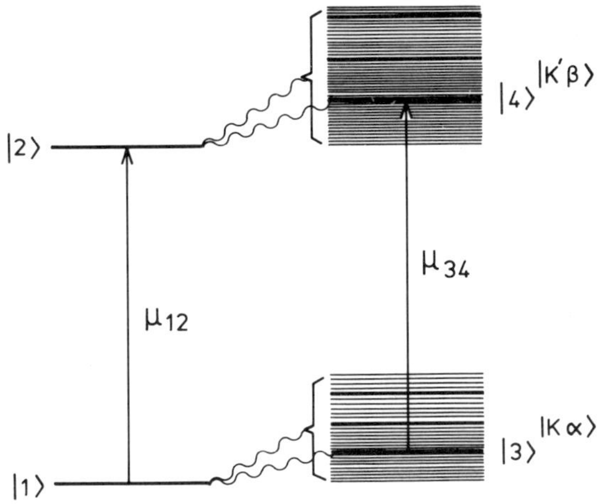

Figure 1. Level structure of an optical transition $|2\rangle \leftarrow |1\rangle$ interacting with pseudolocalized librational levels $|3\rangle$ and $|4\rangle$, and delocalized crystal states $|k\alpha\rangle$ and $|k'\beta\rangle$.

ized phonons (librations) of the guest molecule in the ground and excited state. The other heavily drawn lines in the figure indicate two- and three-quantum excitations of these librations. The levels $|k\alpha\rangle$ are acoustic phonons of branch α with wave vector $\underline{k}$. We note that for clarity purposes we have omitted in the figure higher quantum excitations and other molecular librations. The underlying microscopic model-hamiltonian for this figure has the following form (10):

$$H = H_A + H_R + V_{AR}$$

$$H_A = \sum_f (\varepsilon^f + \sum_\kappa V^f_{\kappa\kappa}(\bar{n}_\kappa + \tfrac{1}{2}) + \hbar(\Omega + \Delta\Omega^f)(B^+B + \tfrac{1}{2}))a_f^+ a_f$$

$$H_R = \sum_\kappa \hbar\omega_\kappa (b^+_\kappa b_\kappa + \tfrac{1}{2}) + \frac{1}{3!}\sum_{\kappa,\kappa,\kappa''} (U_{\kappa\kappa'\kappa''} b_\kappa b^+_{\kappa'} b^+_{\kappa''} + cc) + \ldots$$

$$V_{AR} = V_{l-p} + V_{e-p}$$

$$V_{l-p} = \sum_{f,\kappa} U^f_{\kappa\lambda}(b^+_\kappa B + b_\kappa B^+) + \frac{1}{3!} \sum_{f,\kappa,\kappa'} U^f_{\kappa\kappa'\lambda}(b_\kappa b^+_{\kappa'} B^+ + b_\kappa b_{\kappa'} B^+ + cc)$$

$$V_{e-p} = \sum_{f,\kappa} V^f_\kappa (b_\kappa + b^+_\kappa) a^+_f a_f + \sum_f V^f_\lambda (B + B^+) a^+_f a_f + $$

$$+ \sum_{f,\kappa} V^f_{\kappa\kappa}(b^+_\kappa b_\kappa - \bar{n}_\kappa) a^+_f a_f + \sum_{\substack{f,\kappa,\kappa' \\ (\kappa \neq \kappa')}} V^f_{\kappa\kappa'} b^+_\kappa b_{\kappa'} a^+_f a_f \qquad (1)$$

with $\Delta\Omega^f = V^f_{\lambda\lambda}/\hbar$ and $\Delta\Omega^g = V^g_{\lambda\lambda}/\hbar = 0$. The index λ is used when the pseudolocal phonon is involved.

In eqns. 1 a^+_f (a_f) denotes the creation (annihilation) operator for electronic state f with energy ϵ^f, b^+_κ (b_κ) and B (B) are the creation (annihilation) operators for the band phonons and librations respectively. $\Delta\Omega^f$ represents the frequency change of the libration in going to the excited state. V^f_κ, $V^f_{\kappa\kappa}$ and $V^f_{\kappa\kappa'}$ are the linear, diagonal quadratic and nondiagonal quadratic electron-(acoustic) phonon interaction matrix elements. V^f_λ is the corresponding quantity for the electron-libration coupling. $U^f_{\kappa\lambda}$ represents the harmonic interaction between the libration and the acoustic phonons and results from the fact that the pseudo-localized libration is not an eigenstate of the pure crystal (18). $U_{\kappa\kappa'\kappa''}$ and $U_{\kappa\kappa'\lambda}$ stand for the cubic anharmonic phonon interactions and $\bar{n}$ is the average phonon occupation number: $\bar{n}_\kappa = [\exp(h\bar{\omega}_\kappa/kT) - 1]^{-1}$, where $\bar{\omega}_\kappa = \omega_\kappa + \delta\omega_\kappa$ and $\delta\omega_\kappa$ is a small renormalization factor of the phonon frequency due to anharmonicity, which is neglected.

We note here that the electronic energy has been redefined to include the thermal average of the electron-phonon coupling energy. As is obvious, from this hamiltonian, the electronic states are treated on equal footing with the localized levels, as seems appropriate from a spectroscopic point of view (19). These localized levels (H_A) are interacting with the anharmonic phonon bath (H_R) through the electron-phonon coupling (V_{e-p}) and phonon-phonon interaction effects (V_{l-p}). Let us consider briefly the perturbations contained in V_{AR}. There is an important difference between the phonon-phonon (V_{l-p}) and electron-phonon (V_{e-p}); V_{e-p} is, through the above mentioned transformation, only operative within the excited state manifold, while V_{l-p} acts within both ground and excited states. The terms in V_{e-p} describe the usual Raman scattering processes of the optical excitation with the pseudo-localized and band modes. The first two terms are the linear electron-phonon coupling with bath and pseudo-localized phonons, respectively. These terms are responsible for the phonon side bands in optical absorption and emission spectra. The third and fourth term stand for diagonal and off-diagonal quadratic coupling with the band phonons. In the McCumber-Sturge theory of optical lineshape (15) the diagonal contribution leads to a temperature dependent lineshift, and the off-diagonal to a temperature dependent linebroadening.

The other contribution to V_{AR}, V_{l-p} describes the processes that determine the lifetime of the pseudo-localized phonon. The first term describes the resonant decay of the librational mode. The pseudo-localized librations are not the eigenstates of the (mixed) crystal (18), and their harmonic decay (18) can be envisioned as the spreading out of the librational phonon wave packet into resonant crystal modes. Note that this process is temperature independent. The second term describes the cubic anharmonic libration-band phonon interaction. The physical picture here is that a libration decays into two band phonons, in either a frequency sum or difference generation process. The total effect of these perturbations on the homogeneous linewidth $(\pi T_2)^{-1}$ of the zero-phonon line (the transition $|2\rangle \leftarrow |1\rangle$) can be expressed by the following well-known relation:

$$\frac{1}{T_2} = \frac{1}{T_2^*} + \frac{1}{2T_1} \tag{2}$$

where T_1 represents the population relaxation effects (e.g. excited state fluorescent decay) and T_2^* is the pure dephasing time. Using Redfield theory in a form presented by Cohen-Tannoudji (20) de Bree and Wiersma (10) calculated the effects of the different perturbations on the homogeneous lineshape. For the present discussion we will focus on the dominant contribution, the diagonal quadratic electron-libration coupling, to the lineshape, which in the UPS limit (*vide infra*) becomes:

$$1/T_2^* = \tfrac{1}{2}(\tau_3^{-1}(T)e^{-\hbar\omega_{31}/kT} + \tau_4^{-1}(T)e^{-\hbar\omega_{42}/kT}) \tag{3}$$

where τ_3 (T), τ_4 (T) and ω_{31}, ω_{42} are the temperature-dependent lifetimes and frequencies of the librations in the ground and excited state respectively. Note that in this expression there is no sum over multiple quantum excitations of the libration. In section E we will discuss this point. De Bree and Wiersma also showed that in the special case where the cold (optical) and librationally-hot transition overlap *and* the lifetime of the libration in ground and excited state is identical, the well-known effect of exchange of coherence between these transitions may occur. In this case the contribution to the dephasing parameter becomes:

$$1/T_2^* = \frac{\delta^2\tau}{1 + \delta^2\tau^2} \; e^{-\hbar\bar{\omega}/kT} \qquad (4)$$

where $\bar{\omega} = \frac{1}{2}(\omega_{31} + \omega_{42})$ is the average librational frequency and τ the lifetime of the libration, which is necessarily identical in ground and excited state in this formalism. While this equation strictly only holds (10) for $\delta\tau \lesssim 1$, where δ (in rad/sec) presents the frequency difference between the transitions that undergo exchange, it has also been applied in the limit where $\delta\tau > 1$ ("slow exchange"). The difference between the UPS and exchange limits of the Redfield relaxation theory is clear from a comparison of Eq. 3 and 4. In the UPS limit, the homogeneous lineshape broadens with temperature as a sum of two exponents, while in the exchange limit single exponential activation is required. In the case of a quasi-exponential activation of the linewidth, knowledge of the lifetime of the libration in the ground and/or excited state enables a proper choice of the dephasing mechanism. It is further important to note that, when the librational frequencies and their temperature-dependent lifetimes are known, no freedom is left in predicting the broadening of the homogeneous lineshape arising from the quadratic electron-libration coupling.

In the intermediate case where one of the exchange conditions is not met, the lineshape function is more complex and for this case we refer to Ref. 10, where a detailed discussion of all aspects of optical Redfield relaxation theory is given.

III. EXPERIMENTAL

Single crystals of pentacene in benzoic acid were grown by the Bridgman method from zone-refined benzoic acid doped with pentacene obtained from Fluka. The final concentrations of pentacene in benzoic acid were in the range of 10^{-4}-10^{-5} mole/mole. After cleaving the crystals from the boule, by using a razor blade, they were mounted in a temperature-variable cryostat built by Oxford Instruments.

In all experiments, photosite I of pentacene in benzoic acid was generated by irradiating the mixed crystal with an argon-laser operating at 5145 Å. In the photon echo experiments a steady state concentration of photosite I high enough to perform the experiments, was obtained by irradiating the crystal with 500 μW of argon laser light. The echo-exciting laser beams and the argon laser beam were focussed on the same 100 μm spot of the crystal. In case of photon echo experiments on the phonon sideband, 2 mW of argon laser light was used and it was checked that by lowering the intensity of the argon laser no change in the measured echo decay was observed. A schematic of the set-up used to generate the accumulated photon echo's is shown in Fig. 2. The echo experiments were further performed as described in Ref. 21.

The fluorescence lifetime of pentacene in site I was measured by excitation with a nitrogen-pumped dye laser and display of the fluorescence signal on a Tektronix transient digitizer. Reabsorption of the fluorescence signal was kept to a minimum by using crystals of only 10% absorption.

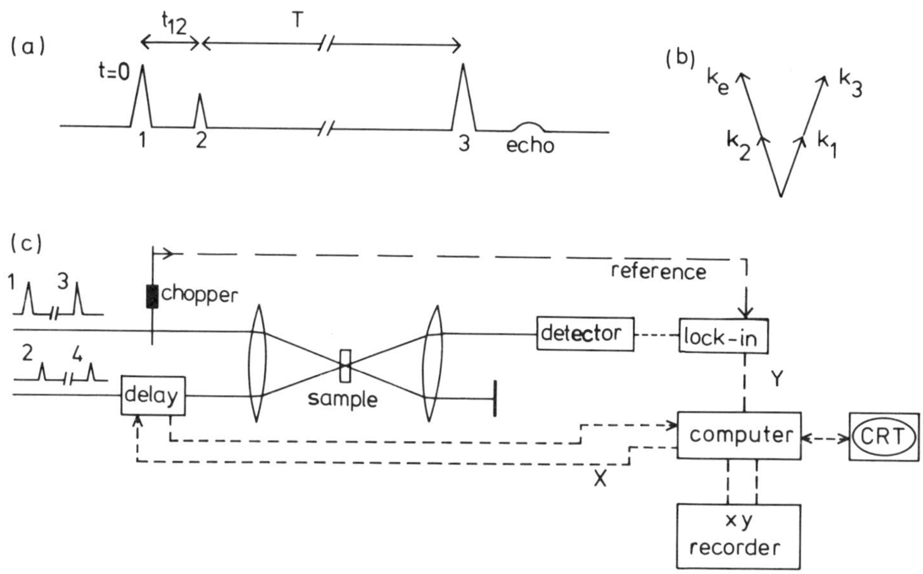

Figure 2. Schematical set-up for the generation and detection of accumulated photon echoes.
(a) excitation pulse cycle; t_{12} is the variable delay time between the excitation pulses and T the fixed cycle time.
(b) phase-matching diagram of the excitation pulses and echo.
(c) set-up for detection of intensity modulations on the probe beam (2 and 4) from modulation of the pump beam (1 and 3). An Apple II computer steps the delay t_{12} between the excitation pulses and also digitizes the incoming signals from the lock-in amplifier.

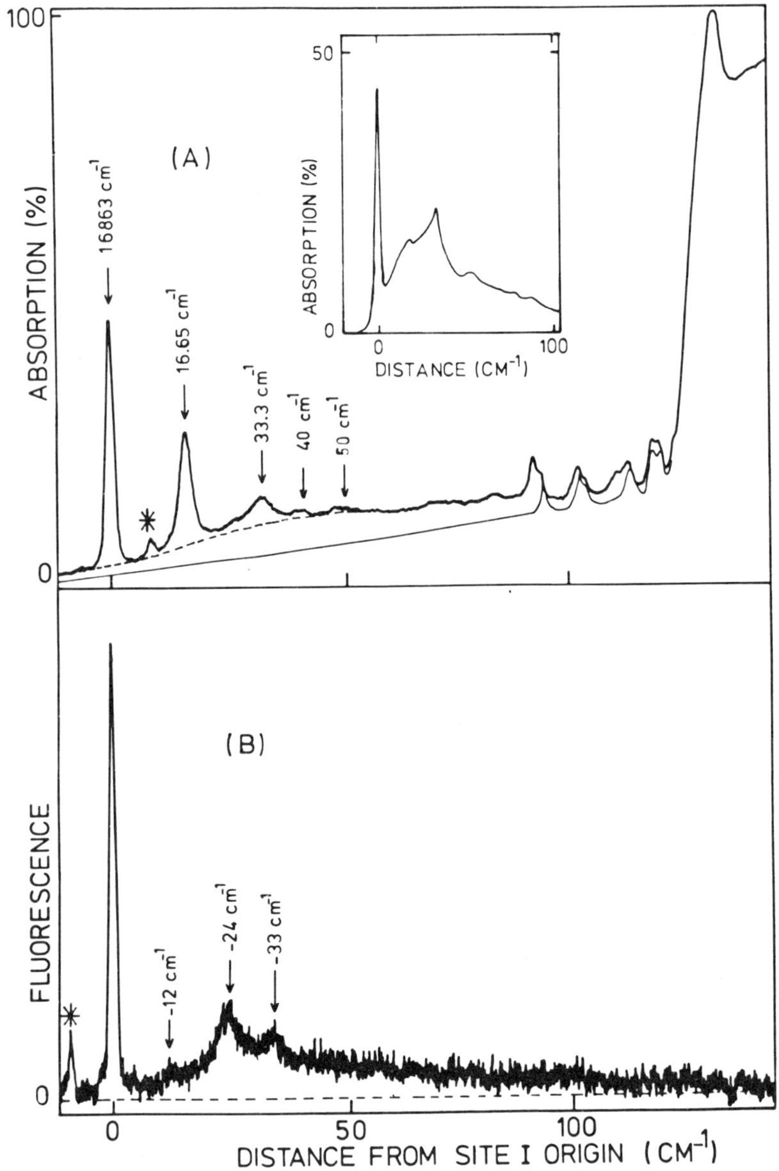

Figure 3. Absorption (A) and emission (B) spectra of photosite I of pentacene in benzoic acid at 1.5 K. Asterisks mark the position of the zero-phonon line of site V (cf. Ref. 16). The intensity of this peak was used in emission to calibrate the intensity of the zero-phonon line of site I. The insert in Fig. 3A shows the absorption spectrum of the unburned origin.

IV RESULTS AND DISCUSSION

A. Absorption and emission spectra

Fig. 3A shows the absorption spectrum of photosite I of pentacene in benzoic acid at 1.5 K. The underlying solid line in the spectrum represents the absorption of the sample, before proton rearrangement has occurred. It is clear that next to the existence of well-defined sites also a broad absorption is generated in the rearrangement process. The dotted line in the spectrum indicates what we believe is the true absorption background of photosite I. The starred absorption is due to photosite V (16). Taking this into account we calculate for photosite I a Debye-Waller factor ($I_{zpl}/(I_{zpl} + I_{psb})$) of 0.49 ± 0.03. Here I_{zpl} is the intensity of the phonon sideband. The insert of Fig. 3A shows the absorption spectrum of the unburned origin at low temperature. Comparison of the two spectra shows a considerable difference in phonon sideband structure indicating a noticeable change in electron-phonon coupling after proton rearrangement. This is also reflected in the Debye-Waller factor for this transition which is only 0.16 (22). Figure 3A further shows that the photosite spectrum exhibits a clear progression in a 16.7 cm^{-1} mode. This mode, previously noted by Olson et al. (16), is assigned by us as the lowest-frequency librational mode of electronically-excited pentacene in this site. From the intensity distribution in the librational progression we calculate (23) a reorientation angle of 1.3° along this coordinate. Note that in the unburned origin the same mode is observed but with a different intensity pattern and riding on a much more intense background absorption. In the photosite spectrum we further discern an absorption at 40 cm^{-1} from the origin which may be due to another librational mode of pentacene. Monte Carlo calculations of the type reported by Koehler (24) are presently in progress to investigate this idea. Fig. 3B shows the emission spectrum of photosite I after excitation with a cw-dye-laser in the librational absorption. Note that the relative intensity of the origin, after correcting for reabsorption, is four times greater than presented. This shows that in emission the intensity in the librational bands is far less than in absorption. However, we clearly observe a progression in a 12 cm^{-1} mode and assign this mode to the lowest-frequency libration of pentacene in the groundstate. The feature at 33 cm^{-1} from the origin may correspond to the 40 cm^{-1} peak observed in the absorption spectrum. As stated before, in emission much of the intensity in the phonon sideband is found in the acoustic phonons at higher frequency; a possible reason for the non-observation of this groundstate mode by Olson et al. (16). It is gratifying to note that the Debye-Waller factor in emission is found to be

0.51 ± 0.03, within experimental error identical to the one in absorption. This fact lends credence to our choice of base line in the absorption spectrum as discussed above.

What remains to be understood is the dramatic difference between the phonon sideband in absorption and emission of this site. While in absorption all the intensity is found in a librational progression, in emission most of the intensity is found in the acoustic modes. An interpretation of this effect in terms of a "Duschinsky"-type mixing, allowed by the loss of inversion symmetry of this site, is underway (25).

The most important conclusion pertinent to this work is that in photosite I the lowest-frequency libration of pentacene is 12 cm^{-1} in the ground state and 16.7 cm^{-1} in the excited state. An attempt to observe, at higher temperature, a hot band at 4.7 cm^{-1} to the blue of the origin was inconclusive due to the fact that at higher temperatures ($\gtrsim$ 4 K) new spectral features arise due to thermally activated proton rearrangements in the host crystal.

B. Librational lifetime

As was pointed out in section II, next to knowledge of the librational frequencies, the lifetimes of these modes need to be known in order to predict theoretically the dephasing characteristics of the pure electronic transition. The first *direct* measurement of a librational lifetime was recently made by the Völker group (26), on the system free-base porphin doped into n-decane, using the photochemical holeburning technique. For the 5.6 cm^{-1} libration in the excited state they obtained from the holewidth a lifetime of 115 ± 20 psec.

To measure the excited state librational lifetime of site I of pentacene in benzoic acid we have performed accumulated photon echo and lineshape experiments. Fig. 4B shows on an expanded scale the lineshape of the fundamental librational transition at 1.5 K. It is clear that the lineshape is very close to being Lorentzian. If we assume that the underlying inhomogeneous width is identical to that of the origin we calculate a homogeneous width of 2.2 cm^{-1} for this librational transition which corresponds to a librational lifetime of 2.4 psec. The dots in Fig. 4B correspond to points of the simulated Voigt lineshape with an inhomogeneous width of 1.0 cm^{-1} and a homogeneous width of 2.2 cm^{-1}. As this librational lifetime is of essential importance in the analysis of our results we decided also to directly attempt to measure this lifetime using the technique of accumulated photon echoes. Fig. 4A shows the result of such an echo experiment together with the auto-correlation of the excitation pulses. By convoluting

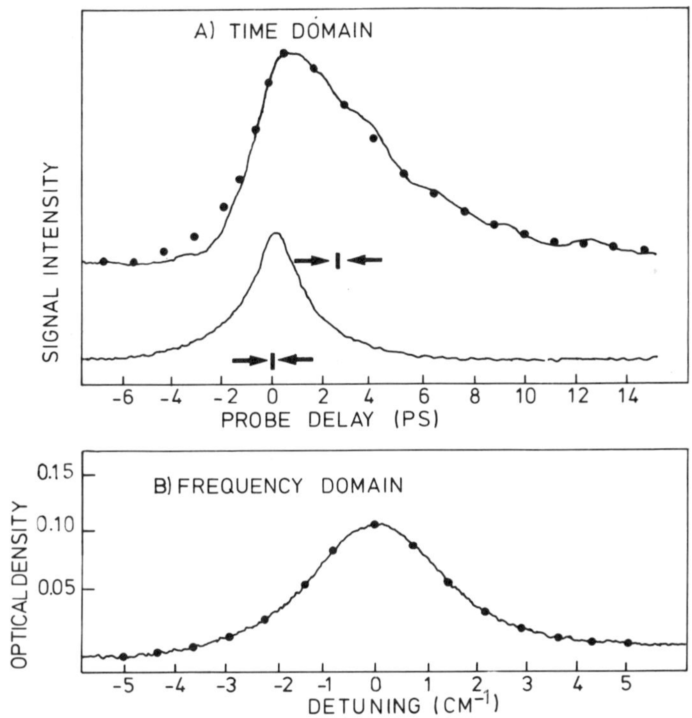

Figure 4. Determination of the librational lifetime.
(a) in the time domain from decay of the photon echo (upper trace). The lower trace represents the auto-correlation of the excitation pulses. The vertical bars represent the position of the intensity-weighted time average of both signals (see text).
(b) in the frequency domain from a lineshape analysis of the libronic transition. The dotted fits in A and B are discussed in the text.

the correlation signal with an exponential decay curve with a decay constant of 2.5 psec the dots in Fig. 4A are generated. The fit is excellent except at the rising edge of the signal. The discrepancy here is ascribed to excited state absorption of pentacene. Another way of determining the decay time is to calculate the intensity-weighted time average of both the auto-correlation and the echo signal. One can show (27) that in case of a symmetrical impulse function and exponential decay of the system's response, the time difference between these averages equals the decay parameter of the response function.

With this method we also calculate a decay time of 2.5 psec. We therefore conclude that the lifetime of the libration in the excited state equals 2.5 ± 0.1 psec.

The excellent agreement between the librational decay time as obtained from a lineshape analysis and photon echo decay also shows that the inhomogeneity in the librational transition is identical to that of the origin. Thus a lineshape analysis of the librational transition at higher temperature should lead to an accurate measurement of the underlying homogeneous linewidth. In section D the temperature dependence of this linewidth will be discussed.

As will be obvious from Fig. 3B it is impossible to measure the librational lifetime in the ground state from the fluorescence spectrum. In principle, it would be possible to measure this lifetime by using multi-resonant time-resolved CARS experiments (28). Because of the weak oscillator strength of the fundamental librational transition in emission, we have not attempted to perform such an experiment. Direct measurement of this lifetime therefore is presently out of reach but an *indirect* determination is possible once the question of the appropriate dephasing mechanism is settled.

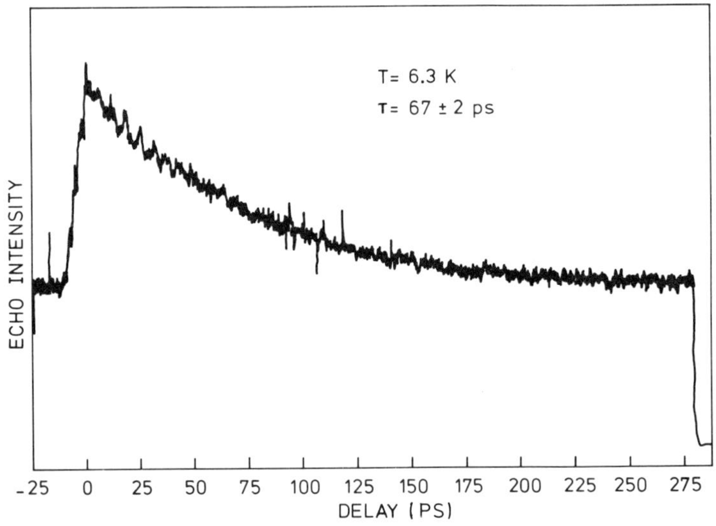

Figure 5. Photon echo decay at 6.3 K of the zero-phonon line of photosite I of pentacene in benzoic acid. At a delay of 280 psec the pump beam was blocked to determine the incoherent background contribution to the signal (Ref. 12).

C. Temperature dependence of the homogeneous linewidth of the zero-phonon line of photosite I.

In order to determine the temperature dependence of the homogeneous linewidth of the zero phonon line of photosite I we have performed photon echo and linewidth measurements. An example of an accumulated photon echo decay is given in Fig. 5

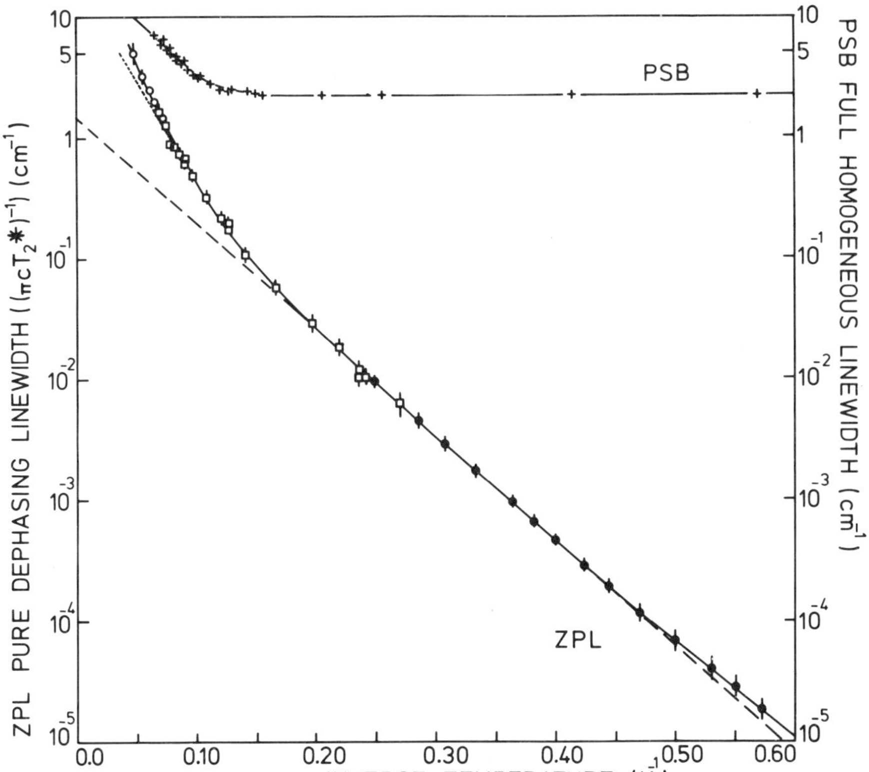

Figure 6. The lower curve presents the pure dephasing contribution to the homogeneous linewidth of the zero-phonon line of photosite I of pentacene in benzoic acid. The data points represented by open circles were obtained from a lineshape analysis. The points indicated by open squares were obtained from photon echo decay measurements at a fixed temperature. The filled circles were deduced from a measurement of the photon echo intensity at a fixed (3.37 nsec) delay between the excitation pulses. The upper curve gives the homogeneous width of the libronic transition 16.7 cm^{-1} to the blue of the zero-phonon line. The fits to the upper and lower curves are discussed in the text.

where it should be noted that the steady state concentration of photosite I in these experiments is such that the absorption strength at the origin corresponds to an optical density of only 0.02. The pure dephasing contribution to the homogeneous linewidth obtained from these echo decays is presented in Fig. 6 by the open squares. These experimental points, $(\pi T_2^*)^{-1}$ in cm^{-1}, were obtained from the photon echo decay time (T_2) by subtraction of the lifetime (T_1) contribution using Eq. 2. At temperatures below 4 K this method becomes impractical because of the long scanning times and other practical problems. At these temperatures we have therefore inferred the echo decay times from a measurement of the echo intensity as a function of temperature at a fixed delay. An example of such a run is given in Fig. 7 together with a fit to be discussed later. The data points obtained from these runs are indicated by filled dots in Fig. 6. Finally at temperatures above 15 K linewidth measurements were performed and the homogeneous linewidth extracted by deconvoluting the known inhomogeneous linewidth from the Voigt lineshape. These data points are indicated by open circles in Fig. 6.

Up to 20 K the fluorescence lifetime of pentacene in site I was measured to be 23 ± 1 nsec and independent of temperature. Note that in the temperature range from 1.5 to 20 K the homogeneous linewidth changes from 7 MHz to 5 cm^{-1} (1.5×10^5 MHz). The upper part of Fig. 6 shows the temperature dependence of the homogeneous width of the librational transition at 16.7 cm^{-1} from the origin.

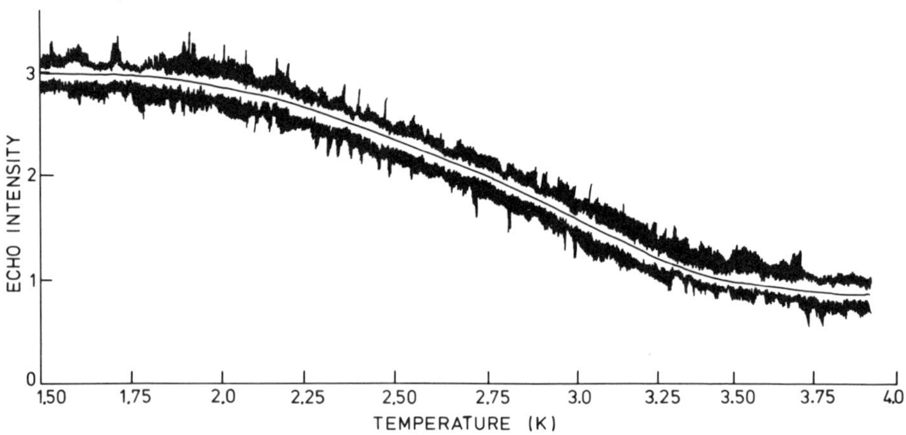

Figure 7. Accumulated echo intensity as a function of temperature at a fixed delay of 3.37 nsec. The solid line, which is the fit to the intensity change, is discussed in the text.

The first thing to note is that as long as the linewidth of the librational transition remains constant, the linewidth of the origin exhibits a quasi-exponential activation, indicated by the dashed line in Fig. 6, with an activation energy of 14.0 cm^{-1} and a pre-exponential constant of 1.5 cm^{-1}. This activation energy is close to the average of the librational frequency in the ground and excited state $(12.0 + 16.7)/2 = 14.3$ cm^{-1}, which indicates that *both* librations are involved in the dephasing process. The second thing to note is that at low temperature the activation energy is distinctly lower than 14.0 cm^{-1} as expected from the UPS mechanism. According to eq. (2), at low temperature the lowest frequency libration will dominate the dephasing process. The solid line going over into the dotted line in Fig. 6 is the best fit to Eq. 3, where $\tau_4(T)$ is obtained from the T_1-contribution of the homogeneous linewidth of the librational transition (*vide infra*) and τ_3^{-1} (T = 0 K) is an adjustable parameter. From this fit the low-temperature lifetime of the pentacene libration in the ground state is found to be 18 ± 2 psec, which is to be compared with the 2.5 psec lifetime of this mode in the excited state. The fit between the experimental results and the theoretical expression in Eq. 3 can even be made perfect, as shown by the solid line, by adding to τ_3^{-1} (T = 0 K) a temperature dependent part, where we assume that the broadening of the ground state libration is similar to that of the excited state.

The solid line in Fig. 7 is also a fit to Eq. 3 with the parameters just mentioned. In this temperature range, however, optical dephasing is dominated by phonon scattering to the ground state librational mode (12 cm^{-1}) and consequently a fit to a single exponential activation with a 12 cm^{-1} activation energy is also possible. The pre-exponential constant of 0.5 cm^{-1} obtained in this latter fit is not directly interpretable in terms of a lifetime as this constant includes effects of the excited state librational mode which was neglected.

To conclude this section, we emphasize that the homogeneous lineshape of the pentacene transition in this site up to 20 K can be explained on the basis of a dephasing mechanism where uncorrelated phonon scattering processes occur in the ground and excited state. The only ad-hoc assumption made is that the ground state libration exhibits a similar temperature dependence as the excited state libration. This point is further discussed in section D.

Finally we note that if the exchange equation 4 is used to fit the data in the region up to 5 K, as given by the dashed line in Fig. 6, we calculate from the pre-exponential constant of 1.5 cm^{-1} and $\delta/2\pi = 4.7$ cm^{-1} a librational lifetime of 6.9 psec.

This number is in disagreement with the measured librational lifetime of 2.5 psec in the excited state and inferred lifetime of 18 psec in the ground state. The 6.9 psec lifetime, however, is close to the *average* librational lifetime of (2.5 + 18)/2 = 10.25 psec.

D. Librational dephasing and relaxation

The homogeneous width of a libronic transition in optical spectra is, as in the case of the pure electronic transition, determined by population relaxation (T_1) and pure dephasing (T_2*) effects. At low temperature the pure dephasing contribution to the homogeneous width will be negligible and it is therefore justified to calculate from this width the low-temperature librational lifetime. With increasing temperature, however, dephasing processes are known to become important and the homogeneous width will be the sum of both. In the dBW optical dephasing model (Eq. 3) the temperature dependent librational lifetimes are important parameters and need to be known. In order to extract from a linewidth measurement on a libronic transition the part only caused by population decay we need information on the dephasing of this transition. In complete analogy to the pure electronic transition we can write for the pure dephasing contribution to the homogeneous width of the libronic transition the following:

$$\frac{1}{T_2^*} = \tfrac{1}{2}(\tau_3^{-1}(T)e^{-\hbar\omega_{31}/kT} + \tau_6^{-1}(T)e^{-\hbar\omega_{64}/kT}) \qquad (5)$$

Here $\tau_3(0)$ is the 18 psec lifetime of the libration in the ground state and $\tau_6(0)$ the 1.2 psec lifetime (c.f. section E) of the doubly-excited librational level in the excited state at low temperature. Furthermore ω_{31} equals the 12 cm^{-1} frequency of the ground state libration and ω_{64} represents the 16.7 cm^{-1} energy difference between the doubly- and singly-excited librational state. We have calculated, using Eq. 5 and the quoted low-temperature librational parameters, the pure dephasing contribution to the homogeneous linewidth of the libronic transition. This turns out, in the temperature region of interest, never to exceed 10% of the homogeneous width. This fact has kept us from trying to include in the calculation the temperature dependence of the librational lifetime (*vide infra*). This omission has little effect on the calculation of the librational lifetime and *no* effect on the best fit in Fig. 6. By subtracting this calculated dephasing width from the measured homogeneous width presented in Fig. 6, the T_1-contribution to the linewidth can be obtained. From this width it is a trivial matter to obtain the upper limit (*vide supra*) of the temperature dependent librational lifetime ($\tau_4(T)$) as displayed in Fig. 8.

We first note that the temperature dependence of the librational lifetime is much steeper than observed for the lowest-frequency libron in pure naphthalene (29). In addition the low-temperature lifetime of this pentacene libration is a factor of 100 shorter than the lowest-frequency naphthalene libration.

The lifetime of a pseudo-localized libration is determined by libration-phonon scattering processes of which the dominant terms are given by (10):

$$V_{l-p} = \sum_{\kappa} U^{f}_{\kappa\lambda}(b^{+}_{\kappa}B + b_{\kappa}B^{+}) + \frac{1}{3!}\sum_{\kappa,\kappa'} U^{f}_{\kappa,\kappa'\lambda}(b_{\kappa}b^{+}_{\kappa'}B^{+} + b_{\kappa}b_{\kappa'}B^{+} + c.c.) \tag{6}$$

where the notation is similar to that used in the theoretical section. Thus, the first term in V_{l-p} describes the harmonic mixing between the crystal (acoustic) normal modes and the pseudo-localized molecular libration and leads to resonant decay of the libration (18). As the acoustic phonon density of states forms a quasi-continuum at the librational frequency we may employ Fermi's Golden Rule to calculate the resulting resonant decay rate Γ^{l}_{res}.

$$\begin{aligned}\Gamma^{l}_{res} &= \frac{2\pi}{\hbar}\sum_{\kappa}\Big(|<(l-1)_{\lambda},(m_{\kappa}+1)|b^{+}_{\kappa}B|l_{\lambda},m_{\kappa}>|^{2} \\ &\quad - |<l_{\lambda},(m_{\kappa}-1)|B^{+}b_{\kappa}|(l-1)_{\lambda},m_{\kappa}>|^{2}\Big)|U^{f}_{\kappa\lambda}|^{2}\rho_{\kappa}(\epsilon_{\lambda}) \\ &= \frac{2\pi}{\hbar}\sum_{\kappa} l\Big(|<0_{\lambda},(m_{\kappa}+1)|b^{+}_{\kappa}B|1_{\lambda},m_{\kappa}>|^{2} \\ &\quad - |<1_{\lambda},(n_{\kappa}-1)|B^{+}b_{\kappa}|0_{\lambda},m_{\kappa}>|^{2}\Big)|U^{f}_{\kappa\lambda}|^{2}\rho_{\kappa}(\epsilon_{\lambda}) \\ &= l\,\frac{2\pi}{\hbar}\sum_{\kappa}|U^{f}_{\kappa\lambda}|^{2}\rho_{\kappa}(\epsilon_{\lambda})\end{aligned} \tag{7}$$

Γ_{res} presents the decay rate from the librational level with quanta to $(l-1)$, and m is the quantum number of a lattice phonon with wave vector κ. $\rho_{\kappa}(\epsilon_{\lambda})$ is the phonon density of states at the energy ϵ_{λ} of the singly-occupied libration. This expression shows that librational decay proceeds in single quantum steps down the librational ladder leading to a resonant decay rate that increases linearly with the librational quantum number l. The second term in Eq. 6 represents the cubic-anharmonic libration-phonon interaction which leads to phonon up- and down-conversion processes. Using again Fermi's golden rule we come to the following contribution to the librational relaxation rate:

$$\Gamma^{l}_{anh} = l\,\frac{2\pi}{\hbar}\sum_{\kappa\kappa'}\frac{18}{(3!)^2}\,|U^{f}_{\kappa\kappa'\lambda}|^2\,\rho_{\kappa\kappa'}(\epsilon_\lambda)\ \times$$

$$\times\left[\{(1 + n(\epsilon_\kappa) + n(\epsilon_{\kappa'}))\,\delta(\epsilon_\lambda - (\epsilon_\kappa + \epsilon_{\kappa'}))\}\right.$$

$$\left. + \{2(n(\epsilon_\kappa) - n(\epsilon_{\kappa'}))\,\delta(\epsilon_\lambda - (\epsilon_{\kappa'} - \epsilon_\kappa))\}\right] \qquad (8)$$

Here $\rho_{\kappa\kappa'}(\epsilon_\lambda)$ stands for the two-phonon density of states at ϵ_λ for up-

$$\left(\rho_{\kappa\kappa'}(\epsilon_\lambda) = \int_0^{\epsilon_\lambda}\rho_\kappa(\epsilon)\rho_{\kappa'}(\epsilon - \epsilon_\lambda)d\epsilon\right)$$

and down-

$$\left(\rho_{\kappa\kappa'}(\epsilon_\lambda) = \int_0^{\epsilon_\lambda}\rho_\kappa(\epsilon)\rho_{\kappa'}(\epsilon_\lambda - \epsilon)d\epsilon\right)$$ conversion respectively.

In contrast to harmonic decay (Γ^{l}_{res}), Γ^{l}_{anh} is temperature dependent through the Bose-Einstein occupation numbers $n(\epsilon_\kappa)$. Note again that with the chosen anharmonic phonon terms, librational decay proceeds in single quantum steps down the librational ladder and also that Γ^{l}_{anh} is proportional to l. The librational lifetime τ_l resulting from these phonon-scattering processes is related to the sum of Γ^{l}_{res} and Γ^{l}_{anh} in the following way:

$$\tau_l^{-1} = \Gamma^{l}_{res} + \Gamma^{l}_{anh} = l(\Gamma^{1}_{res} + \Gamma^{1}_{anh}) \qquad (9)$$

Fischer (30) has shown that other anharmonic phonon interactions give a negligible contribution to the librational lifetime. For example, direct communication between the doubly excited librational level and its ground state (of the type $\langle 2_\lambda, m_\kappa|b_\kappa B^+B^+|0_\lambda,(m+1)_\kappa\rangle$) is not efficient because the phonon density of states involved in the Golden Rule equation for this process is much lower than that for the processes of eq. (6).

We are now in a position to analyze the librational lifetime data, presented in Fig. 8, in terms of different libration-phonon interactions. The solid line in Fig. 8 is a theoretical fit to the sum of the harmonic and anharmonic relaxation rates of the form:

$$\tau_4(T)^{-1} = \Gamma^{1}_{res} + |V_d|^2(2n(\tfrac{1}{2}\epsilon_\lambda) + 1) +$$

$$+ |V_u|^2(n(\epsilon_s - \epsilon_\lambda) - n(\epsilon_s)) \qquad (10)$$

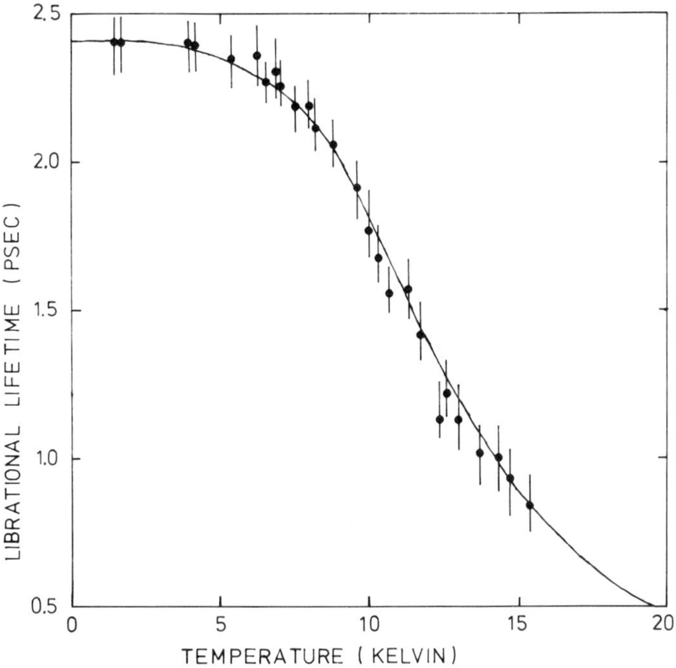

Figure 8. Librational lifetime as a function of temperature. The data points were obtained from the upper curve in Fig. 6 by using Eqs. 2 and 5. The solid line is a fit of the form given by Eq. 10 with parameters mentioned in the text.

with the following parameters

$$\Gamma^{1}_{res} = 1.9\ \mathrm{cm}^{-1} \qquad |V_d|^2 = 0.3\ \mathrm{cm}^{-1} \qquad \epsilon_\lambda = 16.7\ \mathrm{cm}^{-1}$$
$$|V_u|^2 = 80\ \mathrm{cm}^{-1} \qquad \epsilon_s = 50\ \mathrm{cm}^{-1}$$

Here the summations in eqs (7) and (8) are approximated by the leading terms (29). $|V_d|^2$ stands for the effective matrix element in the downconversion process, $|V_u|^2$ for that involved with upconversion. We first note that at zero Kelvin only Γ^1_{res} and $|V_d|^2$ contribute to the linewidth. The relative importance of these terms can be derived from the low-temperature curvature of the linewidth, which makes an accurate determination of the resonant decay and down-conversion processes possible. The resulting parameters used in the fit show that *resonant* decay is the most important one at low temperature. The fit turns out to be rather insensitive to the choice of ϵ_s in the up-conversion process; a larger value of ϵ_s can be compensated by a larger $|V_u|^2$ to maintain a good fit. Irrespective of the precise magnitude of $|V_u|^2$ it is clear that the cross-section for up-conversion far exceeds that of down-conversion as in the case of libron decay in pure naphthalene (29).

With the assumption that the decay of the libration in the ground state of pentacene is affected by similar processes as in the excited state, we can simulate its temperature dependence. The actual best fit to the linewidth data in Fig. 6 was made with the following parameters for $\tau_3(T)^{-1}$:

$$\Gamma^1_{res} = 0.26\ cm^{-1} \qquad |V_d|^2 = 0.04\ cm^{-1} \qquad \epsilon_\lambda = 12.0\ cm^{-1}$$
$$|V_u|^2 = 80\ cm^{-1} \qquad \epsilon_s = 50\ cm^{-1}$$

These parameters were chosen such that the relative ratio of $|V_d|^2$ and Γ_{res} are the same as in the excited state and furthermore that the doorway state in the up-conversion process is the same as in the excited state.

E. Optical dephasing induced by librational overtones

One of the pertinent questions to ask is why at elevated temperature optical dephasing still is describable in terms of a four-level model. First of all there is no a priori reason why other librations should not play a role in the dephasing mechanism. Therefore, in the most general case Eq. 3 should include a sum over all singly-excited librational levels of the molecule. The reason that, in the pentacene case, no other librations seem to play a role in the dephasing process must be that these other librations are of much higher frequency and therefore do not contribute in the temperature range of our studies.

A more serious question concerns the role of librational overtones in the dephasing process. For photosite I of pentacene in benzoic acid the homogeneous width of the optical transition to the librational overtone at 33.3 cm^{-1} from the origin is, again assuming a 0.1 cm inhomogeneous contribution, 4.4 $cm^{-1} \pm 0.2\ cm^{-1}$, which is due to a 1.2 psec librational lifetime. If such a state is dealt with on equal footing as the fundamental libration in Eq. 3 the effect of this state would add substantially to the width of the optical transition at temperatures above 10 K. The reason that these states should not be incorporated in the dephasing model stems from the fact that in the Redfield dephasing theory, as developed by de Bree and Wiersma, for the UPS case (10), the dephasing rate for the optical transition is the sum of the scattering rates *out* of the initial and final states For phonon scattering processes from these levels to the singly-excited libration states these rates are, by the condition of detailed balance, directly connected to the lifetime of the librational levels. This detailed balance argument need not apply to scattering processes that involve higher excited librational levels as they may not directly relax to the librational ground state. In fact Eq. 7 and 8 show that for the libration-phonon interaction term in V_{l-p} librational overtones only relax down the librational ladder, in single quantum steps. A strong argument in favour of this idea is given by the fact that the homogeneous width of the librational overtone, 4.4 ± 0.2 cm^{-1}, is twice the homogeneous width of the fundamental libronic transition in the optical spectrum. In this situation the upward scattering rate to these overtones equals zero and no contribution of these states to optical dephasing is expected. The conclusion of this section therefore is that multiply-excited librational states will have *no effect* on the homogeneous lineshape of the zero-phonon line.

F. Relevance to previous experiments on the unburned origin.

With our present understanding of the mechanism for dephasing in the zero-phonon line of photosite I of pentacene in benzoic acid it seems relevant to return to the problem of dephasing in the unburned origin. As mentioned in the Introduction the situation for the unburned origin is somewhat confusing as different activation energies for the homogeneous width (14, 16, 17) have been reported. When the temperature-activated homogeneous width of the unburned origin previously reported by our group (14) is compared to that of photosite I up to 20 K, hardly any difference between the two is noticeable. We therefore propose that the dephasing mechanism for the two is the same, with a small change in librational parameters. The holeburning results reported by Olson et al. (16), showing an activation of the low-temperature homogeneous width of 11 cm^{-1}, can be interpreted as reflecting primarily phonon scattering in the ground state involving a 11 cm^{-1} libration. The pre-exponential constant of 0.3 cm^{-1} obtained from these low-temperature holeburning data also suggests a long lifetime of the libration in the ground state. As a single exponent was used to fit the data it is obviously incorrect to directly relate this number to a lifetime. Further support for assuming the dephasing mechanism to be identical comes from the fact that the low-temperature photon echo data of the *unburned* origin (fig. 3 of Ref. 14) can also be interpreted using an 11 cm^{-1} activation energy. The transition at 33.8 cm^{-1} from the origin in emission in the unburned origin then should be correlated to the 33 cm^{-1} transition observed for photosite I, rather than being interpreted as the overtone of a 16.9 cm^{-1} ground state libration as suggested previously (14).

Finally the activation energy of 13.8 cm^{-1}, obtained by Andrews and Hochstrasser (17) from a coherent Raman study on the unburned origin, is close to the average librational frequency of (16.4 + 11)/2 = 13.7 cm^{-1} in this transition. We can not, however, reconcile their pre-exponential factor to the librational lifetimes and suspect that this factor must contain contributions from other sources. The interesting conclusion therefore is that the proton rearrangements responsible for photosite I in the benzoic acid crystal have little effect on the frequency and lifetime of the guest libration. The most dramatic effect is seen in the intensities of the phonon sidebands and indicates that the site symmetry has changed in the proton rearrangement process.

V. CONCLUDING REMARKS

The results presented in this paper provide convincing evidence that in the mixed crystal of pentacene in benzoic acid, optical dephasing proceeds via uncorrelated phonon scattering with singly-excited librational levels. In the system pentacene in naphthalene the same mechanism was proposed (21) but a direct measurement of the librational lifetime was impossible. In both cases it was found that the librational lifetime was rather different in the ground and excited state implying a drastic change in librational potential. It is exactly this point which makes exchange of optical coherence in these transitions an unlikely event.

The only well-documented case of coherence exchange in an optical transition presents the system free-base porphin in n-octane. Here Völker et al. (5) found that both the homogeneous linewidth and shift are activated with the same energy which could be related to a pseudo-local mode in the absorption and emission spectrum. In a recent paper from the same group (31) it was concluded that this "exchange model" also could account for the temperature-induced broadening of the homogeneous linewidth, e.g. in site 4, of magnesium porphin in n-octane. We offer a different interpretation for the observations. For site 4, Völker et al. (31) report a local mode frequency of 18.7 cm^{-1} in the ground state and ≈ 27 cm^{-1} in the excited state. If coherence exchange would be the dominant dephasing mechanism then, according to Eq. 4, the homogeneous linebroadening should be activated with the *average* local mode frequency of ≈23 cm^{-1} while experimentally an activation energy of 18.5 ± 1.5 cm^{-1} is observed. This observation is only consistent with Eq. 3 where in the low-temperature limit T_2^* is dominated by the lowest frequency mode of 18.7 cm^{-1} as observed experimentally. According to Eq. 3 the pre-exponential factor in this case should be identified with $1/2\pi\tau_3$. Using the data from Ref. 3 we calculate a librational lifetime of 1.2 psec for the ground state mode. Note that application of the "slow-exchange" limit of Eq. 4 leads to a librational lifetime which is a factor of two longer. In view of the above it would be interesting to attempt to measure the librational lifetime directly in emission from a "line-narrowing" experiment.

Another interesting case is the mixed crystal of triphenylmethyl in triphenylamine where neither coherence exchange nor UPS processes involving a libration are involved in optical dephasing. The obvious reason for this anomaly is the fact that the guest is so similar to the host that the guest librations are amalgamated into the host libron bands. For this system adiabatic coupling to host acoustic phonons, leading to a low-temperature T^7 dependence for the homogeneous linewidth, was found to dominate optical dephasing (32).

It remains interesting to speculate on the reason for the different behaviour of free-base porphin in n-octane compared to all other "aromatic"-type mixed crystals studied sofar (32b). It is clear that in order to observe coherence exchange effects the librational potential must be little affected by optical excitation.

In aromatic-type mixed crystals the guest-host interaction is quite substantial as evidenced by the gas-to-crystal shift which can amount to several hundreds of wavenumbers. In Shpol'skii matrices on the contrary the guest-host interaction leads only to minor spectral shifts of the optical transition. Even and Jortner (33) recently reported for free-base porphin a gas-to-Shpol'skii shift of only tens of wavenumbers. The different behaviour of Shpol'skii matrices compared to the aromatic host crystals must be attributed to the difference in polarizibility of these compounds. We therefore conclude that the likelyhood for the observation of coherence exchange in optical transitions is maximal in those systems where the "cage-polarizibility" is small. Next to Shpol'skii matrices also the noble-gas matrices seem to be good candidates for the observation of optical exchange.

In the case of pure vibrational excitations in either an aromatic or Shpol'skii host the librational potential is little affected and the librational frequency in the ground and excited state is nearly identical. In this extreme limit of exchange theory ($\delta \approx 0$ in Eq. 4), vibrational dephasing by molecular librations is blocked. This effect was recently demonstrated for vibrational transitions of pentacene in naphthalene and benzoic acid (13).

We end by noting that librational decay at low temperature in the mixed crystal of pentacene in benzoic acid is dominated by direct decay into host acoustic modes. It remains to be seen whether this presents an exceptional case or that in other mixed crystals, like in pure crystals (28, 29), cubic libration-phonon anharmonicity determines the librational lifetime.

ACKNOWLEDGEMENTS

The investigations were supported by the Netherlands Foundation for Chemical Research (SON) with financial aid from the Netherlands Organization for the Advancement of Pure Research (ZWO). We wish to thank D.P. Weitekamp and R.W. Olson for helpful discussions, and H.B. Levinsky for useful comments and for participating in the fluorescence lifetime measurements.

REFERENCES

1. D.S. McClure in *Electrical Conductivity in Organic Solids*, edited by H. Kallman and M. Silver (Interscience Publsihers, New York 1961).
2. T.R. Koehler and J. Schmidt, Chem. Phys. Letters 75, 38 (1980).
3. C.A. van't Hof and J. Schmidt, Chem. Phys. Letters 36, 457 (1975).
4. T.J. Aartsma and D.A. Wiersma, Chem. Phys. Letters 42, 520 (1976).
5. S. Völker, R.M. Macfarlane, A.Z. Genack, H.P. Trommsdorff and J.H. van der Waals, J. Chem. Phys. 67, 1759 (1977); S. Völker, R.M. Macfarlane and J.H. van der Waals, Chem. Phys. Letters 53, 8 (1978).
6. C.A. van't Hof and J. Schmidt, Chem. Phys. Letters 36, 460 (1975); 42, 73 (1976).
7. G.J. Small, Chem. Phys. Letters 57, 501 (1978).
8. C.B. Harris, J. Chem. Phys. 67, 5607 (1977).
9. K.E. Jones and A.H. Zewail in *Advances in Laser Chemistry*, edited by A.H. Zewail (Springer, Berlin 1978).
10. Ph. de Bree and D.A. Wiersma, J. Chem. Phys. 70, 790 (1979); Ph. de Bree, Ph.D. Thesis, University of Groningen, The Netherlands, 1981.
11. I.I. Abram, Chem. Phys. 25, 87 (1977).
12. W.H. Hesselink and D.A. Wiersma in *Spectroscopy and Excitation Dynamics of Condensed Molecular Systems*, edited by V.M. Agranovich and R.M. Hochstrasser (North-Holland Publishing Company, Amsterdam 1983).

13. K. Duppen, D.P. Weitekamp and D.A. Wiersma, to be published in J. Chem. Phys.
14. K. Duppen, L.W. Molenkamp, J.B.W. Morsink, D.A. Wiersma and H.P. Trommsdorff, Chem. Phys. Letters 84, 121 (1981).
15. D.E. McCumber and M.D. Sturge, J. Appl. Phys. 34, 1682 (1963).
16. R.W. Olson, H.W.H. Lee, F.G. Patterson, M.D. Fayer, R.M. Shelby, R.M. Macfarlane and D.P. Burum, J. Chem. Phys. 77, 2283 (1982).
17. J.R. Andrews and R.M. Hochstrasser, Chem. Phys. Letters 82, 381 (1981).
18. See, for example:
A.A. Maradudin in *Solid State Physics*, edited by F. Seitz and D. Turnbull (Academic Press, New York 1966), Vol. 18 and 19.
For applications to molecular crystals we refer to G.J. Small, J. Chem. Phys. 58, 2015 (1973); P.N. Prasad and R. Kopelman, J. Chem. Phys. 58, 5031 (1973).
19. T.J. Aartsma and D.A. Wiersma, Chem. Phys. Letters 47, 425 (1977).
20. C. Cohen-Tannoudji in *Frontiers in Laser Spectroscopy (Les Houches, 1975)*, edited by R. Balian, S. Haroche and S. Liberman (North-Holland, Amsterdam 1977), Vol. 1.
21. W.H. Hesselink and D.A. Wiersma, J. Chem. Phys. 73, 648 (1980).
22. H.B. Levinsky and D.A. Wiersma (unpublished results).
23. M. Wagner, Z. Naturforschung 14a, 81 (1959).
24. T.R. Koehler, Mol. Liq. Cryst. 50, 93 (1979); J. Chem. Phys. 72, 3389 (1980).
25. R. Silbey, H.P. Trommsdorff and D.A. Wiersma (unpublished results).
26. A.I.M. Dicker, J. Dobkowski and S. Völker, Chem. Phys. Letters 84, 415 (1981).
27. B.T. Thole, private communication.
28. D.P. Weitekamp, K. Duppen and D.A. Wiersma, Phys. Rev. A27, 3089 (1983).
29. K. Duppen, H.M.M. Hesp and D.A. Wiersma, Chem. Phys. Letters 79, 399 (1981).
30. G. Fischer, Chem. Phys. 16, 181 (1976).
31. A.I.M. Dicker, L.W. Johnson, S. Völker and J.H. van der Waals, Chem. Phys. Letters 100, 8 (1983).
32. W.H. Hesselink and D.A. Wiersma, Chem. Phys. Letters 50, 51 (1977); D.A. Wiersma in *Photoselective Chemistry*, edited by J. Jortner (John Wiley, New York 1981).
33. U. Even and J. Jortner, J. Chem. Phys. 77, 4391 (1982).

PICOSECOND FLUORESCENCE DECAY OF ORGANIC CHROMOPHORES AND BIOLOGICAL PIGMENTS, MONITORED WITH A SYNCHROSCAN STREAK CAMERA

Friedrich Dörr

Institut für Physikalische und Theoretische Chemie,
Technische Universität München, Lichtenbergstrasse 4,
D-8046 Garching, Germany

Abstract

Fluorescence decay curves have been measured for solutions of (I) indigo derivatives and 2-(2 hydroxyphenyl)-benzoxazole (=HBO), and of (II) intact phycobilisomes and biliproteins in various forms of aggregation. The results from group I support the hypothesis that hydrogen bond vibrations can act as effective accepting modes for $S_1 \rightarrow S_o$ internal conversion. Relative stabilities of protomers and conformers in the S_o and S_1 states are discussed. With antenna pigments of photosynthetic algae, both intact antennae and subunits, the energy transfer has been studied. The rather broad emission decays non-exponentially, with the slower decay at longer wavelengths. After polarized axcitation the anisotropic component decays faster than the isotropic one. Structural changes in the protein upon thermal or chemical treatment are clearly reflected in changes of the decay function.

Experimental

A modelocked Ar^+-laser (Spectra Physics 178-18 W plus modelocker 342) pumps a dye laser (either Spectra Physics 375 or a homebuilt ring laser). Without an additional absorber the latter produces pulses of $\tau_{1/2} \approx 0.3$ ps. SHG in 2 mm KDP yields an autocorrelation pattern with clearly resolved fringes, indicating a high transversal coherence (1). The long-time stability is also excellent (6-8 hrs.). The sample cell is positioned immediately in front of the entrance slit of a synchronscan streak camera

K. B. Eisenthal (ed.), Applications of Picosecond Spectroscopy to Chemistry, 127–138.

Headland photonics imacon 675 plus PAR 1205 D SIT vidicon OMA. Operation frequency is 81 MHz. Due to triggering jitter the actual time resolution is, at present, 25 ps. Under favourable conditions, 10 ps were reached with decay curve simulation by forward convolution with the actually recorded pulse profile, and least square fitting. Compounds of group I have been provided by W. Lüttke (indigos) and M. Melzig (HBO). Biochemical samples for group II experiments were prepared by H. Scheer et al. and by W. Wehrmeyer et al.. For details of the preparation and sample characterization see (2) and (3).

Results

I.1. Indigo and derivatives; influence of H-bridges on radiationless relaxation

Table 1 summarizes our former and more recent fluorescence decay measurements in various hydrogen bonding and non-bonding solvents (4).
The decay curves of the indigos are monoexponential within the measuring error. From the Strickler-Berg relation the radiative lifetime is $\gtrapprox$ 3 ns for all derivatives studied. However, compounds (1), (2) and (3) which are capable of forming intraolecular hydrogen bonds, exhibit rather short S_1 lifetimes which do not markedly vary with the solvent, e.g hydrogen bonding ethanol or aprotic acetone or benzene. In comparison, compounds (4), (5) and (6) cannot form intramolecular hydrogen bonds. They show rather long fluorescence decay times which become shortened when H^+ donating or accepting hydrogen bonds to the solvent can be formed. This findings support qualitatively our hypothesis that hydrogen bond vibrations, both intra- and intermolecular , provide effective accepting modes for internal electronic energy conversion.

A strong coupling to the $S_1 \rightarrow S_0$ electronic transition is also indicated by the calculated π-electron charge redistribution: PPP results yield for the S_1 state a marked increase of π -electron density, q, on the O atoms and a decrease on the N atoms. Typical numbers (electron charges) are for indigo $S_0 \rightarrow S_1$: $\Delta q_N = -0.17$, $\Delta q_O = +0.06$. Even if there were no true double minima on the potential curve O... H... N, large FC factors can so arise and a strong coupling will result. In a more detailed model, Heller and Blattmann (5) emphasize the importance of the interaction of H-bridge and skeleton vibrations.

NR.	COMPOUND \ SOLVENT	ABSORPTION MAXIMUM $\tilde{\nu}_{max}/10^3 cm^{-1}$ FLUORESCENCE LIFETIME τ/ps ETHANOL	ACETONE	BENZENE
(1)	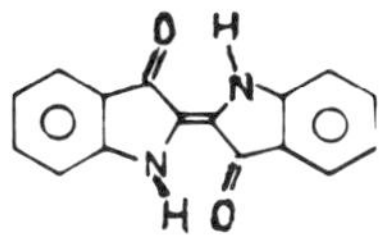	16,4 21 ±3	16,8 67 ±5	16,8 43 ±3
(2)	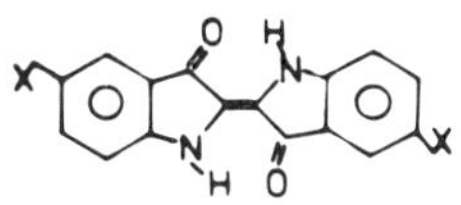	16,0 4 ±3	16,2 30 ±3	16,2 34 ±3
(3)	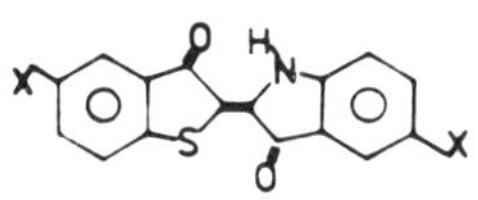	17,0 10 ±3	17,1 15 ±3	17,0 30 ±3
(4)	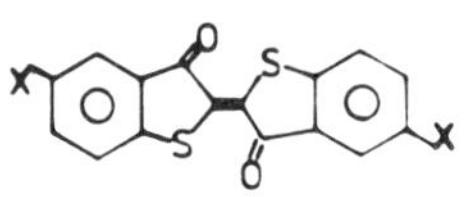	18,0 240 ± 15	18,2 860 ±30	18,0 (9,3ns)+
(5)	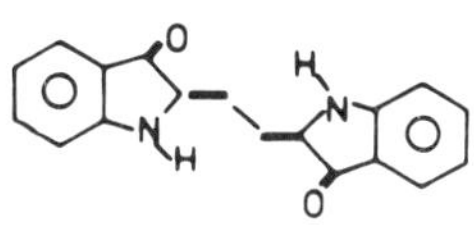	16,9 14 ±3	17,6 150 ±6	17,8 (2,4ns)+
(6)	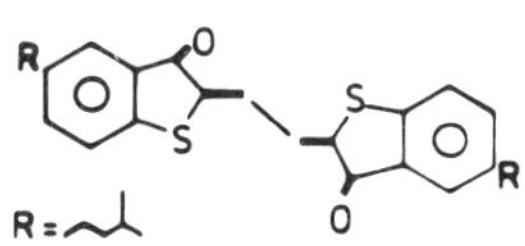	18,9 160 ±15	18,9 (1,25ns)+	18,6 (2,9ns)+

Table 1. Fluorescence decay times τ/ps of indigo derivatives in dilute solutions, T = 300 K. Solvents: E: ethanol; A: acetone; B: benzene. (+) Determined by photon timing.

I. 2. 2-(2 hydroxyphenyl)-benzoxazole; excited state proton transfer

With 2-(2 hydroxyphenyl)-benzoxazole (= "HBO"), the ability of forming intra- and/or intermolecular hydrogen bonds depends on the conformation. In fig. 1, four structures are shown which can be presumed to be relatively stable.

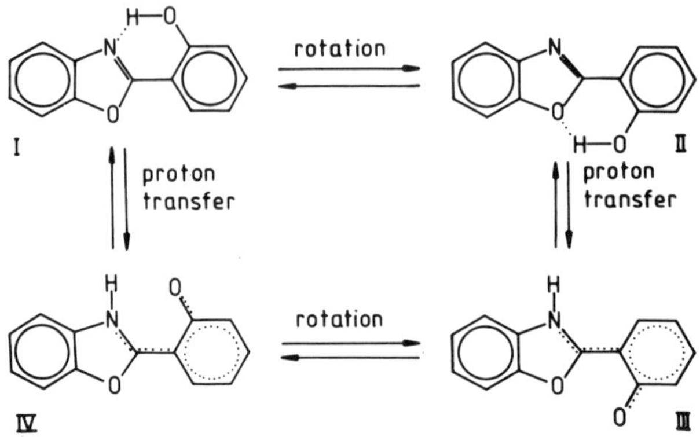

Fig.1 Tautomeric and rotameric forms of HBO

The number of possible forms and the influence upon the respective equilibria from the nature of solvents (aprotic; hydrogen bond donating or accepting; polarity; p_H), from temperature and from deuteration, make the photophysical analysis rather complex. Several authors have already reported on absorption and fluorescence spectra of HBO (6-9). Generally, absorption sets on for $\lambda < 350$ nm, only in ethanol a very weak absorption can be detected around 380 nm. Fig. 2 gives an impression of the drastic changes with the solvent in the fluorescence spectrum.

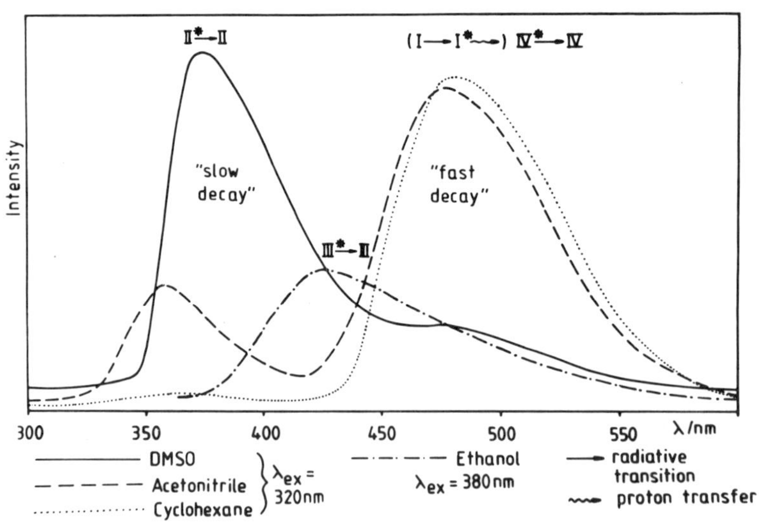

Fig. 2. Emission spectra of 10^{-4} m HBO in various solvents; assignments of transitions correspond to structures of fig. 1.

The complete discussion takes into account absorption, emission and fluorescence excitation spectra in various solvents as well as PPP calculations (10). The decay curves can be fairly well fitted by the convolution of two exponentials which correlate, except for ethanol as solvent, whith the two spectral regions <440 or >440 nm, resp.:

$$I\ (t) = A_1\ \exp.(-t/\tau_1) + A_2\ \exp(-t/\tau_2)$$

For DMSO (a hydrogen bond acceptor) as solvent, $\tau_1 = 68 \pm 8$ ps; $\tau_2 = 1080 \pm 360$ ps. With short pass filters, the following

ratios A_1/A_2 yield the best fit: $400 \cdot\cdot \lambda \cdot\cdot 440$ nm: $A_1/A_2 = 0$; <460: 0.79 ∓ 0.27; <480: 1.89 ± 0.23; <700: 2.84 ± 0.20. (Our streak camera optics cuts off at $\lambda < 400$ nm.)
As an example, fig. 3 shows an experimental decay curve of the total emission of HBO (400 ... 700 nm), the fitted biexponential curve (solid) and the deviation $\chi = (y_{fit} - y_{exp})$ from it; the risetime is not resolved.

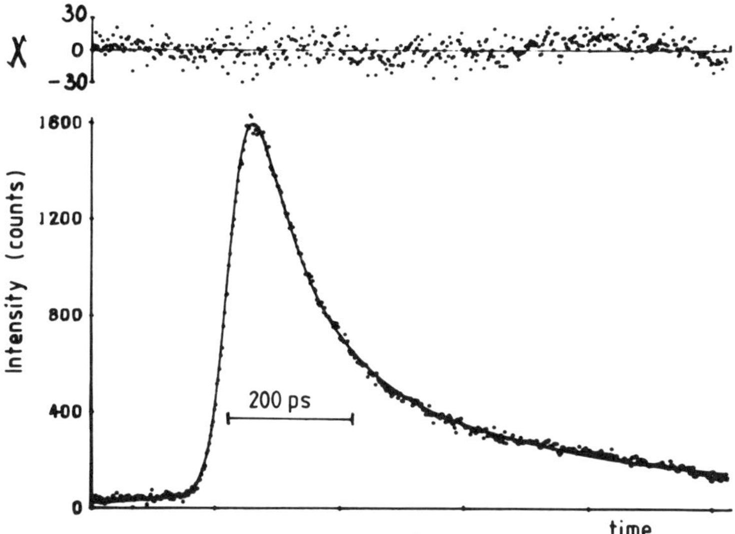

Fig. 3. Fluorescence decay of 10^{-4} m HBO in DMSO. Observed spectral range 400... 700 nm. Solid line: biexponential fit with $A_1 = 0.74$; $A_2 = 0.26$; $\tau_1 = 68$ ps; $\tau_2 = 1080$ ps; scale expanded.

Fig. 4 sums up our model interpretations of the various transitions and conformational changes. "SS" indicates a presumed, not directly observed, "strongly solvated" twisted state in ethanol as solvent.

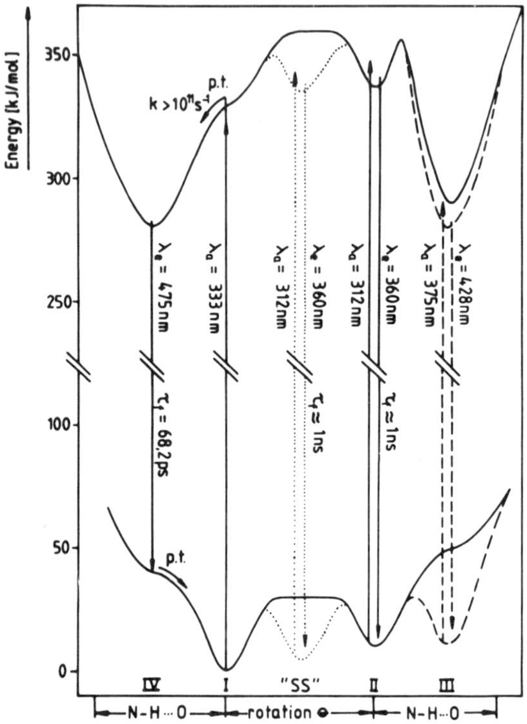

Fig. 4. HBO, estimated potential curves and transitions. Roman figures correspond to structures in fig. 1; p.t.: proton transfer; SS: strongly solvated

II. Energy transfer in antenna pigments of algae Spirulina platensis and Mastigocladus laminosus

Photosynthetic algae and cyanobacteria are equipped with light harvesting antennae, stacks of phycobiliproteins which are organized in complexes, the phycobilisomes. Their chromophores are tetrapyrrole dyes bound to various proteins. For detailed structures see refs. (2) and (3). The arrangement is such that within a stack of biliproteins the different chromophores are ordered according to their S_1 energies in a manner such, that excitation energy can be transferred downhill from the outer surface of the antenna complex to the reaction center, chlorophyll a_{II}. This is represented in a simplified way by the term scheme of fig. 5.

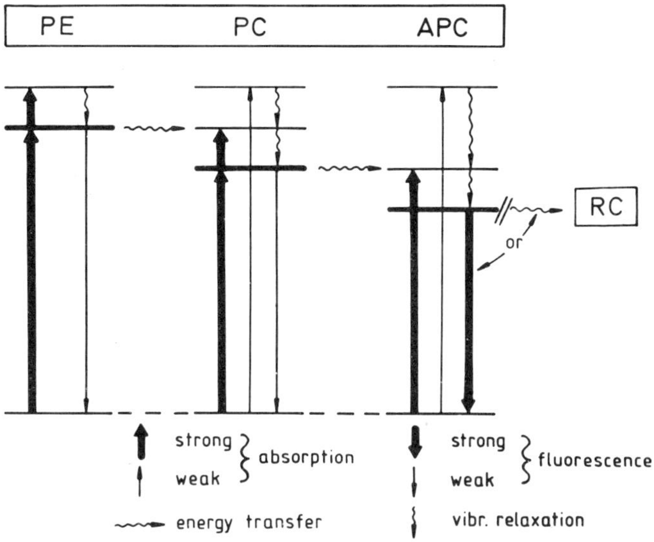

Fig. 5. Excitation energy cascade in the antenna, schematic.

In this figure, PE (570) is phycoerythrin, PC (620) C-phycocyanin, and APC (650) allophycocyanin which provides for the energy transfer to the reaction center, RC. The numbers in parentheses are the longest wavelength absorption maxima in nm of the isolated pigments. They are supposed not to change markedly upon aggregation since the absorption spectrum of an appropriate mixture is very similar to that of the undissociated phycobilisomes. In order to obtain information on the kinetics of energy transfer we measured decay curves, including polarization, of intact phycobilisomes as well as of the constituent biliproteins in various states of aggregation. By the same method the progress of thermally and of chemically induced denaturation was recorded. The optical setup was the same as described above. Excitation wavelength was 600 nm in all cases, thus exciting predominantly PC. The photon flux was $\leq 10^{13}$ photons/cm^2· pulse so that $S_1 - S_1$ annihilation should be negligible (11). Filters and a polarizer sheet (at fixed orientation) could be inserted between the sample cell and the streak camera.

Fig. 6 shows fluorescence decay curves of the isolated intact antenna of Mastigocladus laminosus, after excitation mainly in the PC component, as monitored through different interference filters.

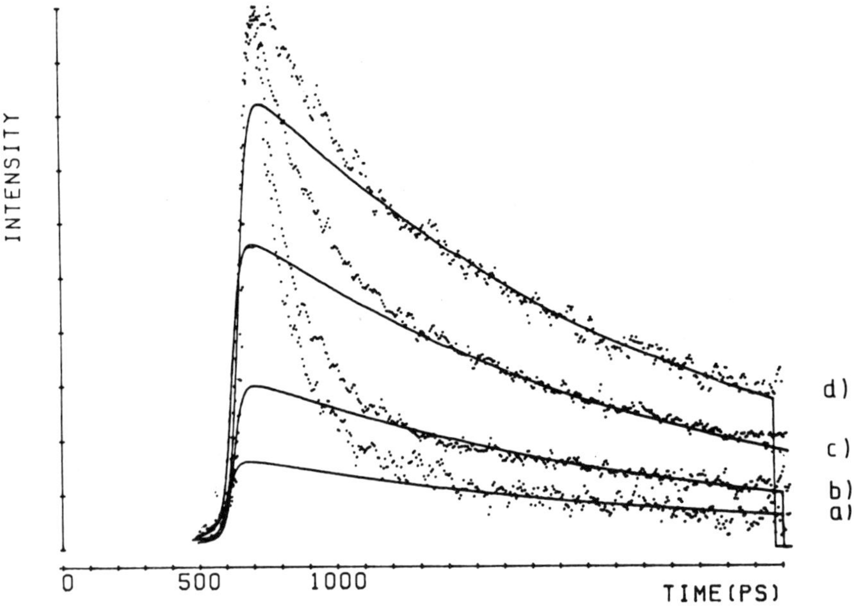

Fig. 6. Antenna of Mastigocladus laminosus, detached from the reaction center: fluorescence decay (λ_{exc} = 600 nm) monitored through different interference filters: a) λ = 614 nm; b) 630 nm; c) 654 nm; d) 668 nm; $\Delta\lambda_{1/2}$ = 12 nm.

From fig. 6 it can be seen that with increasing wavelength a long lived component ($\tau \approx$ 1,4 ns in an exponential fit, solid line) becomes more and more dominant, the lower the observed quantum energy is. This component is presumed to be the emission of the final acceptor, APC. It is quenched in the living cell in which APC transfers the energy to the reaction center. The shortlived part (70... 150 ps) can also be fitted fairly well by one exponential, though model considerations rather would suggest a more complex, e.g. multiexponential or Förster type, behaviour.

This fluorescence is attributed to the overlap of the leakage emission of various PC chromophores in the transfer chain. In fact, the experimental data are still too poor to permit detailed conclusions on the energy flow in this system. Also it has turned out in the fitting attempts that the absolute values of the optimized parameters differ considerably from sample to sample, depending sensibly on the preparation and on aging. Despite this variation the general trends reported here remain the same for all samples.

The depolarization of an anisotropically excited fluorescence yields information on energy migration to differently oriented acceptors and/or orientational realignment of the chromophores on the decay time scale. Geometric factors and other parameters controlling the energy flux are influenced by changes in the protein structure which may be initiated by thermal or chemical treatment. The fluorescence emission was recorded through an analyser of fixed orientation. The excitation pulse polarization was altered by means of a $\lambda/2$ rotator. The measured signals were corrected for the sensitivity of different recording channels. Polarization ratios were tested for artifacts by comparison of $I + 2I$ as computed from measured $I_{\parallel}(t)$ and $I_{\perp}(t)$, with the magic angle ($\approx 54°$) value of $I(t)$; $I_{\parallel(\perp)}$: intensity component with polarization parallel (perpendicular) to that of the exciting light. The buffered solutions were kept at constant temperature, usually 295 K. Only a few typical results are given here as examples in figs. 7-9. For more details, data, and arguments, see the references cited below.

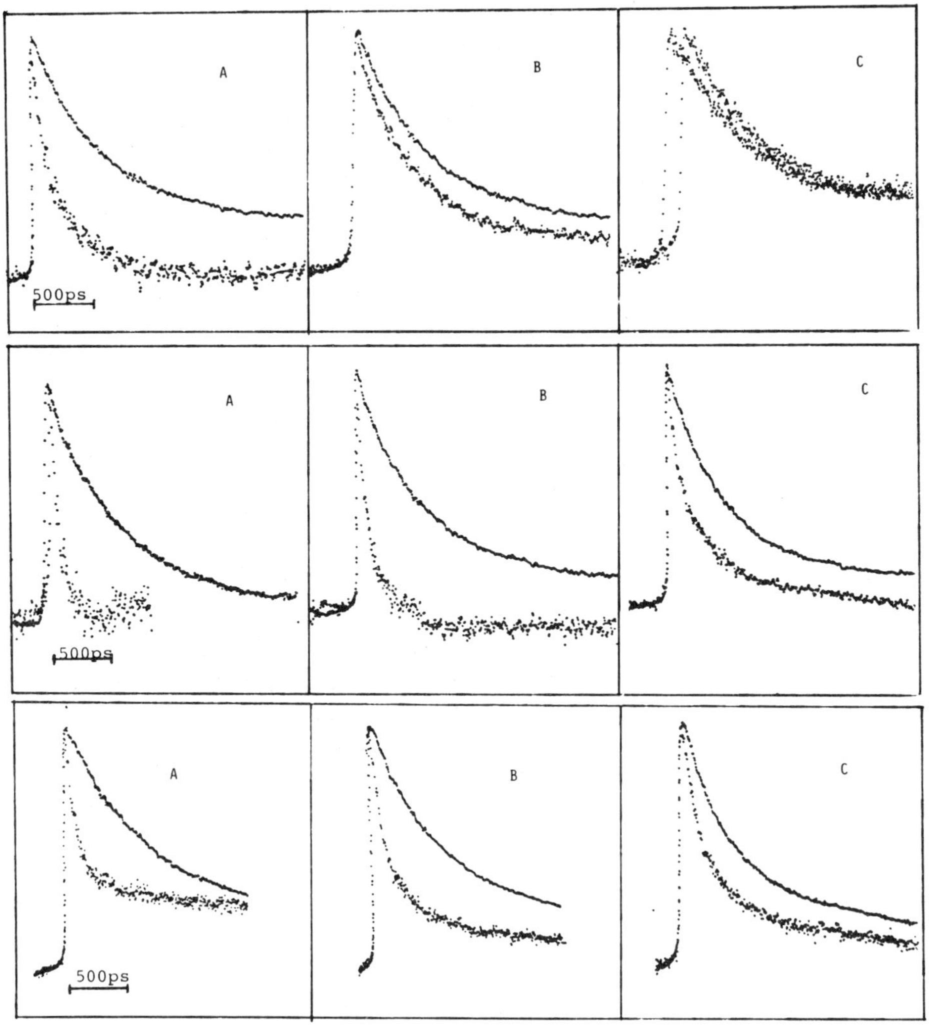

Fig. 7 Spirulina platensis, monomer
Fig. 8 Spirulina platensis, trimer
Fig. 9 Mastigocladus laminosus, PC trimer

Figs. 7-9: Influence of temperature on the decay of anisotropy after photoselective excitation. Upper curves, isotropic, lower curves, anisotropic decay (see text). Sample temperatures: A 18°C; B 39°C; C 52°C; λ_{exc} = 600 nm, $\lambda_{obs} \approx$ 620 - 650 nm.

Figs. 7 and 8 correspond to the monomer and trimer units, respectively, of phycocyanines from alga Spirulina platensis. The freshly prepared buffered solutions (see ref. (2)) were stocked at 0°C. The decay curves of the isotropic emission ($I_{\parallel} + 2 I_{\perp}$, upper curves) and of the anisotropic component ($I_{\parallel} - I_{\perp}$, lower curves) were constructed from consecutive measurements of $I_{\parallel}(t)$ and $I_{\perp}(t)$, taken a few minutes after the indicated temperature had been reached. For visual comparison, the initial intensities have been normalized to the same heights; the general lowering of emission yields at higher temperatures is not represented.

Discussion

In fig. 7A, PC monomer of Spirulina platensis at 18°C, the photoselected anisotropy decays much faster than does the isotropic emission. It is concluded that energy migration to differently oriented acceptors is rather effective. At 31°C, however (fig. 7 B, C), the anisotropy is conserved over a longer time, and at 51°C there is no significant difference between the curves. We attribute this to less effective energy transfer due to changes in the protein structure at higher temperatures. Accepting this interpretation, the trimer of the same biliproteins must have greater thermal stability than the monomer since, in fig. 8, even at 51°C the anisotropy decays still considerably faster than in the monomer. Fig. 9 refers to the PC trimer of the (thermophylic) alga Mastigocladus laminosus. Here, only slight changes occur within the applied temperatur range. The isotropic part of the emission of PC from Mastigocladus (fig. 9, upper curve) differs from that of PC from Spirulina: in figs. 7 and 8 it decays a little slower, in fig. 9 a little faster at higher temperatures. This effect, though small, is reproducible; different preparations (subunits) of the same proteins show the same trends, see refs. (12,13). Upon chemical denaturation of the proteins (with urea), marked changes are observed in the decay functions. These studies are still in progress and will be reported later.

Acknowledgement: Financial support by the Deutsche Forschungsgemeinschaft is highly appreciated.

References

(1) Mindl, T., Hefferle, P., Schneider, S., Dörr, F. 1983, Appl. Phys. B 31, pp. 201-207

(2) Kufer, W., Scheer, H. 1979, Hoppe-Seyler's Z. Physiol. Chem. 360 pp. 935-956; 1979 Z. Naturforschg. 34 c pp. 776-781

(3) Nies, M., Wehrmeyer, W. 1980, Planta 150, pp. 330-337; 1981, Arch. Microbiol. 129, pp. 374-379

(4) Schneider, S., Lill, E., Hefferle, P., Dörr, F. 1981, Nuovo Cimento 63 B, pp. 411-419

(5) Heller, H.J., Blattmann, H.R. 1972, Pure and Appl. Chem. 30, pp. 145-165

(6) Cohen, M.D., Flavian, S. 1967, J. Chem. Soc. B, pp. 317-321

(7) Williams, D.L., Heller, A. 1970, J. Phys. Chem. 74 pp. 4473-4480

(8) Betin, I.I., Nuvumukhametov, R.N., Shigonon, D.N., Chernova, N.I. 1976, Dokl. Akad. Nauk SSSR 227, pp. 126-129

(9) Mordzinsky, A., Grabowska, A. 1982, Chem. Phys. Letters 90, pp. 122-127

(10) Woolfe, G.J., Melzig, M., Schneider, S., Dörr, F. 1983, Chemical Physics 77, pp. 213-221

(11) Searle, G.F.W., Barber, J., Porter, G., Tredwell, C.J. 1978, Biochim. Biophys. Acta 501, pp. 246-256

(12) Hefferle, P., John, W., Scheer, H., Schneider, S. 1983, Photochem. Photobiol, in print

(13) Hefferle, P., Nies, M., Wehrmeyer, W., Schneider, S. 1983, Photobiochem. Photobiophys. 5 pp. 41-51 and 325-334

NON-LINEAR OPTICAL SPECTROSCOPY AND MOLECULAR DYNAMICS IN LIQUIDS

G.A. Kenney-Wallace

Lash Miller Laboratories, University of Toronto, Toronto
Canada M5S 1A1

I INTRODUCTION

Selected for discussion in this paper are some recent studies we have undertaken to investigate nonlinear optical spectroscopy, in both the time and frequency domain, as a tool for exploring molecular liquids and binary systems. The issues we address are important to both the development of a microscopic picture of molecular dynamics and the influence of those motions of the dynamics of photophysical and photochemical processes, as well as on chemical reaction which take place from the ground electronic states. In some respects, given our long standing research interests on the picosecond dynamics of electron solvation, one might subtitle this paper "solvation revisited", since an underlying theme of this work is the nature of solvation dynamics and solvation structures in chemical systems. Ultrashort laser pulses(1) now offer *direct* access to the picosecond (ps) and femtosecond (fs) time domain. What are some of the challenges that we meet at this frontier?

In discussing picosecond relaxation processes in liquids several years ago (2) I noted the parallelism between the exciting developments in the generation of ultrashort laser pulses in the 1970's and the resurgence of interest in understanding liquids, in particular the dynamics of rotational and vibrational motion in liquids *at the microscopic level*. The transitory short-range order and correlated molecular motions believed to describe the short-time behaviour of liquids have been the focus of much of the theoretical attention for which molecular or generalized hydrodynamic models have been developed (3). This regime presents a major challenge to the experimentalist and theorist alike. Inserted into the usual direct link between experiment and theory is the

K. B. Eisenthal (ed.), Applications of Picosecond Spectroscopy to Chemistry, 139–162.

role of computer simulations. The rapidly increasing importance of simulations as an "experimental" testing ground arises because one may observe the response of equilibrium thermodynamic and transport properties of computer liquids to the shape of the intermolecular potential. The latter governs the intermolecular forces and torques of real fluids, and as a multidimensional function of distance and angle it cannot always be clearly divided into attractive and repulsive parts. Consequently, collisional or interaction-induced effects, energy and momentum transfer, rotation, translation and vibration, which collectively govern molecular motion in Nature's liquids, should be experimentally examined for any given system from several experimental techniques. Chosen for their complementary nature, each technique can monitor a different property of the molecule, eg. a labelled spin or dipole moment, which is explicity related to a different correlation function of the motion. The link between spectroscopy and theory is through the formalism of time-correlation functions(4). This interplay between picosecond phenomena and laser techniques is also a major motivation in our picosecond research projects and thus lends itself naturally to a study of liquids.

Nonlinear optical phenomena provide a novel approach to the study of electromagnetic properties of atoms and molecules whose behaviour can be probed spectroscopically through their time or frequency-resolved responses to the applied laser field E (ω,t) (5,6). While the nonlinear susceptibilities $\chi^{(n)}$, n^{th} order in the applied field, characterise a wide range of *molecule-field* interactions, we wish to emphasise that, in high density media, the nonlinear responses could be significantly modified by the *molecule-molecule* interactions, namely n-body intermolecular correlations. These and local electric fields arising from them determine the *effective* intermolecular forces and torques in the liquid and could modulate the actual $\chi^{(n)}$. In the absence of a strong applied field or flow in the field, the molecular orientation is random and so angular distributions are uniform. This is not necessarily the case in strong fields. To investigate these ideas more closely, a series of experiments were designed based on the transient responses to the Optical Kerr effect and on four wave mixing (4WM) interactions in phase conjugation, both of which occur through $\chi^{(3)}_{ijkl}$, third order in the applied field. The primary goals were (a) to examine the "architecture" of $\chi^{(3)}$, given that from a Born-Oppenheimer standpoint the nonlinear susceptibility should have electronic, vibrational and reorientational components, (b) to examine the time-dependent responses of a liquid subjected to both weak and strong fields, through the $\chi^{(3)}$, where *a priori* expectations of statistical mechanics would indicate significant orientational correlation contributions, and (c) to investigate the molecular interactions of liquids under circumstances similar to selective solvation effects. We focus on (c) in this paper.

If we are probing the orientational dynamics of the molecules then we need a reference coordinate scheme as shown in Figure 1. A polarised laser pulse ($\vec{E}_z$) of energy $\hbar\omega$ can undergo a resonant or nonresonant interaction with an ensemble of molecules, one of

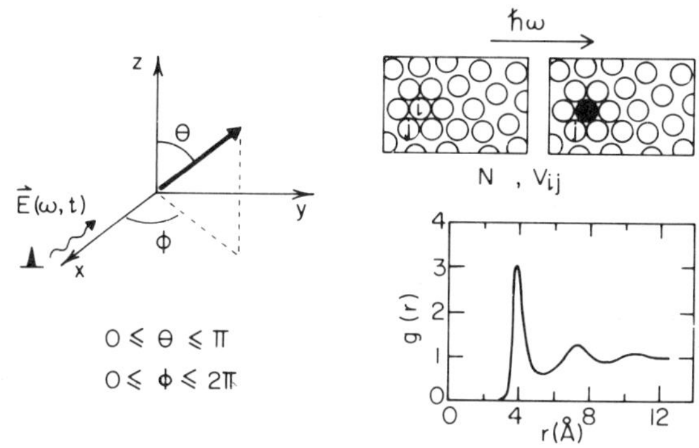

Figure 1(a) Coordinate system for molecular rotation. (b) Volume element of N particles, modelled by V_{ij} pair potential, where i th particle is to be labelled spectroscopically (c) Pair correlation function g(r) of a simple liquid.

which is shown with orientation defined by θ and ϕ. Either molecules are *selected* since the transition dipole moment vector lies coincidentally in an orientation parallel to $\vec{E}(z)$ or have such an alignment *imposed* during passage of the laser pulse, as a consequence of the induced torque experienced by the molecule. The motion of the spectroscopically labelled (black, i^{th}) molecule will be described by a time-correlation function derived from the time-evolution of the probability orientation density, which describes the initial and final orientations about the reference z axis. The nature of the optical interaction will determine the rank of the tensor involved and hence, in the small-step diffusion limit, the type of relaxation time observed. If we dilute the system with solvent molecules, and alter the pair interaction potential V_{ij}, then the new solvation shell should lead to shifts in the pair correlation function g(r) shown in Fig. 1 and the new environment accordingly be reflected in changes in the reorientational dynamics of the labelled molecule.

This paper is presented in four interrelated parts. In section II appear some contextual remarks on solvation phenomena. Since the studies reported here focus on CS_2 as a reference molecule in a range of polar and non polar liquids, section III outlines the reasons for these choices and presents some background material from a molecular dynamics perspectives. In Section IV we develope the nonlinear optical spectroscopy of the Kerr effect and describe some novel phase conjugation experiments. The influence of selective solvation on the dynamics of CS_2 is observed and subsequent changes in the orientational relaxation times are revealed with subpicosecond resolution.

II SOLVATION STUDIES: FROM CLUSTER TO CONTINUUM

The study of solvation phenomena is fundamental to the microscopic structure of liquids and solutions and to the details of the energetics and dynamics of chemical reactions that take place in those solutions. While the simple picture of a few molecules or ions coordinating a central species in water is an image readily reproduced in elementary texts, the quantitative details of even aqueous systems are yet to be established. Neutron, and xray scattering have combined with theory and computer simulations to focus on the determination of g(r), which usually shows density oscillations over distances corresponding to a few molecular diameters before $g(r) \rightarrow 1$ and the onset of the continuum occurs. In theories of solvation, of molecular rotation or local field effects in liquids, it is the boundary conditions describing the transition from the first and second solvation shells into the continuum (or structureless medium) that continue to provide a major challenge. In this respect, experimental evidence that can better define the shape of g(r) as a function of changes in bulk density and temperature also provides a stringent test of the intermolecular pair potentials for that system and the weighted significance of the short and long-range interactions. It is of great importance to evaluate both the static (g) and dynamic (j) pair correlation function, in particular since the latter is expected to be frequency-dependent. Evidence that $g(r) \sim 1$ must be evaluated knowing there are both positive and negative contributions to g(r), and the length scale (or k vector) of the measurement determines the observation. Thus g,j are important factors in shaping the time and frequency evolution of the class of ultra-fast chemical phenomena that depend on particles *escaping* from or *creating* solvation structures. In optical studies of molecular orientation, we can investigate the single particle and many-particle orientation effects, through which information on the (g/j) ratios can be deduced.

What solvation structures do we envisage? Consider electron transport processes in liquids. We studied extensively the picosecond localization and solvation dynamics of excess electrons injected into polar and nonpolar fluids, via laser two photon ionization or ps $\bar{e}$ beam spectroscopy (PEBS) (7,8). In alcohols and alcohol-alkane systems, we observe correlations between the time-dependent IR and visible spectral shifts accompanying the localization and then solvation of the electron and the orientational relaxation times of the solvating alcohol molecules. The data clearly establish that initially the scattered electron samples the statistical structure of the "frozen" fluid at times $<<10^{-12}$s and that at later times the fluid responds to this perturbation in a manner intrinsic to the dynamics of the system. A discrete solvation cluster about the localized electron represents the final, configurationally relaxed quantum state of e^-_s or $(ROH)^-_n$, in which short-range interactions govern the transition and binding energies. We conjecture that $6<n<12$. A quantitative and predictive model for electron solvation demands a more detailed understanding of these solvation structures. While qualitative agreement exists between the earliest ps laser report of e^-_s solvation dynamics in methanol and ethanol (9) and the initial PEBS study, (7a) consideration of all the ps data now available (7-11) also suggests that quantitative interpretations of solvation times cannot be separated from arguments on the photoionization process, energetics and lifetime of the ionizing state of the parent molecule (or ion) in its coordination shell (7b).

These e^-_s chemical species are prototype anions. While relatively short lived ($<10^{-6}$s, typically) they are nevertheless among the most rapidly reacting species; indeed the hydrated electron is now the most kinetically documented of chemical species, matching the H atom its conjugate partner. Since the discrete solvation cluster in which the electron is dressed collapses back into independent solvent molecules upon electron transfer to an acceptor species, this class of simple e^-_s reactions has fundamental appeal to those interested in the microscopic theories of thermal or optically activated electron transfer processes. Thus motivated, we have also explored laser-induced electron transfer (LIET), using e^-_s in alcohol clusters as a donor species. (12) Unexpectedly, a photochemically active channel was discovered to occur within the e^-_s cluster. Above a threshold, the electronically excited state of e^-_s, instead of relaxing non-radiatively to the ground state, coupled to a nearbye potential energy surface from which vibrational dissociation and electron transfer took place to give RO^- + H and (n-1) solvent molecules. This LIET process only occurs under fairly intense laser excitation, and is not a spurious or nonlinear optical phenomenon. Deuterium isotope effect studies pinpoint the LIET mechanism as operating through the OH moiety of an inner ROH molecule in the solvation cluster, and selective solvation studies confirm the intracluster nature of LIET. We are still investigating the origins of the intensity threshold and the role played by the laser field in promoting LIET, which evidence suggests occurs on a subpicosecond time-scale. But once again we are reminded of the pivotal role of the cluster of molecules surrounding the reactive species in determining both the dynamics and energetics of a simple electron transfer process. In studying the formation of e^-_s and in LIET we were also drawn to the question of strong field effects (13,14). In reflecting how a molecule responds to the sudden, strong field perturbation, presented by a bare charged particle or a ps laser pulse, one naturally is led to consider nonlinear optical phenomena in liquids. As will become clear in this paper, nonlinear optical phenomena can not only lead to insights on this question of strong field effects but also present novel routes to the study of the molecular dynamics in liquids.

We have focused so far on problems of past or current interest to us but naturally solvation and nucleation phenomena are ubiquitous, from solution to atmospheric chemistry. Solvation structures play a major role in organometallic and inorganic chemical kinetics, and detailed ps studies of (e.g.) metal-ligand exchange where one solvent molecule substitutes for a ligand, will contribute to our understanding of the differences in dynamics and reaction barriers through such changes in solvation energy on a local basis.

It is also worth noting the possibilities of looking at the solvation structure in its isolated form, that is, short-range order without the continuum contribution. This problem has attracted keen interest in supersonic molecular beam studies of van der Waals complexes (15-19). The successive addition of n solvating molecules to form a solvated species AB_n, where spectral shifts and lineshape changes in the absorption of A as it becomes AB_n, or the fluorescence decay of those consecutively solvated species, can be monitored. These changes in spectroscopic properties can be viewed as building up "condensed-phase" behaviour, but above n=2 often the structures are a matter of conjecture. For example, were successive solvating atoms adding to the central molecule or were they forming a linear chain? It is only very recently that both detailed laser spectroscopic and mass determinations have *simultaneously* been possible (20), thus unequivocally identifying both the energy levels and the structure of each species in the beam expansion for a given set of vibrational, rotational and translational temperatures. From this work it will be possible to examine the specific signatures of liquid or bulk behaviour (collective effects, band structure), so the onset may be correlated with structural information. While one could argue that liquid behaviour is statistical, it is really *a question of timescales*, that is as to whether or not sufficient time elapses during the event under study for all phase space to be sampled. It is also a question of surface area to volume, for over some narrow range of n surely the cluster is mimicing a cold surface rather than a liquid? While these beam data undoubtedly will assist in the quantitative development of ideas of solvation structures, they represent a special case with respect to studies of dynamics in liquids at warmer temperatures, since there the clusters of molecules can exercise many more degrees of motional freedom and high frequency fluctuations play a major role in microscopic behaviour.

III PERSPECTIVES ON MOLECULAR DYNAMICS IN LIQUIDS

The ultimate goal of studying molecular motions is to produce a *model-free theory* of liquid dynamics. By choosing molecules to study whose symmetry properties and minimal number of intermolecular interactions simplify intepretations of the spectral data, one hopes to be able to build up a microscopic theory of molecular vibration, rotation, translational diffusion and rotational-translational and rotational-vibrational coupling. Time-correlation functions (TCF) are the core of understanding time-dependent processes in dense media and linking the spectroscopy to the statistical physics (3,4,21). But relating the orientational time correlations to molecular properties, such as angular velocities, torques, and dynamic pair correlation parameters, in a *model-free* manner presents the most vitally important area of research. Experimental information on the 10^{-13} to 10^{-11} second time regime can thus play a crucial role. In the following paragraphs we briefly elaborate on some of the above statements and conclude by linking these perspectives to the particular nonlinear laser spectroscopy experiments discussed here for molecular liquids. A more detailed discussion of these points appears elsewhere (22).

The calculated response of an equilibrium system to a weak force field is the result of *linear response theory*, which states that when the probe field is *weakly* coupled to the system, knowledge of how the system behaves in the absence of the molecule-field coupling suffices to predict how the system behaves in response to the perturbation. Thus the perturbed system can be treated in terms of TCF of the *equilibrium* system. The fluctuation-dissipation theorem enables one to determine which TCF is "sampled" by a given spectroscopic probe (4,21). For example, the velocity TCF shows the link between the frictional and random forces, and the self-diffusion coefficient is the time integral of the velocity autocorrelation function. The memory function K(t) is the time correlation function of the random forces, and if K(t) is known at short times only, the behaviour of the corresponding TCF can be calculated over the full time regime (4,21) or K(t) deducted from scattering data in order to evaluate *models* of rotational diffusion at early times.

In order to treat molecular reorientation, the reorientational probability density $P(\theta,\phi;t)$ is expanded in spherical harmonics $Y_{\ell,m}(\Omega)$, which from symmetry arguments can be reduced to Legendre polynomials P_ℓ. Wigner rotation matrices $D_{mn}^{(\ell)}(\Omega)$ are used to transform $Y_{\ell,m}(\Omega)$ from the laboratory frame to the molecular frame, ensuring of course that upon rotation energy is conserved. For example, if $\Omega_u(t)$ specifies the molecular orientation unit vector, project u(t) along the dipole in Fig. 1. Any orientational distribution can be completely specified if all of its moments $M^{(n)}$ are known. The spectral moments, calculated from the normalised experimental

$$P_\ell(u(o).u(t)) = \frac{4\pi}{(2\ell+1)} \sum_{m=-\ell}^{\ell} Y^*_{\ell,m}(\Omega_u(o))\, Y_{\ell,m}(\Omega_u(t)) \qquad [1]$$

intensities, can in turn generate the appropriate TCF (4). Since the time-correlation function $C_\ell(t)$ is the Fourier transform of the power spectral density, the moments are the time derivatives:

$$M_\ell^{(n)} = i^n \left\{ \frac{d^n}{dt^n} C_\ell(t) \right\}_{t=o} \qquad [2]$$

While $M^{(2)}$ is invariant to the nature of the interaction since it carries no information on torques, $M^{(4)}$ in principle yields the mean-square torque experienced by the molecule.

The orientational TCF $\langle P_\ell(u(o).u(t)\rangle$ is thus accessible spectroscopically and now may be related to molecular motion, although the interpretations at this stage are not necessarily model-free. In IR for example (ℓ=1) the rotational diffusion of point dipoles leads to the Debye relaxation, where D_r is the rotational diffusion constant (4).

$$\langle P_\ell\ (u(o).u(t))\rangle = \exp\ [-\ell(\ell+1)D_r t] = \exp\ -\ 2D_r t \qquad [3]$$

Interactions with the dipole moment of the molecule via IR spectroscopy, or dielectric relaxation are ℓ=1 or first rank tensor interactions and the correlation function for $T_o^{(1)}$ is $P_1(\cos\theta)$. NMR measurements on nuclei with quadrupole moments are proportional to a second rank (ℓ=2) rotational tensor. Depolarised Rayleigh scattering involving the linear polarizability yields information on ℓ=2 as does Raman. The angular momentum correlation times are thus predicted to be different by factors of $\ell(\ell+1)$ and the TCF from higher ℓ values decay more rapidly. Figure 2a illustrates this in the decay of the P_1 and P_2 correlation function corresponding to ℓ=1 and ℓ=2 rank "experiments" in a computer simulation of a diatomic liquid CO(4). Note the clear differences, the initial t^2 dependence and the eventual *long time* exponential decay. Processes occuring via hyperpolarizabilities or nonlinear interactions will behave as ℓ>2 if diffusion, not jumps, dominate.

In some experiments one probes *single particle* correlation functions $\langle A_i(0)\ A_i(t)\rangle$ such as in NMR and Raman, while in others such as Rayleigh, Kerr and dielectric relaxation the *many-particle* function $\sum_i \langle A_i(0)\ A_j^*(t)\rangle$ and cross terms $\sum_{i\neq j} \langle A_i(0)\ A_j(t)\rangle$ are obtained. Once again, the relaxation times observed in each experiment are *not* anticipated to be the same, collective angular correlation effects generally leading to longer relaxation times. However, it is precisely these differences that lead to important information for fine tuning our understanding of molecular dynamics. Finally, the boundary conditions describing the tangential angular momentum coupling to one molecule to another may not be hydrodynamic *stick* at early times and require further modification of the Debye-Stokes-Einstein approach to rotational diffusion (23). Models invoking non-diffusional processes, such as instantaneous random angular jumps or impulsive randomising collisions, have been proposed too (3,4).

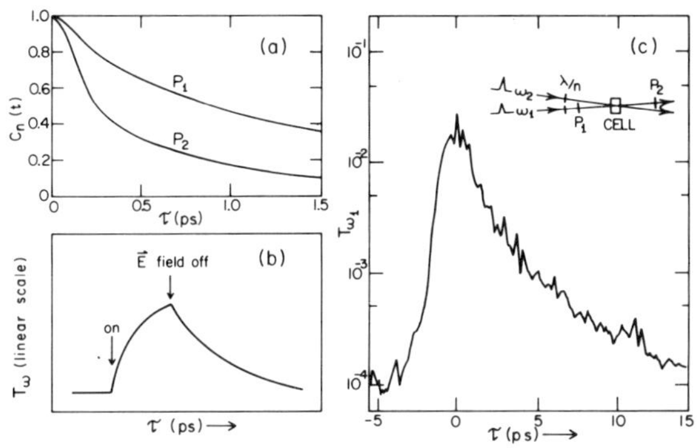

Figure 2a) Computer simulation of TCF for a diatomic liquid showing P_1 and P_2 behaviour; b) Optical Kerr transients; c) Ps OKE effect in chlorobenzene

The above perspectives are really glimpses of the enormous depth and range of theoretical and experimental work that have led to the important questions in liquids under discussion today. In concluding, I want to emphasise three points which are important to examine and to answer for interpretations of ps laser spectroscopy measurements, in particular the nonlinear effects we are studying. First, while linear response theory is a well-established route from which TCF and K(t) linking theory and experiment can be derived, one must be aware of the *weak coupling* condition for the *molecule-field* interaction upon which its valid application rests. *At what point is the system-perturbed far from equilibrium in strong laser field?* Second, simultaneous interactions arising through different $T^{(\ell)}$ terms can occur in some experiments and thus the extent to which it is possible to experimentally distinguish between them may depend critically on the choice of molecular system. Third, comparison with the other experimental or theoretical observations must be made critically, since it is not always the case that all the parameters of the problem are the same, in particular when comparing measurement of the slope versus the integral of a correlation function, such as exemplified by Fig. 2a.

Fourier transform (FT) pervades molecular dynamics theory. From purely a mathematical perspective, the time and frequency domains are rigorously linked through a FT operation and hence information gained in one domain can be readily interconverted into the other, as shown below for the polarization and light intensity in the medium.

$$P(r,t) = \int P(\omega,k)\, e^{\,i(k\cdot r-\omega t)} d\omega dk; \quad I(t) = \int I(\omega) e^{i\omega t} \qquad [4]$$

However, when the dynamical properties of a particular molecular system are under discussion, then experimental, intepretative and even cultural considerations are folded into this perspective to reveal a much more complex issue than initially appeared. Since this topic repeatedly arose during this NATO workshop in the guise of a competition between the time domain and the frequency domain, it seems appropriate to make a few general comments here. The topics chosen for discussion here are selective rather than exhaustive, but focus on the spectroscopy and the statistical mechanics involved in systems under discussion here.

In studying a given molecular system, it is self-evident that one needs to know the positions of the energy levels before the dynamics between any two coupled levels can be studied spectroscopically. Hence the frequency domain plays a hierarchical role in that a *high resolution spectrum* is first required to evaluate the zeroth order description of the system. This zeroth-order picture often breaks down due to overlapping levels or perturbed levels, but these are revealed as extra spectral components in the frequency domain picture. However, once a system exhibits coupling between three or more states such as often happens in larger molecules, and compensating effects mask the perturbations, then *both* time and frequency-domain measurements are necessary to solve the problem. An interesting case is N_2O^+, where spectral line positions do not change but the radiative lifetimes vary significantly between different rotational levels (24). Thus without prior knowledge of the excitation conditions, the FT operation cannot be meaningful. Quantum beats (25,26) provide another example where both time and frequency domain approaches are necessary. Ultimately it is preferable to measure the excitation spectrum with subdoppler resolution to find the coupled levels leading to this coherent phenomenon, since only those levels *within* the $\Delta\omega$ spectral linewidth of the laser give rise to the quantum beats. In the picosecond time domain one must be continually alert to the fact that at some point $\Delta\omega$ carried on the laser pulse will couple in so many levels that such beat phenomena will be washed out. A visible laser pulse of 10ps duration and 0.2 cm^{-1} bandwidth would not reveal quantum beats even in the small molecule SO_2, where a relatively sparse density of states occurs, and yet a 5ns pulse of 0.01 cm^{-1} linewidth revealed dramatically modulated fluorescence decays (26).

In liquids and dense gases where collisions, intramolecular, molecular motions and energy relaxation occur on the picosecond timescale, spectroscopic lineshape studies in the frequency domain were for a long time the principle source of dynamical information on the *equilibrium state* of manybody systems. Lineshapes analyses, although "indirect", are very accurate sources of dynamical information for those motions contributing towards the central spectral components. The information from the spectral wings is more complex. If indeed one observes that a single Lorenztian line shape fits all the data, and thus a single mechanism and exponential relaxation time is deduced, then the spectral wings present no problems. The experimental limits on the precision of such measurements are similar to signal:noise problems in ultrahigh resolution spectroscopy. If however there are several components to be fitted to the lineshape, then the non-Lorenztian lineshape has to be fitted by an algorithm that is not model-free. In other words, mechanistic assumptions built in to the lineshape fitting parameters are a fundamental limitation in complex systems. The wings are particularly sensitive to such effects, since the rotational wings and interaction-induced wings often display comparable spectral density in small molecules. While separation of timescale arguments are frequently used to resolve two components unambiguously, it is also clear that cancellation effects (e.g. between 2,3 and 4 body terms) can be masked by the fact the information is not directly time-resolved (23a, 27).

Direct laser measurements in the picosecond and femtosecond time domain can reveal uniquely this separation of time-scale argument, if valid for a given atomic or molecular liquid. In principle the "spectral wings" component will appear first in a time-resolved experiment, before the evolution of the central components, thus removing model-based fitting parameters that are often invoked in frequency domain analyses. Furthermore, the possibility of observing strong field effects and the time-evolution of dynamical variables of a system perturbed significantly from equilibrium are major goals not only of our studies but for nonlinear response theory, without which the most comprehensive and general dynamical description of molecular liquids will remain incomplete. Both domains are important experimental windows on very fast relaxations processes and, if the system is weakly coupled to the probe beam, they can provide useful complementary information on the dynamical properties of the equilibrium ensemble-averaged state.

We now return to the experiments presented in this paper and the reasons why CS_2 was selected as a reference molecule in the solvents studied. Usually one considers that any two or more particles interact through a field that involves anisotropic overlap, dispersion, and permanent and induced multipole interactions, including those via hyper-polarizabilities (β,γ) when nonlinear terms are included. It is of paramount importance to minimize the number of interactions probed in a given experiment, otherwise there will be too many contributions to the spectrum, and time-dependence of the signal will be difficult to deconvolute with any realism based in the physics of the system. If the molecule possess elements of symmetry, the number of independent coefficients required to describe the mean orientation in space is accordingly reduced. Hence selecting a linear triatomic such as CS_2 or CO_2 of $D_{\infty h}$ symmetry, with only polarizability and quadrupole contributions to the interaction tensor, limits the terms involved. (Of course collision-induced effects are always present in the liquid phase.) Furthermore, in the case of CS_2 it appears as though the *hard core model*, emphasing short range effects, effectively reproduces those features of the potential that also seem to be important to the local equilibrium liquid structures (28). Liquid CS_2 has no strong electrostatic multipole moments, so this is perhaps not surprising. Only at lower temperatures is the quadrupole moment expected to become structually influential in a major way. The theory and simulation reproduces the equilibrium g(r) observed in neutron scattering experiments satisfactorily (28). In computer simulations CS_2 has been studied (28b) using a hard core potential to calculate g_2, which has been deduced from light scattering and Cotton-Mouton experiments (29). CS_2 is most familiar to ps laser studies as a nonlinear optical system, whose fast relaxation responses have been extensively used in optical Kerr shutters since it has one of the highest nonlinear refractive indices n_2 known, 2×10^{-11} esu (30). Thus the *equilibrium* properties of pure CS_2 liquid are reasonably well documented for a polyatomic molecule.

IV NONLINEAR LASER SPECTROSCOPY AND PHASE CONJUGATION

We designed nonlinear laser spectroscopy experiments to probe the *time-dependent* responses of CS_2 as it was driven far from equilibrium and as the nature of its nearest neighbour interaction was altered. Many novel optical effects can be induced by the interaction of intense ps laser pulses through the third-order nonlinear susceptibility $\chi^{(3)}$ of an isotropic medium (5,6). The nonlinear polarization $P^{(3)}$ thus generated can be used to study dynamical effects in *molecular* systems where vibrational and

$$P_i^{(3)}(\omega) = \sum_j 6[\chi^{(3)}_{1122}(\omega{=}\omega{+}\omega'{-}\omega')E_i(\omega)E_j(\omega')E_j^*(\omega') \quad [5]$$

$$+\chi^{(3)}_{1212}(\omega{=}\omega{+}\omega'{-}\omega')E_j(\omega)E_i(\omega')E_j^*(\omega')$$

$$+\chi^{(3)}_{1221}(\omega{=}\omega{+}\omega'{-}\omega')E_j(\omega)E_j(\omega')E_i^*(\omega')]$$

orientational terms as well as electronic can contribute to the overall optical anisotropy. We have been studying phase conjugation and degenerate four wave mixing (4WM), the optical Kerr effect (OKE), and self-induced ellipse rotation (SIER) through both intracavity measurements (31) and in the more conventional external pump-probe configuration (32). The passage of an intense picosecond polarised laser pulse (ω_2) pulse through a nonlinear medium induces a transient birefringence $\delta n(t)=(\delta n_{\parallel}-\delta n_{\perp})$ due to field-induced alignment, which in a liquid will ultimately relax over a timescale (τ_r) correlated to the molecular reorientational dynamics in the fluid. In OKE a second, polarized, but much attenuated laser probe pulse (ω_1) will experience a polarization rotation as it passes through the induced birefringence, and the transmission (T_ω) as a function of the time delay of the probe, through

$$\delta n(t) = n_2^{elec} \langle E^2_{\omega_2} \rangle + \frac{n_2^{mol}}{\tau_r} \int_{-\infty}^{t} \langle E_{\omega_2}(t') \rangle e^{-(t-t')/\tau_r} \qquad [6]$$

crossed polarizers, is shown in Fig. 2b. The birefringence has both an instantaneous and a slower response to the picosecond pump pulse, since n_2^{elec} as the electronic component of the nonlinear susceptibility is expected to follow the laser pulse profile but the molecular component n_2^{mol}, a combination of vibrational and orientational parts, carries with it a time-dependence characteristic of each molecular system (30).

The goals of these continuing projects have been to examine the rise and decay transients in the Kerr effect (Fig. 2b) and thus to examine the limiting cases of the *rise* as a strong field and the *decay* as a weak field relaxation; furthermore, by comparing $T_{(\omega)}$ for SIER and OKE, and selecting different pump and probe polarizations, it should be possible to deduce the relationships between $\chi^{(3)}_{ijk\ell}$ components of the nonlinear susceptibility tensor (32,33). The OKE in chlorobenzene is shown in Fig. 2c (13).

Phase conjugation is a new and exciting nonlinear effect whose rapid development over the past five years owes much to its demonstrated potential for optical image reconstruction. The major thrust of the research has been on the wide range of laser and nonlinear optical device-oriented applications (34). The approach via 4WM based on $\chi^{(3)}$ interactions is a very useful way to produce conjugate waves. Only a few but nevertheless important applications of 4WM have appeared for the investigation of chemical and molecular dynamics (35,36). We focus on our own studies of CS_2 in various binary systems (37) as an example of the power and versatility of the 4WM approach to reveal detailed microscopic information from both frequency and time domain studies. Figure 3 illustrates some of the basic theoretical concepts and experimental 4WM configuration on which our experiments are based. (Theoretical analyses of phase conjugation effects are beyond the length of this paper but have been recently reviewed from many different approaches in ref. (34)). Consider two counter propagating waves originating from the same laser (E_1,ω_1; E_2,ω_2) interacting with a third (E_3,ω_3) to generate a nonlinear polarization $P^{(3)}$ in the medium (dotted circle). The conjugate wave is ω_4, a time-reversed image of the ω_3 input wave, and the field $E_4(r,t) = E_3^*(r,t)$ for all distances $z<z_o$. A macroscopic nonlinear polarizability $P^{(3)}$ is generated at frequency ω_4. Phase matching constraints lead

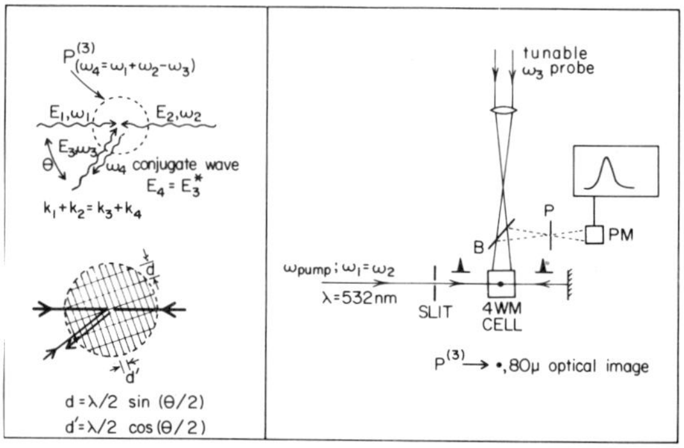

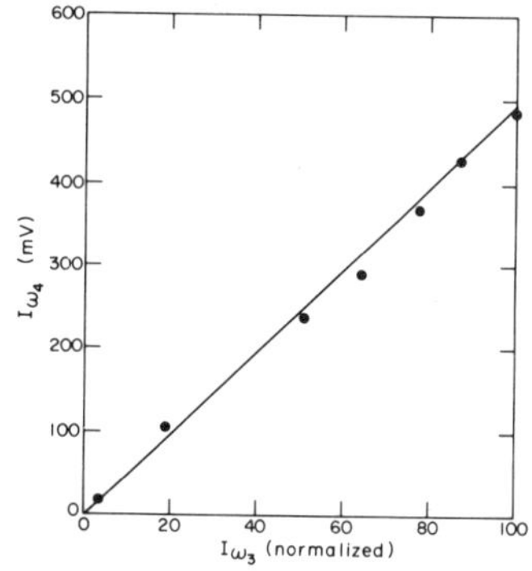

to $k_1+k_2 = k_2+k_4$. Figure 3 and 4. See Text for discussion.

$$P^{(3)}_{(\omega_4=\omega_1+\omega_2-\omega_3)} = \chi^{(3)}E_1(\omega_1)E_2(\omega_2)E_3^*(\omega_3) \qquad [7]$$

If $P^{(3)}$ is the vehicle for generating a conjugate wave, then we predict that the molecular motions of the nonlinear medium leading to loss of phase memory will modulate $P^{(3)}$. Thus the intensity, spectral linewidth, centre frequency and polarization of ω_4 should carry information on those molecular motions. Monitored in a pump-probe configuration, the properties of the conjugate wave can be studied while the "*coherence*" or "*phase*" grating is initially established and subsequently *destroyed by these intermolecular or intramolecular dynamics of the molecules which lead to loss of coherence in the ensemble.*

Actually, in degenerate 4WM, interference of the waves leads to several "phase gratings" being established in the liquid with spacing proportional to θ the angle between the beams. Several gratings can be simultaneously established with various permutations of ω_1 to ω_4, but those shown in Fig. 3 illustrate the grating (spacing d) between ω_1 and ω_3 from which ω_2 scatters into ω_4, and the more narrowly spaced grating (d') from interference between ω_2 and ω_3, from which ω_1 scatters into ω_4. Each grating will have a different polarization criterion and scattering efficiency, or reflectivity R for the conjugate wave. In our experiments we are studying ω_4 in transparent (lossless) systems and so only the real part of $\chi^{(3)}$ is involved and no thermal gratings appear. Several studies of "amplitude" gratings involving absorption to excited electronic states, from which photochemistry (38), radiationless relaxation, energy transfer and acoustic effects (39-41) can follow, have been recently reported and illustrate the versatility of grating experiments.

In Figure 3b the experimental configurations for phase conjugation studies in both time and frequency domain are shown. For the quasi degenerate 4WM tunable laser studies, from which we recorded the *spectrum* of ω_4, we used a Quanta-Ray DCR-1A Q-switched, frequency-trebled Nd:YAG laser to pump a coumarin 500 dye laser at 355 nm, and then chose pump frequencies $\omega_1 = \omega_2$ at 532 nm and probe ω_3= tunable, to generate the counter propagating conjugate wave at ω_4. A full description of this study is published elsewhere (37) and we will note only salient details here. The pump and probe beams were linearly polarized and, in order to minimise transverse transit time effects, the pump beams were focused to a line 80μ wide in the 4WM cell using a cylindrical lens. Transit time effects here refer to the temporal uncertainty imposed by the fact the conjugate wave has components generated from across the full 80μ segment. In n-hexane, for example, with a refractive index of n_D^{20}= 1.37506, visible light travelling right across this "slit" image would correspond to a tranverse uncertainty of 194fs. In fact this is an upper limit and the actual uncertainty is averaged over smaller distances and depends on the phase grating prepared, but transit time effects will put a *lower* limit on the determination of the molecular motions involved. A smaller optical image would lead to smaller transit times, so this is not a general limitation and 80μ was merely a design choice in the optical 4WM experiments here. The 400μ pinhole (P), which acts as a coarse filter to remove scattered light from the backwards propagating ω_4 signal, is at least ten times larger in diameter than the beam waist of the conjugate wave; thus it does not block the wave. Typical power densities in the focused ω_1, ω_2 beams were 30 MW cm^{-2} while the tunable ω_3 probe wave was 50 $kWcm^{-2}$. The linewidth of ω_3 was 0.1 cm^{-1}, and folding in all instrumental and (conservative) transit time corrections, the precision on the conjugate wave spectra shown in Figure 5 is 0.2 cm^{-1}. The degenerate 4WM measurements of I_{ω_4} were also carried out using a Q-switched, 20ns pulsed ruby laser at 694nm and $\omega_1=\omega_2=\omega_3=\omega_4$. Full details will appear elsewhere (42). Again pump (ω_1,ω_2) power densities were typically 30MW cm^{-2} in the unfocussed beams and up to 100 MW cm^{-2} when focused. The reflectivity (R) of the conjugate wave fixed pump intensities in CS_2 over a given interaction length ℓ is given as [8].

$$I_{\omega_4}=[n_{CS_2}\ \chi^{(3)}]^2\ I_{\omega_1}\ I_{\omega_2}\ I_{\omega_3}\ ;\ R=\left\{\frac{\omega\chi^{(3)}}{\varepsilon_o n^2 c^2}\right\}^2 \ell^2\ I_{\omega_1}\ I_{\omega_2} \qquad [8]$$

We calculated R ∿ 5 x 10^{-3} for comparison with the observed R ∿ 4 x 10^{-3}. Figure 4 shows the predicted linearity of the response in the amplitude of ω_4 for varied probe pulse intensity in ω_3 at constant (I_{ω_1} I_{ω_2}); the data are from the ruby experiments but of course the behaviour is general for all such 4WM schemes. The ω_4 signal disappears completely if any of ω_1,ω_2 or ω_3 are blocked.

We will now discuss data from both frequency and time domain studies of CS_2 to investigate selective solvation effects with various classes of solute-solvent interactions. Figure 5 shows three conjugate wave spectra. The middle spectrum is that generated in pure CS_2 liquid by tuning ω_3 in the Nd:YAG experiments. The line width of ω_4 is broadened in comparison to the ω_3 wave to reveal a spectrum with a Lorenztian lineshape well down into the spectral wings, where ultimately the resolution in the far spectral wing is masked by the transit time effects. However, there appears to be no hint of a second wing component. The dashed line is a best fit to the data from which the 2.3 cm^{-1} (hωhm) linewidth was calculated to give a *single exponential relaxation time* of 2.2 ± 0.2 ps, where the 0.2 ps includes *all* experimental uncertainties. This is the *dephasing time* of the phase grating in the pure liquid at room temperature. Also in Fig. 5 is the ω_4 spectrum for CS_2 in alkane (RH) n-pentane (20% volume fraction), a liquid chosen for low $n_2 = 4 \times 10^{-13}$esu, its weak C-H dipole interactions and quite contrasting liquid structure to CS_2. It is clear that while the spectrum has broadened, it still maintains a Lorenztian line shape, and thus the CS_2 molecules, *which are the predominant source of nonlinear polarization $P^{(3)}$ in the fluid*, are rotating faster. In contrast, in a more strongly dipolar interacting system, CS_2 in an alcohol (ROH), such as phenol at a 50% mole fraction (η~1.5 cp) the conjugate wave displays quite different behaviour. The narrow central line and much broader wings indicate that at least two relaxation mechanisms are responsible for the loss of coherence. These examples illustrate clearly the difference between CS_2 in a solvation structure where the interactions change from primarily induced-dipole and quadrupole to dipolar, in which correlations from the torques of the permanent dipoles are involved too. The structure of the solvation shell and immediate environment will govern the dynamics and strongly interacting molecules will rotate more slowly.

What are the molecular dynamics leading to loss of coherence as manifested by ω_4? Translational diffusion is important only at long (μs) times. Collective orientational relaxation is the mechanism commonly assigned to previously observed ~2ps relaxation times in CS_2 and so we conclude that the principal mechanism leading to loss of coherence is indeed rotational motion of the CS_2 molecules.

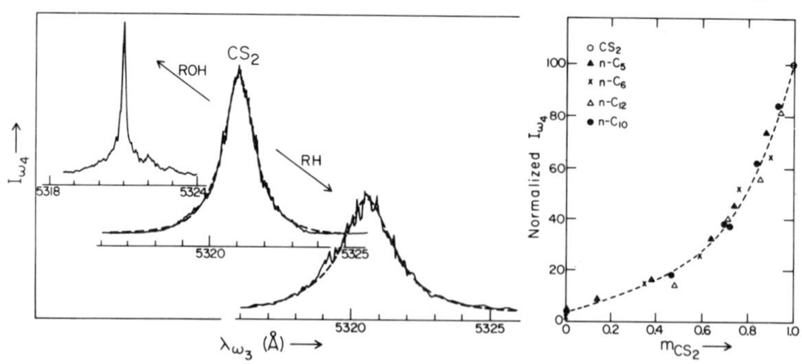

Figure 5 Conjugate Wave Spectra in CS_2.

This is also a quantitative and direct confirmation of the *reorientation* origin of the short component in the CSRS Raman dephasing experiments reported earlier in CS_2 (43). There the dephasing time of the ν_1 Raman mode at 656.5 cm^{-1} was seen to be 20 ± 1 ps, but a temporally unresolved ps component accounted for a substantial fraction (>80%) of the loss of coherence. Temperature-dependent Raman studies (44) show vibrational dephasing is also very long in CS_2 compared to correlation times of the density fluctuations in the medium, indicating that the mean interaction potential is rather weak. The evidence on dephasing from both these 4WM rotational and the cited vibrational studies suggest that elastic collisions are not very effective at disrupting the imposed phase memory. *If this were the case, we would not expect to see collision-induced effects in CS_2 (such as observed as spectral wings in the depolarized Rayleigh experiments) in these 4WM coherently driven phase conjugation experiments.* Thus we can describe the mechanism probed via phase conjugation in CS_2 as a pure reorientational dephasing time, which at high power densities will evolve and decay in time as a *coherently driven* Kerr effect. That is, the initially prepared phase conherence in the ensemble will be accompanied by a high degree of orientational alignment. Hence the relaxation times may change as a function of pump intensity if the molecule-molecule relaxation-interactions are shape-dependent, e.g., 2-body and higher angular correlations are present. Such experiments are in progress (42).

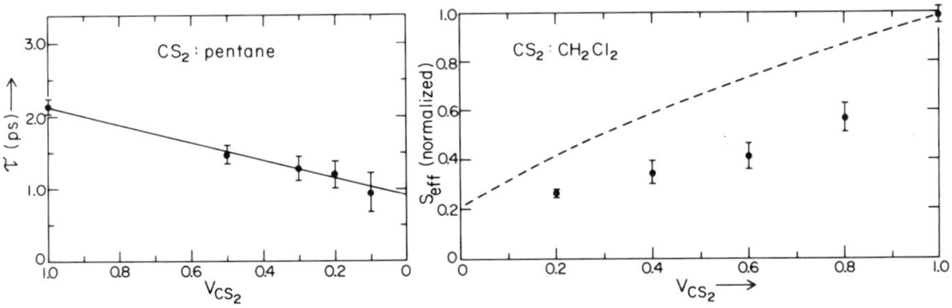

Figure 6 Relaxation times and Interactions of CS_2 systems.

When CS_2 is dilute in a nonpolar liquid of vanishingly small $\chi^{(3)}$ such as an alkane (RH), the spectrum of ω_4 changes lineshape to reflect the concomitant changes in the microscopic environment, namely the local intermolecular interaction and thus the reorientational motions. The viscosities of CS_2 (0.36 cp) and n-pentane (0.24 cp) are not very dissimilar at room temperature, and the orientational relaxation time of n-pentane was recently measured to be ~2.2ps (45), similar to pure CS_2. The CS_2-alkane interactions and structure however are quite different from the pure liquids. The relaxation time (derived from the spectral linewidths of ω_4) is plotted as a function of volume fraction (V) of CS_2 in figure 6a. Even in the most dilute solutions, less than 10% of the signal is solvent background and the dominant source of $P^{(3)}$ are the now isolated CS_2 molecules. Upon substitution of alkane for CS_2 molecules in the solvation shell the local density, volume and g(r) parameters must shift, which is why all data are normalized to a volume fraction rather than number density. *For each dilution ($0.8<V<0.2$) there will be a statistical distribution of n-pentane molecules in the *inner* solvation shell, and this could lead to a distribution of relaxation times within the full ensemble; thus these data represent the most probable times associated with rotation in the first shell solvation structure. Since CS_2 is governed by hardcore, short range interactions, changes in and beyond the second shell can be neglected as a primary influence on the dynamics. At infinite dilution, $\tau \sim 0.9 \pm .1$ ps.**We conclude from Figure 6a that the *changing microscopic interactions around CS_2 and the subsequent changes in the orientational relaxation times can be revealed through these 4WM phase conjugation experiments with subpicosecond resolution.* Furthermore the nature of the correlation function changes from the initial many body CS_2-CS_2 correlation function $\sum_{i,j} \langle A_i(0) A_i(t)\rangle$ or *collective* orientational time to the ensemble averaged *single particle* time $\sum_i \langle A_i(0)\, A_i(t)$, corresponding to orientational dephasing of CS_2 in n-pentane.

How can one separate, at least to a first approximation, dynamical effects from structural effects? In ns times, a *steady-state* phase grating is established, which reflects a conjugate

*In pure liquid CS_2, the neutron scattering data are consistent with a loose hexagonal arrangement of CS_2 molecules about a reference molecule, with the first very sharp intermolecular peak at 3.9 Å (S-S) and no significant structure beyond ~10 Å.

**Upon dilution, viscosity changes (following Debye-Stokes-Einstein predictions) would predict 1.45 ps for single rotation in n-pentane, assuming spherical geometry for the volume swept out by the particles. Not surprisingly, the τ recorded here while linear in viscosity do not yield the predicted slope.

wave that carries information on the interactions of the *orientationally-averaged* ensemble. Thus the amplitudes of the ns $I_{\omega 4}$ signal in the ruby degenerate 4WM experiments are plotted as a function of the molefraction of CS_2 (m) in a range of alkanes of varying chain length, from n-pentane to n-dodecane; the data all fall on a similar curve as Fig. 5a shows. Given that the monomer orientational times for these alkanes increases with chainlength but are nevertheless all $<<10^{-9}$s, the intensity of ω_4 becomes a measure of $P^{(3)}$ *modulated by the time-averaged, many-body interactions of* CS_2 *embedded in the alkane structures.* Thus the data show that the linear alkanes exhibit comparable CS_2 alkane steady-state interactions. If we investigate different classes of molecules, whose symmetry, structure and interactions with CS_2 very significantly from class to class but are very similar *within* a class, then we find a separate curve analagous to that for alkanes in Fig. 5 for dipolar alcohols, for aromatic hydrocarbons, and for halomethanes (42).

In a *non-interacting model*, the nonlinear polarizability of CS_2 will not be influenced by its solvation structures and thus no amplitude modulation of $\chi^{(3)}$ is expected. *Strong interactions* will enhance or diminish $\chi^{(3)}$. The assumption of pair-additivity underlies all calculations of intermolecular potentials. Similarly, if the dynamical interactions probed by 4WM are merely angle-dependent pair interactions leading to an effective $\chi^{(3)}$, which is the algebraic sum of the individual $\chi^{(3)}$ (CS_2) and $\chi^{(3)}$ (solvent) components, then a $\chi^{(3)}$ can be calculated as shown (dotted line) in Fig. 6b for CS_2 in CH_2Cl_2. The actual 4WM signals observed (S_{eff}) are also plotted for the various solutions, where $\chi^{(3)}$ is calculated from $(I\omega_4)^{\frac{1}{2}}$ and $S_{eff}=\chi^{(3)}/n^{-2}$, normalized for V=0, 1 to the pure liquid values and corrected for refractive index (n) differences in the solutions (42). The data quite clearly deviate from the non-interacting model and demonstrate that in the CS_2-CH_2Cl_2 systems the local dipolar interactions and orientational correlations have substantially modified the nonlinear polarizability of CS_2 and thus the *molecule-field* interactions. Local field corrections, when incorporated following several different prescriptions (3), do not alter the degree of deviation of the data from the non-interacting case, because both the calculated and observed $\chi^{(3)}$ are similarly scaled (42). Once again, picosecond 4WM experiments currently in progress are determining the time evolution and decay of the conjugate wave and the amplitude of different contributions to $\chi^{(3)}$, namely electronic, vibrational and orientational, to determine which part of the "architecture" of $\chi^{(3)}$ has been modified by the *molecule-molecule* CS_2-CH_2Cl_2 interactions. These data will also reveal the dephasing tunes for specific dipole, induced-dipole and quadrupole interactions for TCF analysis (42).

We have accomplished the first steps in all the primary goals as outlined in the introduction and shown how nonlinear laser spectroscopy is a powerful tool for dynamics. In summarising these ongoing phase conjugation experiments, let us reiterate three points. First, the lineshape of ω_4 is sensitive to all processes intimately linked to the intermolecular (and intramolecular) interaction which result in loss of the initially imposed coherence through $P^{(3)}$. Second, 4WM can reveal significant microscopic changes in the dynamics of these CS_2 binary systems, in which the solvent is essentially invisible to the laser field but the solute: solvent interactions and thus g(r) and angular correlations are systematically changing. Third, that the TCF appropriate to describe interactions through a 4th rank tensor such as $\chi^{(3)}$ should be based on a P_4 description of the orientational probability density, which should decay 3.3 times faster than P_2 *in a small step diffusive model*. Since the relative amplitudes of the P_4 and P_2 components and the role of 3 and 4-particle correlations (to which P_4 is more sensitive) have also to be considered in evaluating the TCF, comparisons between Kerr and 4WM data with those τ_{or} from other spectroscopic techniques must be done carefully (for the reasons discussed in section III) before firm conclusions regarding the *model* of molecule motion can be established (i.e. diffusive versus large step changes in strong collisions). How do the Kerr and Rayleigh times compare from an empirical standpoint? In OKE the collective orientation density probed corresponds to only forward scattering (or zero scattering vector) in contrast to depolarized Rayleigh scattering, which contains Fourier components for all scattering angles. While consistently longer Kerr times (2.2 ± .2 ps) had been recorded for CS_2 than those seen in depolarized Rayleigh light scattering (1.7-1.9 ps), evidence has just been presented to suggest power broadening effects may have influenced some high power density Kerr measurements (46b). In fact, that observation is consistent with our argement that the more orientational alignment prepared in the ensemble, the more likely the relaxation time will *not* correspond to the equilibrium case, and assumptions built into linear response theory may well break down too. The OKE low power limit of 1.56 ps (46b) is now somewhat faster than the most precise Rayleigh measurement of 1.7 ps (27a), but similar to the Raman deduced single-particle time of 1.5 ps. If small step diffusive motion dominated CS_2 relaxation-interactions, and if the P_4 component were rather small, then in the limit of low power the Kerr time should converge to the Rayleigh, since both probe many-particle times. But this is not a model-free conclusion. Thus we believe more precision studies of CS_2 orientation are needed to evaluate both the if's and the physics of the strong field and weak field relaxations, via interactions of different tensorial rank.

CS_2 poses further challenges in femtosecond (fs) spectroscopy. In work recently published onthe ultrafast responses of CS_2, using fs pump-probe studies with $\omega_1=\omega_2$ (λ=615nm or 620nm), two components have been resolved of lifetime ~300 fs and ~2ps respectively (32, 46-48). OKE studies (32, 46, 48) measure $\delta n(t)$; the phase retardation measured in interferometry is δn_{11} (47). While in our initial experiments the pulse duration (t_p) was 300 fs and the ultrafast response 300 fs, if of purely of electronic origin, it could not be unambiguously distinguished from t_p (32). The shortest pulses used to date of t_p ~70fs clearly indicate that this very fast signal is of longer duration (46,47). Times now reported are 200 fs (48), 240fs (46) and 330 fs (47). The longer component ranges from 2.2 ps (30b, 32, 34), 1.7 ps from our unpublished OKE data, 2ps (47), 2.17ps (46a) 1.56ps (46b) and 1.4ps (48). While curve-fitting and deconvolution procedures are sensitive to lack of knowledge of the skewedness of the laser pulses unless third-order correlation function pulse diagnostics are made, the *fast* component in CS_2 undoubtedly corresponds to a real relaxation mechanism, whose origin could be electronic or molecular in the sense that it is an interaction-induced process governed by molecular encounters. The latter could modulate the total electronic polarisability at very high frequencies. In depolarized Rayleigh studies (27,44) the spectral wings have long been attributed to interaction-induced effects of a collisional nature, and this is the most immediate explanation to consider (46). (It is perhaps a coincidence to note that the fast component in CS_2 is indeed of the correct order for the P4 response) The real question arises as to how these higher-order relaxation times can be extracted from interaction-induced effects, when they all can occur for small molecules on comparable times scales? This includes effects leading to "over-shoot" or oscillatory librational behaviour for a particle in a cage of solvating molecules. Librational motions are extremely fast and of the order of the 100-300 fs responses observed here. If the origin of the fast response were purely total electronic polarizability, the signal should follow the laser pulse profile. The decay appears longer in CS_2. If vibronic coupling or any electronic (e.g. 2-photon) *resonance* contributions were present, it would introduce a damping term into the susceptibility expression for $\chi^{(3)}$ and we would anticipate a real linewidth or finite relaxation time to be observed for the ultrafast response. Such two-photon absorption in liquid CS_2 has been observed in this spectral region (49). While some have concluded no electronic effects in CS_2 exist (50), it is obvious that in the limit of ultrashort fs pulses there must be electronic hyperpolarizability and thus it can be sampled in nonlinear laser experiments. The *real* question is how important a contribution $\chi^{(3)}$ makes to the magnitude and dynamics of the fs response. Thus electronic resonant enhancement (and the coherent transient 4WM argument (30a,32,51)) can only be properly investigated by *tunable* fs studies, as we have stated earlier (32), in both molecular and atomic liquids. The amplitude of the collision-induced effects should be density-dependent and contain n-body terms. There are time-scale arguments to consider as well in predicting the relative amplitudes of the electronic to molecular (orientation and vibration) contributions to $\chi^{(3)}$. With laser pulses $t_p \gg \tau_{or}$, all degrees of motional freedom in the molecule could develope a full polarization response. If $t_p << \tau_{or}$, the question arises as to how far the orientation polarization response to $P^{(3)}$ can be developed *before* the optical field is switched off. These arguments predict a ratio of electronic orientational contributions that is time-dependent. The torque experienced by the molecule and subsequent relaxation will be both $\vec{E}$ and shape-dependent. When the optical field is switched off, the field-induced angular momentum does not immediately vanish and the torque could persist for a finite time, which may be longer than the correlation time of density fluctuations in the liquid. Finally (but not necessarily the end of this list) it is important to recall that the degree of orientational alignment attained in a Kerr type (high power) experiment may be well beyond the normal, small thermal excursions in an equilibrium fluid, and thus the ensemble *prepared* in such an experiment will experience more extensive ordering and hence undergo different relaxation times to the isotropic state, described by field-dependent (g/j) ratios.

Given all these considerations, one might be tempted to conclude that continual observations of simple Debye relaxation for CS_2 under such a range of spectroscopic experimental conditions were serendipitous. On the contrary; femtosecond and picosecond nonlinear laser spectroscopy have stimulated discussion and questions whose resolution penetrates the very core of molecular dynamics and intermolecular forces in liquids, as we have tried to illustrate.

ACKNOWLEDGEMENTS The author wishes to acknowledge the various and important contributions over the years of her coworkers cited in the references, most recently M. Golombok and C. Kalpouzos who continue the nonlinear experiments, and in particular valuable discussions with Professor Stephen C. Wallace. Financial support from the Natural Sciences and Engineering Council of Canada, the U.S. Office of Naval Research and the Guggenheim Foundation is also gratefully acknowledged.

REFERENCES

1a. *Picosecond Phenomena III*, K.B. Eisenthal, R.M. Hochstrasser, W. Kaiser, and A. Laubereau, Eds., (Springer-Verlag, 1982).
b. R.L. Fork, C.V. Shank, R. Yen and C. Hirlimann, IEEE J. QE-19, 500 (1983).
2. G.A. Kenney-Wallace, Phil. Trans. R. Soc. Lond. A. 299, 309-319 (1980).
3a. *Newer Aspects of Molecular Relaxation Processes*, Faraday Symposium No. 11, Chem. Soc., London (1977). b. *Structure and Motion in Molecular Liquids*, Faraday Disc. No. 66, Chem. Soc., London (1978).
4a. B. Berne in *Advanced Treatise on Physical Chemistry*, vol. VIIIB, ed. D. Henderson (Academic Press 1971); b. R.G. Gordon Adv. Mag. Res. 3, 1 (1968).
5. N. Bloembergen *Nonlinear Optics*, (Benjamin 2nd ed. 1982)

6. R. Hellwarth, Prog. In Quantum Elec. 5, 1-68 (1977).
7a. G.A. Kenney-Wallace and C.D. Jonah, Chem. Phys. Lett., 39, 596 (1976) b. G.A. Kenney-Wallace and C.D. Jonah, J. Phys. Chem. 86, 2572 (1982), and references cited therein on the ps solvation studies to date.
8. D. Huppert, G.A. Kenney-Wallace and P.M. Rentzepis, J. Chem. Phys. 75, 2265 (1981).
9. Y. Wang, M.H. Crawford, M.J. McAuliffe and K.B. Eisenthal Chem. Phys. Lett. 74, 160 (1980).
10. H. Miyasaka, H. Masuhara and N. Mataga, Chem. Phys. Lett. 98, 277 (1983).
11. J. Wiesenfeld and E.P. Ippen, Chem. Phys. Lett., 73, 47 (1980).
12. L.A.H. Billard, F.C. James and G.A. Kenney-Wallace, J. Phys. Chem. 0000 (1983) and references therein to earlier work.
13. E. Quitevis, P.M. Kroger, C. Kalpouzos and G.A. Kenney-Wallace, Can. J. Chem. 61, 975 (1983).
14b. W.T. Coffey, C. Rybarsch and W. Schröer, Chem. Phys. 99, 31 (1983) b. M.W. Evans, J. Chem. Phys. 76, 5480 (1983).
15. *van der Waals Molecules*, Faraday Disc. No. 73 (1982).
16. A. Amirav, U. Even and J. Jortner, J. Chem. Phys. 75, 2489 (1981).
17. T.R. Hays, W. Henke, H.L. Selzle and E.W. Schlag, Chem. Phys. Lett. 77, 19 (1981).
18. P.M. Felker and A. Zewail, Chem. Phys. Lett. 94, 454 (1983).
19. J. Hager and S.C. Wallace, J. Phys. Chem. 87, 2121 (1983).
20. M. Smith, J. Hager and S.C. Wallace, J. Chem. Phys. submitted for publication (1983).
21. W.A. Steele, Adv. Chem. Physics 34, 1 (1976).
22. G.A. Kenney-Wallace, J. Phys. Chem. submitted for publication.
23a. D. Kivelson and P.A. Madden, Ann. Rev. Phys. Chem. 31, 523 (1980). b. J.T. Hynes, R. Kapral and M. Weinberg, J. Chem. Phys. 69, 2725 (1978).
24. S. Abed, M. Broyer, M. Carre, M.L. Gaillard and M. Larzilliere, Chem. Phys. 74, 97 (1983).
25. W.R. Lambert, P. Felker and A. Zewail, J. Chem. Phys. 75, 5958 (1981).
26. M. Ivanco, W. Sharfin and S.C. Wallace, J. Chem. Phys. 76, 2095 (1982).
27a. T.I. Cox and P.A. Madden, Mol. Phys. 39, 1487 (1980).
28. W.B. Street and D.J. Tildesley, in ref. 36,P 27; b. R.W. Impey, P.A. Madden and D.J. Tildesley, Mol. Phys. 44, 1319 (1981).
29a. M.R. Battaglia, T.I. Cox and P.A. Madden, Mol. Phys. 38, 1539 (1979) b. Y. Higashigaki, S.L. Whittenberg, and C.H. Wang, J. Chem. Phys. 69 3297 (1978) c. S.L. Shapiro and H.P. Broida, Phys. Rev. 154 129 (1967). These are selective rather than exhaustive references on the CS_2 light scattering literature.
30a. See E. Ippen and C.V. Shank in *Ultrashort light Pulses*, ed. S. Shapiro (Springer-Verlag 1977) b. D.H. Auston, in ref. 30a; c.K. Sala and M.C. Richardson, Phys. Rev. A12, 1036 (1975) d. E.P. Ippen and C.V. Shank, Appl. Phys. Lett. 26, 92 (1975).

31. C. Kalpouzos, G.A. Kenney-Wallace, P.M. Kroger and E. Quitevis in L. Goldberg, Ed., "*Picosecond lasers and applications*" SPIE, vol. 322 (1982).
32. J. Etchepare, G.A. Kenney-Wallace, G. Grillon, A. Migus and J-P Chambaret, IEEE J.Q. Elec-18, 1826; and to be published
33. B.A. Garetz and G.A. Kenney-Wallace, and in *Picosecond Phenomena* ed. C.V. Shank, E.P. Ippen and S. Shapiro (Springer-Verlag 1978) p208.
34. *Optical Phase Conjugation*, ed. R.A. Fisher (Academic Press 1983).
35. H. Souma, E.J. Heilweil and R.M. Hochstrasser, J. Chem. Phys. 76, 5693 (1982) and references therein.
36. A bibliography of all phase conjugation papers reported up to 1981 is in ref. (34).
37. G.A. Kenney-Wallace and S.C. Wallace, IEEE J. QE-19,719 (1983) and in ref. 1, p. 221 (1982).
38. G.C. Bjorklund, D.M. Burland and D.C. Alvarez, J. Chem. Phys. 73, 4321 (1980).
39. A.E. Siegman, Appl. Phys. Lett. 30, 21 (1977); see also this volume, p. 325.
40. K.A. Nelson, R. Cosalegno, R.J.D. Miller and M.D. Fayer, J. Chem. Phys. 77, 1144 (1982).
41. J.R. Andrews and R.M. Hochstrasser, Chem. Phys. Lett. 76, 213 (1982).
42. M. Golombok, G.A. Kenney-Wallace and S.C. Wallace, submitted for publication; and unpublished data.
43. J.P. Heritage, Appl. Phys. Lett. 34, 470 (1979).
44. T.I. Cox, M.R. Battaglia and P.A. Madden, Mol. Phys. 38, 1539 (1979); ibid, 37, 1413 (1979).
45. G.D. Patterson and P.J. Carroll, J. Chem. Phys. 76, 4356(1982).
46a. B.I. Greene and R.C. Farrow, J. Chem. Phys. 77, 4779 (1982).
b ibid, Chem. Phys. Lett. 98, 273 (1983); and in ref. 1, p. 209 (1982).
47. J.M. Halbout and C.L. Tang, App. Phys. Lett. 40, 765 (1982), and in ref. 1 p212.
48. J. Etchepare, G. Grillon, R. Asher, J.L. Martin, C. Bruneau and Antonetti, in ref. 1, p. 217 (1982).
49. P.P. Ho, P.Y. Yu and R.R. Alfano, J. Chem. Phys. 74, 1605 (1981)
50. J. Etchepare, G. Grillon, R. Muller and A. Orzag, Optics Comm. 34, 269 (1980).
51. J.L. Oudar, IEEE JQE-19, 713 (1983).

PART III VIBRATIONAL RELAXATION

PICOSECOND VIBRATIONAL DEPHASING EXPERIMENTS IN LIQUIDS UNDER HIGH LASER DEPLETION CONDITIONS

D. Ben-Amotz, M. Berg, S. M. George, A. L. Harris, and C. B. Harris

Department of Chemistry, University of California, Berkeley, California 94720

In the presence of a sufficiently high intensity photon field (>1GW/cm^2) a myriad of nonlinear optical processes occur in a molecular system. Transient Stimulated Raman Scattering (TSRS) pumping of vibrational degrees of freedom is often found to dominate the interaction of a nonresonant picosecond photon field with a molecular liquid medium. This process has several potentially powerful applications for investigating ultrafast vibrational dynamics [1,2,3,12] and can be described semiclassically in terms of the coupling of a coherent vibrational oscillator equation,

$$\ddot{q} + (2/T_2)\dot{q} + \omega^2 q = F(t) = -\partial H_{int}/\partial q = N/2(\partial\alpha/\partial q)_0 E^2 \quad (1)$$

with Maxwell's equations for the electromagnetic field,

$$\nabla^2\vec{E} - 1/c^2\partial^2\vec{E}/\partial t^2 = 4\pi/c^2\partial^2\vec{P}/\partial t^2 \quad (2)$$

where q is the normal mode coordinate, ω is the vibrational frequency, and T_2 is the vibrational dephasing time. H_{int} is the interaction term in the coupled photon-phonon Hamiltonian which gives rise to a time dependent force F(t) on the coherent vibrational field. α is the molecular polarizability and N is the molecular number density. $\vec{E}$ is the electric field vector, $\vec{P}$ = NαE is the resulting material polarization and c is the speed of light. In the stationary wave approximation, ($\partial^2 E_L/\partial t^2 << \omega_L \partial E_L/\partial t$), these equations can be reduced to three coupled differential equations [4,5],

K. B. Eisenthal (ed.), Applications of Picosecond Spectroscopy to Chemistry, 165–172.

$$\partial E_L/\partial z = -K_1\omega_L Q E_S$$
$$\partial E_S/\partial z = K_1\omega_S Q^* E_L \qquad (3)$$
$$(\partial/\partial t + 1/T_2)Q = K_2 E_S{}^* E_L$$

where E_L,ω_L and E_S,ω_S are the laser and Stokes field amplitudes and frequencies respectively, Q is the coherent vibrational field amplitude, and K_1 and K_2 are coupling constants which are proportional to $\partial\alpha/\partial Q$.

Often a low laser depletion condition ($\partial E_L/\partial z=0$) is assumed in order to simplify the equations [1,6]. Recently, however, it has become apparent that the low laser depletion approximation may not be made in liquids when modeling the TSRS vibrational excitation process in picosecond excite-and-probe vibrational dephasing experiments, except under very special circumstances. Careful experimental characterization of TSRS under conditions relevant to these experiments has revealed that high laser depletion in the excitation process is the rule rather than the exception [7]. Extensive numerical solutions of the complete set of TSRS equations in the presence of an inhomogeneously broadened vibrational lineshape has led to some new and surprising conclusions concerning the role of high laser depletion in vibrational dephasing (T_2) experiments [5].

The results of these calculations are exemplified in figure 1, which shows the temporal profiles of a normalized laser pulse e_L, Stokes pulse e_S and vibrational amplitudes q_0, q_1, q_3 in a symmetrical five frequency component line shape after substantial depletion of the laser pulse has occurred. The parameters which give rise to these solutions are, laser pulse width t_p = 7psec, T_2 = 7psec, $\Delta\omega_1 = -1.5\text{cm}^{-1}$, $\Delta\omega_2 = +1.5\text{cm}^{-1}$, $\Delta\omega_3 = -3.0\text{cm}^{-1}$, $\Delta\omega_4 = +3.0\text{cm}^{-1}$. More generally, this is equivalent to $t_p/T_2=1$ with $t_p\Delta\omega = T_2\Delta\omega = \pm1.98$ and ±3.96 radians.

These time profiles clearly demonstrate the new effects which occur in TSRS in high laser depletion. The onset of this behavior is evident in the trailing half of the laser pulse when the laser photons have been substantially depleted via their conversion into the coherent Stokes field and a coherent vibrational amplitude. At this point, the coherent Stokes and vibrational fields regenerate a new laser field which is 180° phase shifted (opposite in sign) relative to the original laser field. This process is what causes the laser field amplitude e_L in figure 1 to go through zero and then change sign. This sign change in the laser field is in turn experienced by the coherent molecular vibrational ensemble as a 180° phase shift in the driving force, F(t) in equation 1. The important point is that the different frequency components in the five component lineshape respond to this phase shift differently. The central vibrational component, which is resonant with the driving force ($\omega_0=\omega_L-\omega_S$), must be driven through zero in order to change its phase by 180°, and is left with a relatively small phase reversed amplitude. The off-resonant components, on the other hand, can precess at their natural frequency while the driving force is changing sign. They are able to maintain a nonzero amplitude at all times and are thus left with a relatively large phase shifted amplitude. The final phase of the off-resonant vibrators at any particular point in the sample cell depends on the extent of laser depletion at that point and is thus a function of position in the sample cell.

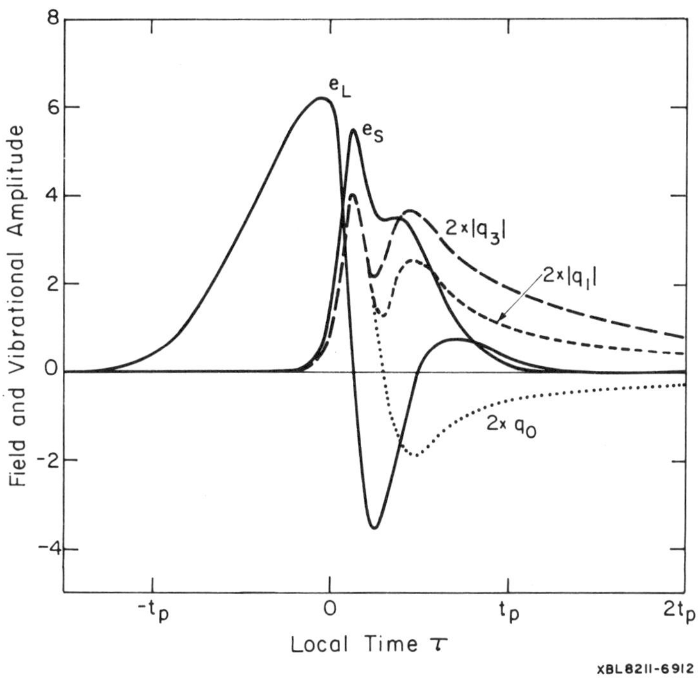

Figure 1. Calculated temporal profiles of the light fields (e_L,e_S) and vibrational amplitudes (q_0,q_1,q_3) for a five component vibrational lineshape with a 0.1:0.2:0.4:0.2:0.1 number density distribution, 1.5cm^{-1} frequency spacing, and t_p=T_2=7psec.

Much of this complex high depletion behavior can be understood by recalling the behavior of a forced, damped, harmonic oscillator at steady state which at some time t=0 experiences a 180° phase shift in its driving field. For example, prior to t=0, the forced harmonic oscillator at steady state has a solution which can be expressed in the usual way as $Z(t)=A\exp(i(\omega_0 t-\delta))$ where ω_0 is the driving field frequency. If the driving field undergoes a 180° phase shift at t=0, the solution for times greater than zero becomes $Z(t)=A[2\exp\times(-t/t_d)\exp(i(\omega_j t-\delta))-\exp(i(\omega_0 t-\delta))]$ where t_d is the damping time and ω_j is the oscillator's natural frequency. When the oscillator is driven on resonance, $\omega_j=\omega_0$, the solution for times greater that zero can be expressed as $Z(t)=R(t)\exp(i(\omega_0 t-\delta))$, where R(t) is real. The oscillator driven on resonance continues to rotate at frequency ω_0 and the movement of the resonant vector R(t) is confined to the real axis; consequently, the real amplitude R(t) must go through zero when the oscillator adjusts to the 180° phase shift. This behavior is analogous to the real resonant vibrational amplitude q_0 illustrated in figure 1. On the other hand when the oscillator is driven off-resonance, $\omega_j\neq\omega_0$, the solution is given by $Z(t)=C(t)\times\exp(i(\omega_0 t-\delta))$. In this solution, however, C(t) is a complex amplitude containing a transient term which oscillates with the frequency $\omega_0-\omega_j=\Delta\omega_j$; thus, the off-resonant vector C(t) rotates in the complex plane. This rotating term allows the off-resonant oscillator to sweep into proper phase with the phase-shifted driving force without being driven through zero. This behavior is analogous to that of the complex off-resonant vibrational amplitudes q_1 and q_3 shown in Figure 1. This ability of the off-resonant vibrational amplitudes to sweep through 180° without being driven through zero is of central importance in the interpretation of the results of picosecond dephasing (T_2) experiments.

The fundamental question which these calculations attempt to address is that of the selectivity of excite-and-probe vibrational dephasing experiments, i.e. the ability of these experiments to measure a homogeneous dephasing time, T_2, in an inhomogeneously broadened vibrational linshape. This question has arisen as a result of recent experimental measurements of vibrational dephasing times under reported low laser depletion conditions [8]. These measurements have produced non-selective results in contrast with earlier, apparently selective, results also obtained under reported low laser depletion conditions [9-11]. Although longer pulses and higher Stokes conversion were suggested to have been responsible for the earlier results, the conclusion was drawn that the homogeneous dephasing time T_2 in an inhomogeneously broadened lineshape could not be measured using TSRS techniques [8]. The present calculations reveal that while little selectivity is possible under low laser depletion conditions, high laser depletion conditions enhance selectivity. Moreover, this enhanced selectivity allows the homogeneous vibrational dephasing time T_2 to be measured with a high degree of accuracy in an inhomogeneously broadened transition.

In order to understand the physical basis for selectivity under high laser depletion conditions, one must consider the coherent probing process. In the probing process a low intensity picosecond laser pulse interrogates the liquid sample containing a coherent vibrational ensemble which was prepared using TSRS. The resulting probe signal consists of coherent Stokes radiation which is proportional to the product of the coherent vibrational amplitude and the probe laser field integrated over the sample cell length. It is in the generation of this probe signal that the effect of high laser depletion on the central and off-resonant frequency components in the vibrational lineshape manifests itself. The 180° phase shift of the driving field in the vibrational excitation process leads to corresponding phase shifts in the vibrational fields within the liquid cell. These large phase shifts in the vibrational fields give rise to destructive interference of the probe Stokes signal as the probe pulse travels through the sample cell. The degree of destructive interference of the probe Stokes signal depends of the amplitudes of the coherent vibrational fields involved. The central frequency component gives rise to the least destructive interference because q_0 is left with relatively little 180° phase shifted amplitude. The off resonant components which are better able to maintain their amplitude during the 180° phase shift lead to more severe destructive interference and thus a greater cancellation of the off resonant contribution to the probe Stokes signal. This difference in behavior between the central and off-resonant vibrational components results in a probe signal which represents the phase decay of a substantially narrowed vibrational linewidth. The dramatic difference in selectivity in the high laser depletion regime is shown in figure 2. The small amount of selectivity seen in low laser depletion is consistent with previous results calculated in the low laser depletion limit [1]. The enhanced selectivity in high laser depletion is a direct consequence of the 180° phase shift in the excitation laser pulse amplitude which occurs when the laser is depleted by more than about 10%.

These calculations also reveal that the theoretical selectivity is strongly dependent upon the laser pulse width, but only weakly dependent upon the dephasing time T_2. Selectivity is enhanced with an increase in the $t_p\Delta\omega$ product and to a lesser extent by an increase in $T_2\Delta\omega$. In addition, the calculation reveals that the accuracy of the experimental measurement of the homogeneous dephasing time T_2 is dependent upon a competition between the homogeneous and effective inhomogeneous dephasing times. The greater the ratio of the homogeneous linewidth to the effective inhomogeneous linewidth, the more closely does the decay of the coherent probe Stokes signal approach a pure ($T_2/2$) exponential [5]. Using a laser pulse width t_p=7psec, these calculations indicate that dephasing times $T_2 \leqslant 4.4$psec can be measured with an accuracy of greater than 60%. The experimental accuracy increases to near 100% as the dephasing time decreases to $T_2 \leqslant 1.4$psec.

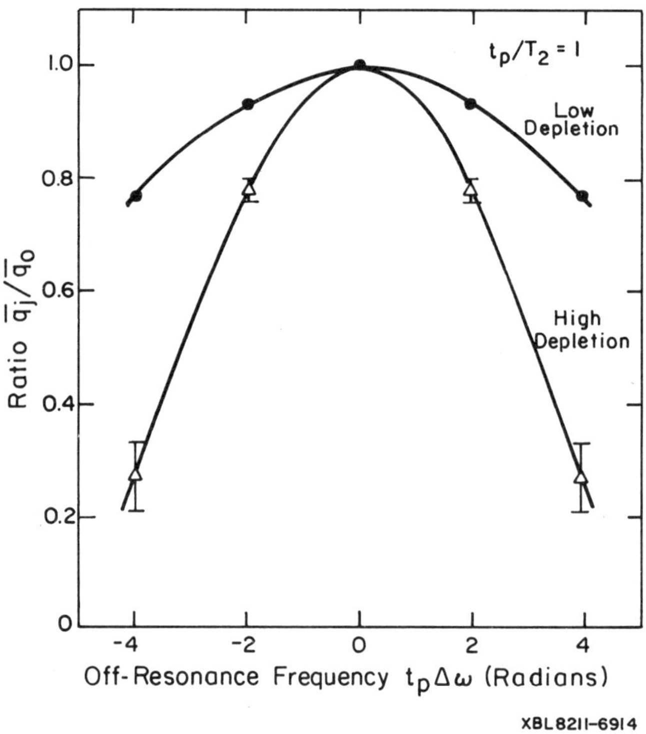

Figure 2. Dependence of selectivity (not including relative number densities) on the off-resonant magnitude $\Delta\omega$ for the five-component vibrational linewidth. The off-resonant magnitude is measured in units of $t_p\Delta\omega$ where the pulse width t_p is constant.

Because of the substantial selectivity predicted by these calculations, it is essential that high laser depletion be considered when interpreting vibrational dephasing experiments. These theoretical results are corroborated to some extent by recent vibrational dephasing experiments conducted in high laser depletion with laser pulse widths of $t_p \simeq 7$psec [12,13]. Briefly, the vibrational dephasing times $T_2/2$ were measured for the symmetric CH_3-stretching vibration in liquid acetonitrile at various temperatures over acetonitrile's entire liquid range at P=1atm. Over the same temperature range at P=1atm, the corresponding isotropic Raman linewidths for the symmetric CH_3-stretching vibration are nearly invariant with temperature [13,14]. A nonselective measurement of the phase decay time due to the total linewidth would then represent the inverse of the total vibrational linewidth and would therefore be invariant with temperature. However the experimental phase decay times obtained at different temperatures were reproducibly different [12,13], and longer than the linewidth transform. These long, temperature dependent decay times indicate the experiment's ability to measure the phase decay of a narrowed subunit within an inhomogeneously broadened vibrational lineshape. The changes of these measured phase decay times with temperature are in qualitative agreement with vibrational dephasing models [12,13]. These experimental findings partially confirm the theoretical results discussed above and suggest that these conclusions may be generalized to the case of a continuous inhomogeneous distribution. In addition, one could, in principle, include in the theory the effects of the radial intensity distribution of the laser pulses and any phase modulation which might be present. However, these extensions of the theory should not alter its basic qualitative conclusions [5]. In the light of these results, contradictions in previous vibrational dephasing studies [1-3,8-11] might well be explained by the selectivity differences associated with the high and low laser depletion regimes.

Acknowledgments

This work has been supported by a grant from the National Science Foundation. A.L.Harris and M.Berg acknowledge the support from the National Science Foundation Graduate Fellowship.

References

1) Laubereau, A., and Kaiser, W., 1978, Rev.Mod.Phys. 50, pp. 607.
2) George, S.M., Auweter, H., and Harris, C.B., 1980, J.Chem.Phys. 73 pp. 5573.
3) Harris, C.B., Auweter, H., and George, S.M., 1980, Phys.Rev.Lett. 44 pp. 737.
4) Kachen, G.I., 1975, Ph.D. Thesis, UCRL-51753, Lawrence

Livermore Laboratory, University of California, Livermore.
5) George, S.M., and Harris, C.B., 1983, Phys.Rev.A 28, pp.863.
6) Carman, R.L., Shimizu F., Wang, C.S., and Bloembergen, N., 1970, Phys.Rev.A 2 pp. 60.
7) Ben-Amotz, D., George, S.M., and Harris, C.B., 1983, Chem.Phys.Lett. 97 pp. 533.
8) Zinth, W., Polland, H.-J., Laubereau, A. and Kaiser, W., 1981, Appl.Phys.B26 pp. 77.
9) Laubereau, A., Wochner, G. and Kaiser, W., 1978, Chem.Phys. 28 pp. 363.
10) Laubereau, A., Wochner, G., and Kaiser, W., 1976, Phys.Rev.A 13 pp. 2212;1976, Opt.Commun. 17 pp. 91.
11) Zinth, W., Laubereau, A. and Kaiser, W., 1978, Opt.Commun. 26 pp. 457.
12) George, S.M., Harris, A.L., Berg, M. and Harris, C.B., in: Picosecond phenomena, Vol.3, eds. K.B. Eisenthal, R.M. Hochstrasser, W. Kaiser and A. Laubereau (Springer Berlin, 1982) pp. 196.
13) George, S.M., Harris, A.L., Berg, M. and Harris, C.B., unpublished results.
14) Yarwood, J., Arnt, R., and Doge, G., 1977, Chem.Phys. 25 pp. 387.

ULTRAFAST COHERENT INFRARED AND RAMAN SPECTROSCOPY OF GASES AND LIQUIDS

H.-J. Hartmann, H. Graener, H.R. Telle and A. Laubereau

Physikalisches Institut, Universität Bayreuth
Bayreuth, W. Germany

Coherent interaction between light and matter has been intensively studied in recent years.[1] Using ultrashort laser pulses and novel experimental techniques we are in the position to derive information from these investigations on vibrational and rotational dynamics in the electronic ground state of simple molecules. In this brief summary three topics will be discussed.

COHERENT PULSE PROPAGATION IN THE INFRARED ON THE PICOSECOND TIME SCALE

In the past coherent propagation of optical pulses has received considerable attention.[2-4] The general situation of self-induced transparency was investigated for molecular systems with small number densities where the phenomenon could be observed in microsecond or sub-microsecond time domain.[3] Little experimental work was concerned with the special case of small area pulses reporting qualitative accordance with theoretical discussions.[4,5] In our investigation quantitative agreement between theory and experiment is obtained for the propagation of low-intensity pulses allowing the determination of molecular time constants. Our experimental data represent the first observation of coherent pulse reshaping on the time scale of a few picoseconds.[6]

We have carried out a detailed theoretical investigation treating the common two-level model and more general situations. Some numerical examples are depicted in Fig. 1. For near-resonant input pulses of Gaussian shape and duration t_p (FWHH) a specific experimental situation with $T_2/t_p = 6$ and propagation length of $\alpha\ell = 10$ is considered. T_2 denotes the dephasing time of the molecular

K. B. Eisenthal (ed.), Applications of Picosecond Spectroscopy to Chemistry, 173–184.

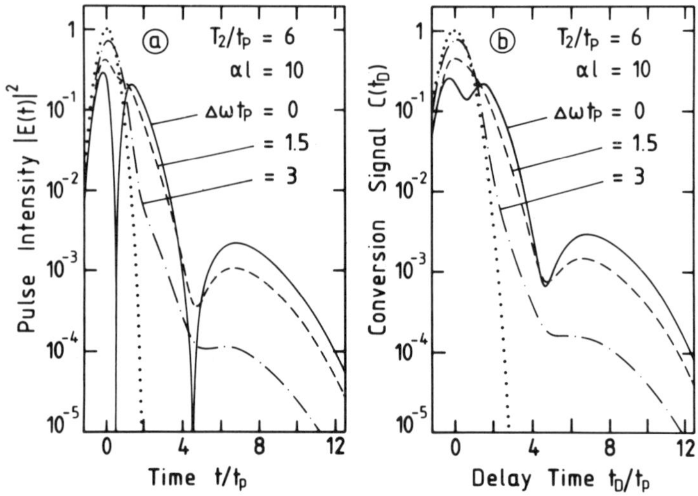

Figure 1:
a) Calculated intensity versus time of the transmitted pulse for three values of frequency setting $\Delta\omega t_p = (\omega_L-\omega_o)t_p = 0$, 1.5 and 3, respectively (solid, broken and dash-dotted lines). A special situation is considered with $T_2/t_p = 6$ and $\alpha\ell = 10$; Gaussian input pulse (dotted curve).
b) Calculated conversion signal $C(t_D)$ for the pulses of Fig. a) versus delay time t_D. The signal decay of the transmitted pulse (solid, broken and dash-dotted lines) allows the determination of T_2. Dotted curve: signal transient of input pulse.

transition of interest. α is the absorption coefficient at the maximum of the absorption line, which may be determined from conventional transmission spectroscopy. The pulse intensity (proportional to the absolute value of the electric field amplitude squared) is plotted in Fig. 1 a) on a logarithmic scale versus time. The peak value of the input pulse (dotted line) is normalized to unity. The transmitted pulse is shown for three values of frequency setting $\Delta\omega$ (solid, broken and dash-dotted curve, respectively). The striking changes of the trailing pulse wing should be noted, while the pulse maxima have decreased only by factors $\lesssim 2$. This behaviour originates from the finite lifetime T_2 of the induced transition dipoles, which produces delayed coherent reemission.

Available infrared detectors do not allow the time-resolved observation of ultrashort pulses. The picosecond detection of the pulse intensity, however, can be accomplished by an ultrafast light gate applying parametric up-conversion.[7,8] This technique provides the convolution of the IR pulse of interest with short probing pulses operating the parametric device. The resulting conversion signal $C(t_D)$ for the pulses of Fig. 1 a) is depicted in Fig. 1 b). Gaussian shape and duration t_p is assumed for the probe pulse. $C(t_D)$ is plotted on a logarithmic scale versus delay time t_D between infrared and probing pulses. It is interesting to see the extended wings of the transmitted pulses which exceed the input pulse wing by several orders of magnitude ($t_D/t_p \gtrsim 3$). This behaviour can be used to determine the dephasing time T_2 of the molecular transition. For quantitative work the effect of the frequency setting $\Delta\omega$ should be noted, which is illustrated by Fig. 1. Our calculations show that the frequency of the incident pulse has to be precisely known, $\Delta\omega t_p \lesssim .5$, in order to determine T_2 accurately.

It is emphasized that intensity variations of the input pulse do not change the shape of the signal transient $C(t_D)$. This property favours experimental applications of the coherent propagation effect investigated here.

We have substantiated our theoretical results measuring pulse propagation in HCl at moderate pressure values. In our experiments, single picosecond pulses are generated by a Nd:glass laser system which consists of a passively mode-locked laser oscillator, pulse selector and amplifier. The laser pulse passes through a saturable absorber for minor pulse compression and is directed into a double-pass parametric generator.[8,9] Tunable infrared pulses are efficiently produced at the "signal" (ω_S) and "idler" (ω_{IR}) frequency positions by stimulated parametric amplification in a 3 cm $LiNbO_3$ specimen. Pulse duration is measured to be $t_p = 4.0 \pm .5$ ps. The "idler" pulse around 3000 cm^{-1} serves for the study of the coherent propagation effect; it passes the sample cell of 1.3 cm or 10 cm length and enters the picosecond light gate. The "signal" pulse is used as probing pulse operating the gate device, which consists of a carefully aligned $LiNbO_3$ crystal of 1 mm length. Parametric sum frequency generation of the transmitted infrared pulse and the properly delayed probing pulse is detected by the help of suitable filters and a S1-photomultiplier. Sub-picosecond time resolution and high sensitivity with a detection range of 10^5 are achieved.[6,8] The shape and amplitude of the input infrared pulse is simultaneously measured in a reference experiment, using a second infrared gate set-up.

Highly purified HCl samples (99.995 %) were investigated with natural isotope abundances. In our measurements the picosecond pulse at the idler frequency is tuned to the maximum of the R(3) line of HCl^{35} situated at 2963.3 cm^{-1}.[10] The corresponding line of the HCl^{37} component is shifted by $\simeq 2$ cm^{-1} to smaller frequencies and also interacts notably with the infrared pulse.

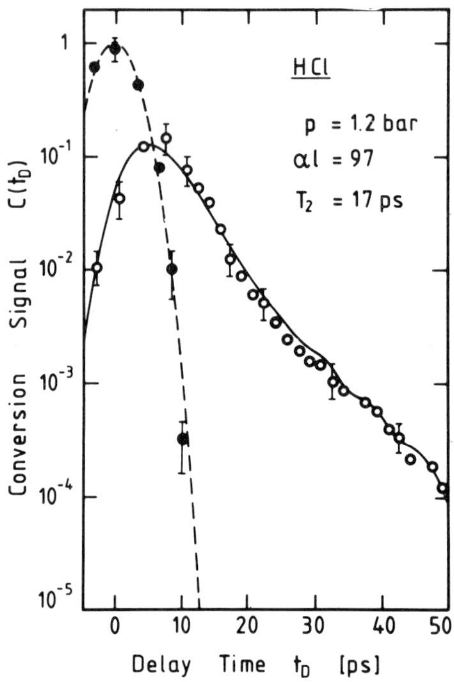

Figure 2:
Coherent pulse propagation in HCl at 1.2 bar; the conversion signal $C(t_D)$ of the resonant input pulse (full points) and transmitted pulse (open circles) is plotted versus delay time; theoretical curves; the drastic change of the pulse wing allows the determination of $T_2(R)$ = 17 ps for the R(3) line.

Some experimental data are presented in Fig. 2.[6] The conversion signal $C(t_D)$ is plotted on a logarithmic scale versus delay time t_D between the infrared and probing pulses. $t_D = 0$ marks the maximum signal observed with an evacuated sample cell. The ordinate scale states experimental transmission values as measured with the light gate technique. The open circles in the Fig. represent the measured conversion signals for a 10 cm sample cell filled with HCl at a pressure of p = 1.2 bar. It is interesting to note the drastic change of the transmitted pulse shape as compared with the input pulse (full points). The trailing wing of the transmitted pulse declines over several orders of magnitude for a time interval of ≃ 40 ps.

The solid line in Fig. 2 is a theoretical curve which accounts quantitatively for the position of the maximum and the decay of the signal. The $C(t_D)$-curve is computed taking explicitly into account the two isotopic R(3) species. Two fitting parameters exist in the computation: the dephasing time $T_2(R)$ of the R(3) line and the frequency difference $\Delta\omega_{is}$ of the two isotopic species. The value of $\alpha\ell = 97$ is taken from conventional spectroscopic data.[11] The results for Fig. 2 are $T_2(R) = 17 \pm 1$ ps and $\Delta\omega_{is}/2\pi c = 2.2 \pm .2\ cm^{-1}$. Additional measurements in the pressure range 1-3 bar verify a linear dependence on 1/p with $T_2(R) \times p = const = 21.5 \pm 1$ ps × bar.

The exponential time constant T_2 corresponds to a Lorentzian frequency width, $\Gamma = 1/(2\pi c\ T_2)$ (HWHH). Our measurements, therefore, yield a line broadening coefficient $\Gamma/p = 0.247 \pm .011\ cm^{-1}$ bar of the R(3) line in excellent agreement with spectroscopic data. The results for the isotopic line splitting are also in full accordance with frequency domain observations.[10]

Our data[6] represent the first time-domain measurement of the dephasing time $T_2(R)$ of a vibration-rotation transition in gases at medium pressure values. The signal decay extending over several orders of magnitude gives strong support to the exponential time behaviour predicted by impact theory.[12]

Vibrational relaxation is believed to contribute little to the measured dephasing time, which is significantly shorter than the time between elastic collisions. A significant contribution from rotational relaxation is expected which originates from the large permanent dipole moment of HCl.[13] The question remains if population decay or pure dephasing represent the dominant rotational relaxation mechanism. Some light is shed on this point by the following discussion.

TIME-RESOLVED COHERENT RAMAN SCATTERING

Picosecond CARS experiments have been performed for a large number of liquids in recent years.[14] Similar techniques have been applied to gases only in few cases.[15] We have studied picosecond Raman scattering of several diatomics and methane in the pressure range 0.1 to 10 bar.[16] Some results on the Q-branch of HCl will be discussed here.

Excitation of the molecular transitions is achieved by low-gain stimulated Raman amplification of a suitable input Stokes pulse, using single pulses of a mode-locked Nd:YAG laser system as pump pulses. The Stokes input is derived from a multiple step parametric generator-amplifier set-up producing intense pulses of 10 ps in the wavelength range 1.2 - 2 µm.[17] Frequency width is ≈ 6 cm^{-1} around 1.5 µm. The molecular excitation is monitored by delayed probing pulses at the second harmonic frequency of the laser pulse

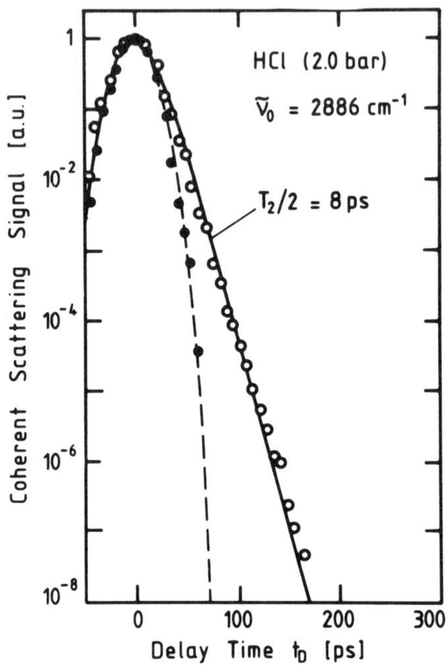

Figure 3:
Coherent Raman Scattering of the Q-band of HCl at 2.0 bar versus delay time between the infrared excitation pulses and the green probing pulse (open circles); full theoretical curve. Cross correlation between input Stokes and probing pulses (full points, broken line) representing an upper limit to the instrumental response function.

generating coherent anti-Stokes scattering in forward direction. This advanced measuring scheme has important advantages. Due to the separated frequencies of the excitation and probing process, background signals are very weak; sensitive detection of probe scattering signals achieves large signal to background ratios up to 10^9.

Experimental data for the Q-branch of HCl ($\tilde{\nu}_o$ = 2886 cm^{-1}) at 2 bar are depicted in Fig. 3.[16] The coherent scattering signal S^{coh} (t_D) of the green probing pulse is plotted on a logarithmic scale versus delay time. Exponential decay with time constant $T_2(Q)$ = 16 $\pm$ 2 ps is observed over 8 orders of magnitude. Principally speaking, this time constant represents a mean value of three prominent lines Q(2) - Q(4) which give major contributions to the spectroscopic Q-band. The observed exponential decay over a large dynamic range suggests very similar values for these Q-lines in accordance with theoretical expectations. A beating phenomenon between the individual Q-transitions is not resolved in Fig. 3 because of time-resolution arguments (probe pulse duration ≃ 16 ps). Additional measurements at various pressure values indicate $T_2(Q) \times p = const = 38 \pm 3$ ps × bar.

It is interesting to compare this result with our T_2 data on the R(3) line discussed above. We note a longer relaxation time of the Q-band by a factor of ≃ 1.8. The molecular transitions involved in the time-resolved infrared and Raman experiments are illustrated in Fig. 4. The measured $T_2(R)$ contains the full contribution of pure dephasing and population relaxation of the rotational transition. The effect of vibrational relaxation (T_2(vib)) may be neglected as is obvious from a comparison of the spectral linewidths of rotation (FIR)[18] and vibration-rotation bands (IR). The same conclusion results from a comparison with the elastic collision time $\tau_c \times p = 190$ ps × bar, since one or more collisions are required for vibrational relaxation via the repulsive part of the intermolecular interaction, $T_2(\text{vib}) \gtrsim \tau_c$. The $T_2(Q)$ time is effected by the rotational relaxation only in part via rotation-vibration coupling leading also to the well-known frequency shifts of the Q-lines. Application of the Anderson-Kubo model[19] for coupled adjacent lines suggests a situation close to the slow modulation

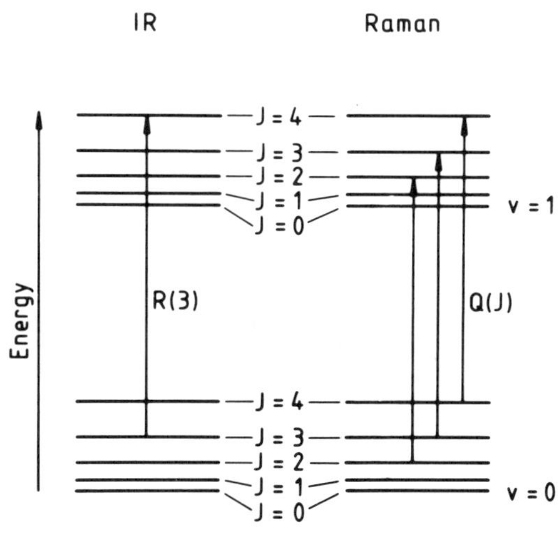

Figure 4:
Energy level diagram (schematic) and molecular transitions observed in the experiments of Fig. 2 and 3. The frequency spacings between the Q(2), Q(3) and Q(4) lines are 1.8 cm^{-1} and 2.4 cm^{-1}, respectively.

limit; consequently population decay (time constant $T_1(rot)$) of rotational sublevels fully contributes to the measured dephasing time $T_2(Q)$; on the other hand, frequency fluctuations (pure dephasing) of the J-sublevels are of minor importance for $T_2(Q)$.

We conclude that $T_1(rot) \approx T_2(Q) = 38$ ps ($p = 1$ bar) and that pure rotational dephasing is comparable in magnitude to $T_1(rot)$ processes to explain the shorter value of $T_2(R)$. The latter mechanism has been often ignored in theoretical discussions of vibration-rotation transitions.

NONEXPONENTIAL DEPHASING OF VIBRATIONAL MODES IN LIQUIDS ON THE SUBPICOSECOND TIME SCALE

In the past exponential decay of the coherent excitation of molecular vibrations has been clearly established on the time scale of several ps or longer.[14] No direct information has been available on the short time behaviour, $t \lesssim 1$ ps. Similarly, for theoretical work on stimulated Raman scattering purely exponential time behaviour was considered.[20] It is, however, known from theory of relaxation processes that this approximation does not hold for very short times which compare with the time scale of the dephasing mechanism.[21]

We have generalized the theory of transient stimulated Raman scattering to include non-exponential dephasing.[22] Our approach introduces the vibrational relaxation in terms of the vibrational autocorrelation function $\phi(t)$.[23] Using stochastic theory the following expression may be derived for this function[24]

$$\phi(t) = \exp\left\{-\frac{|t|}{T_2} - \frac{\tau_c}{T_2}\left[\exp\left(-\frac{|t|}{\tau_c}\right)-1\right]\right\} \qquad (1)$$

Two time constants, τ_c and T_2, describe the vibrational relaxation. The correlation time τ_c characterizes the time scale of the individual dephasing events. For $t \lesssim \tau_c$ the time dependence of ϕ is non-exponential with horizontal slope at $t=0$. For large t, Eq. (1) yields the expected exponential decay with dephasing time T_2. Simple exponential dephasing considered previously is contained in our present treatment for $\tau_c \to 0$.

Experimental information on τ_c is highly desirable for the identification of the dephasing mechanism. τ_c can be determined measuring the vibrational excitation by coherent Raman scattering of delayed probing pulses.

We have carried out numerical calculations of the molecular excitation by stimulated Raman scattering and of the coherent Raman probe scattering using the more general description of vibrational dynamics connected with Eq. (1). Our results are briefly summarized as follows:

(i) Keeping the spectroscopic linewidth $\delta\nu$ constant (FWHM of the Fourier transform of Eq. (1)) we note only little effect of τ_c on the stimulated amplification of the Stokes pulse of the pumping process.

(ii) The time evolution of the vibrational excitation and of the coherent probe scattering signal, however, are notably effected by τ_c.

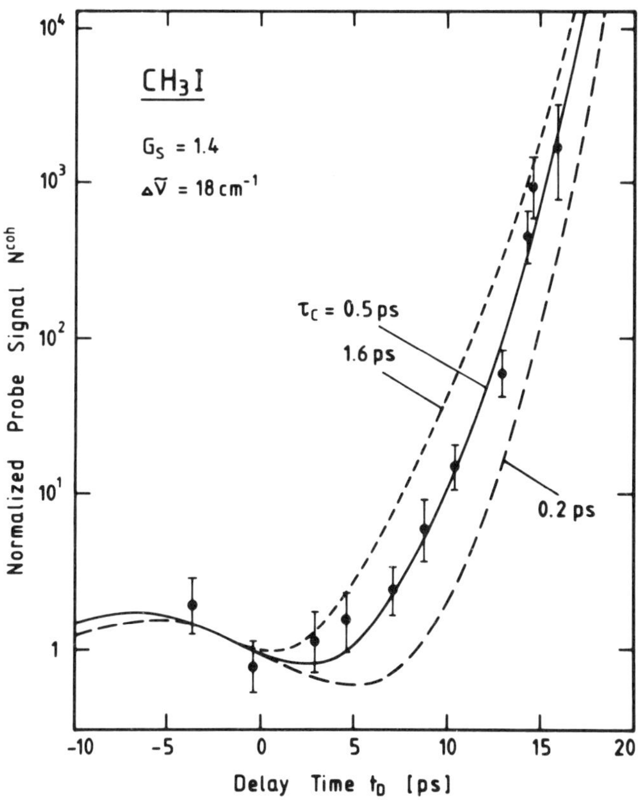

Figure 5:
Normalized probe signal $N^{coh}(t_D)$ versus delay time t_D for the ν_1 CH_3-mode of neat CH_3I; experimental points; theoretical curves.

(iii) For experimental investigations special conditions appear to be advantageous: near-resonant excitation by stimulated Raman scattering at low pump intensity with small amplification factor $G \lesssim 2$ and distinct frequency shift of the stimulated scattering process, $\Delta\nu = \nu_L - \nu_S - \nu_o \simeq 3\ \delta\nu$.

We have devised a double-channel picosecond system for the direct determination of τ_c, measuring the coherent probe scattering signal $S^{coh}(t_D)$ of the vibrational transition and the instrumental response curve $R(t_D)$ under identical geometrical conditions. With this system we have investigated several CH_2- and CH_3-stretching modes in the neat liquid at room temperature.[22]

Data on the symmetric ν_1-mode at 2948 cm^{-1} of CH_3I are presented in Fig. 5. A small amplification of the incident Stokes pump pulse by a factor of $G = 1.4 \pm .3$ has been adjusted for the excitation process. The normalized probe scattering signal, $N^{coh}(t_D)$, is plotted versus delay time t_D; $N^{coh} = S^{coh}/R$ denotes the ratio of the coherent probe scattering signal to the reference signal measured at the same delay setting. Values of $N^{coh} \neq 1$ reflect the delayed response of the vibrational transition via time constants τ_c and T_2.

The solid curve in Fig. 5 has been calculated for $\tau_c = 0.5$ ps and $T_2 = 2.0$ ps. The latter value is consistent with the observed asymptotic time dependence of the scattering signal $S^{coh}(t_D)$. The good agreement with the experimental points should be noted. For comparison, two broken lines are also shown in the Fig. which are calculated for constant $\delta\nu$ and values of $\tau_c = 0.2$ ps and 1.6 ps, respectively; these curves notably deviate from the experimental points. Our data of Fig. 5 present clear evidence for a finite time scale, $\tau_c > 0$, of the dephasing interaction. The results for τ_c and T_2 are in accordance with data on the spontaneous Raman line shape.[25]

Similar data were obtained for the ν_1-stretching modes of CH_3CN and CH_2BrCl yielding values of $\tau_c = 0.7$ ps and 0.4 ps, respectively.[22] The time constants in the subpicosecond range indicate that the relaxation mechanism occurs close to the limit of homogeneous broadening. Assuming dephasing via the repulsive part of the intermolecular potential to be predominant, the constant τ_c represents the time scale of translational motion; i.e. the elastic collision time. Simple models of the liquid state suggest this time to be 10^{-12} to 10^{-13} s in satisfactory agreement with our experimental data. Because of vibration-rotation coupling, τ_c may be also effected by the rotational relaxation time. Further work is required to elucidate this point.

In conclusion we point out that we have investigated the dynamics of vibration-rotation transitions of gases at medium pressure necessitating picosecond time resolution. In this pressure range molecular dephasing may be studied without being effected by Doppler broadening. Two techniques, a novel transient IR spectroscopy and picosecond coherent Raman scattering have been applied yielding different experimental information because of different selection rules.

We have also observed the time constant τ_c of vibrational relaxation in liquids. Our results represent the first time domain observations of the early non-exponential time dependence and demonstrate that the dephasing interaction at room temperature proceeds fast within a fraction of a picosecond.

REFERENCES

1. See, for example, "Picosecond Phenomena III", eds. Eisenthal, K.B., Hochstrasser, R.M., Kaiser, W., and Laubereau, A., 1982, Springer, Berlin.
2. McCall, S.L., and Hahn, E.L., 1967, Phys. Rev. Lett. 18, 908; 1969, Phys. Rev. 183, 457.
3. Patel, C.K.N., and Slusher, R.E., 1967, Phys. Rev. Lett. 19, 1019;
   Rhodes, C.K., and Szöke, A., 1969, Phys. Rev. 184, 25;
   Gibbs, H.M., and Slusher, R.E., 1970, Phys. Rev. Lett. 24, 638;
   Zembrod, A., and Gruhl, Th., 1971, Phys. Rev. Lett. 27, 287.
4. Grieneisen, H.P., Goldhar, J., Kurnit, N.A., Javan, A., and Schlossberg, H.R., 1972, Appl. Phys. Lett. 21, 559;
   Hamadani, S.M., Goldhar, J., Kurnit, N.A., and Javan, A., 1974, Appl. Phys. Lett. 25, 160.
5. Crisp, M.D., 1970, Phys. Rev. A1 1604; 1970, A2 2172; 1972, A5 1365.
6. Hartmann, H.-J., and Laubereau, A., 1983, Opt. Commun., in press.
7. Mahr, H., and Hirsch, M.D., 1975, Opt. Commun. 13, 96;
   Halliday, L.A., and Topp, M.P., 1977, Chem. Phys. Lett. 46, 8;
   Kopainsky, B., and Kaiser, W., 1978, Opt. Commun. 26, 219.
8. Hartmann, H.-J., and Laubereau, A., 1981, Appl. Opt. 20, 4259.
9. Laubereau, A., Greiter, L., and Kaiser, W., 1974, Appl. Phys. Lett. 25, 87;
   Seilmeier, A., Spanner, K., Laubereau, A., and Kaiser, W., 1978, Opt. Commun. 24, 237.
10. Meyer, C.F., and Levin, A.A., 1929, Phys. Rev. 34, 44;
    Herzberg, G., 1950, Molecular Spectra and Molecular Structure, Vol. I, van Nostrand, Princeton, New York.
11. Toth, R.A., Hunt, R.H., and Plyler, E.K., 1970, J. Mol. Spectr. 35, 110.
12. Anderson, P.W., 1949, Phys. Rev. 76; 1952, 86, 809.

13. Williams, D., Wenstrand, D.C., Brockman, R.J., and Curnutte, B., 1971, Mol. Phys. 20, 769.
14. von der Linde, D., Laubereau, A., and Kaiser, W., 1971, Phys. Rev. Lett. 26, 954; Laubereau, A., 1974, Chem. Phys. Lett. 27, 600; for further references, see Laubereau, A., and Kaiser, W., 1978, Rev. Mod. Phys. 50, 607.
15. Greene, B.I., Weisman, R.B., and Hochstrasser, R.M., 1978, Chem. Phys. Lett. 59, 5.
16. Graener, H., and Laubereau, A., to be published.
17. Graener, H., and Laubereau, A., 1982, Appl. Phys. B 29, 213.
18. Smith, E.W., and Girand, M., 1977, J. Chem. Phys. 66, 1762.
19. Anderson, P.W., 1954, J. Phys. Soc. Japan 9, 316; Kubo, R., and Tomita, K., 1954, J. Phys. Soc. Japan 9, 888.
20. Penzkofer, A., Laubereau, A., and Kaiser, W., 1979, Prog. Quant. Electr. 6, 55.
21. See, for example, Martin, P.C., "Measurements and Correlation Functions" (Gordon and Breach, New York, 1968).
22. Telle, H.R., and Laubereau, A., 1983, Chem. Phys. Lett. 94, 467.
23. Kubo, R., "Fluctuations, Relaxation and Resonance in Magnetic Systems", 1962, ed. Ter Haar, D., Plenum, New York.
24. Rothschild, W.G., 1976, J. Chem. Phys. 65, 455.
25. Yarwood, J., Arndt, R., Döge, G., 1977, Chem. Phys. 25, 387; Döge, G., Arndt, R., Khuen, A., 1977, Chem. Phys. 21, 53.

PICOSECOND PROCESSES IN CHEMICAL SYSTEMS: VIBRATIONAL RELAXATION

E. J. Heilweil, R. Moore, G. Rothenberger,
S. Velsko, and R. M. Hochstrasser
University of Pennsylvania

1. INTRODUCTION

Studies of picosecond processes have now been underway for one and one half decades. The level of detail exposed in experiments has roughly paralleled the advances in laser technology that have taken place during that period. The first applications of picosecond lasers to chemical problems used the solid state lasers neodymium-glass and ruby. The slow repetition rate and poor reproducibility of these systems allowed only rather coarse grain experiments. The attempts to measure statistical limit radiationless processes in azulene (1-3) and benzophenone (3-5), and studies of nearest neighbor energy transfer between (6) and within (7) molecules are but few of the processes that were studied with these early picosecond lasers. The important contributions of Kaiser and coworkers on vibrational relaxation in liquids (8,9) involved measurements of coherent Raman scattering and were made possible by more careful control of the Nd:glass laser to generate transform limited TEM_{oo} pulses. A subsequent development involved the picosecond generated continuum (10), which has allowed accurate absorption spectra to be recorded, even at low laser repetition rates, by incorporating extensive multiplexing capabilities (11).

The current generation of picosecond lasers permits a new level of exploration of molecular processes. The synchronous pumping of dye lasers with CW modelocked gas and solid state lasers provides high repetition rate, high stability picosecond or subpicosecond (12) pulses that were already used in many types of experiments. The stability of these lasers has resulted in them being used effectively in coherence decay experiments on

K. B. Eisenthal (ed.), Applications of Picosecond Spectroscopy to Chemistry, 185–203.

surfaces (13) and in molecular solids (14, 15). The CW-pumped passively modelocked dye lasers pioneered by Ippen and Shank (16) have also found wide applicability in the chemical area by providing opportunities for studies in the 0.2 to 10 ps regime. More recently the regime less than 0.1 ps was opened up (17) now making it possible for chemists to study directly processes occurring on the timescale of liquid state collisions.

Applications of picosecond laser methods to biological questions has occurred throughout the whole period of development described above. The low repetition solid state devices were used in pump-probe experiments to study the primary processes in vision (18) and photosynthesis (19). Not unexpectedly, based on the experiences with simpler photophysical processes, the new subpicosecond lasers exposed fast primary events (20-22) not seen in earlier experiments. In addition to photobiological processes, picosecond pulses have been used to perturb biological systems from equilibrium by means of photochemistry, in order that subsequent structural modifications could be followed. One example involves the use of a picosecond pulse to dissociate diatomic molecules O_2, CO or NO from hemoglobin (23-26). The resulting protein structure change and diatom dynamics in the protein are expected to provide vital clues to the mechanism of hemoglobin cooperativity.

In the present paper a number of experiments are presented which illustrate the state of the art in picosecond (as opposed to subpicosecond) methods applied to molecular problems. The theme of the article involves vibrational relaxation.

It is of great interest to generate experimental information on the relaxation dynamics of vibrational states of molecules. In the gas phase, and in the absence of collisions, intramolecular rotational-vibrational energy redistribution cannot onset until the rovibrational levels merge into a quasi-continuum (27). On the other hand in the presence of collisions the relaxation of a particular level resulting in the redistribution of the energy amongst other translational, rotational and vibrational motions often occurs as rapidly as the hard sphere collision frequency. The situation in a solid is different because of the existence of a quasicontinuum of 'external', or lattice modes that can accommodate any amount of the internal vibrational energy. Thus even when excited in the low internal energy region of the vibrational fundamentals the crystal can undergo vibrational energy redistribution. The decay of an internal mode excitation into other internal and external modes is analogous to the relaxation of molecular states excited in the molecular quasicontinuum. We will discuss both these extremes in this article as they occur in preliminary studies of p-difluorobenzene vapor (28) and benzene crystals (14, 29-31).

In the solid, just as with the levels of a molecule in regions of high state density, one must distinguish population and phase relaxation processes. The separation of these processes depends on the definition of the 'bath' or quasicontinuum states. In the case of the solid there is apparently a clear distinction between lattice and internal mode excitations. Thus we must distinguish experimentally between relaxation involving no alteration in the internal mode, and the relaxation of the internal mode population. In one part of this paper we will deal with a crystalline solid - benzene - so that the excitations are excitons. Each internal vibrational level forms a band of exciton states corresponding to excitation amplitude waves having wavelengths from infinity to the lattice spacing. Since the optical excitation generates only a few of these exciton states for each level, the relaxation of the system into the remaining ones is clearly not a relaxation of the internal mode population. Nevertheless certain experimental techniques measure this relaxation which is analogous to the pure dephasing part of the decay of the coherence of a two-level system (32, 33).

Vibrational energy redistribution and relaxation processes occurring in solutions have quite similar physics to the isolated molecule and low temperature crystal cases. Rather well-defined molecular vibrational states can be excited in dilute solution and the relaxation occurs both by internal to external mode energy transfer, and by internal energy redistribution. The competition between these two pathways thus determines the nature of the process. In the case of simple diatomic species in liquids the vibrational excitation might couple to local rotations or to external modes. Our recent studies (34) of CN^- in H_2O exemplify this case and are discussed in this article.

2. INTRAMOLECULAR VIBRATIONAL RELAXATION IN ABSENCE OF COLLISIONS

This section describes the status of our efforts to measure directly the rovibrational energy redistribution in a large molecule. There are many spectral evidences of the nonstationarity of optically accessible rovibrational levels in molecules (36-40) so it seemed necessary to establish the occurrence of these implied relaxations by direct measurements in the time domain. Such measurements are important because there is a regime near to the statistical limit of radiationless transitions where it would not be possible to determine the pathways of vibrational energy redistribution by spectral analysis alone. The study of this 'true dynamical region', like that of the statistical limit, will require time resolved spectroscopy to establish the various dynamical pathways.

Our preliminary experiments have involved p-difluorobenzene

vapor. Even at a few torr, picosecond excitation occurs thousands of times faster than hard sphere collisions so that collision free data can be obtained under bulb conditions if picosecond gating is used. The first experiments to report the time domain identification of intramolecular vibrational relaxation involved the excitation of levels at 263.2 nm and the subsequent evolution of the fluorescence in the 281.9 nm region.

In the experiment (28) the $3^1 30^3$ level (41) of p-difluorobenzene at its 294 K vapor pressure (ca. 20 torr) was excited by the fourth harmonic (263.2 nm) of a glass laser. The $3^1 30^3$ level is 1616 cm^{-1} above the S_1 zero point level. Gated fluorescence spectra were obtained in the spectral region where the transition $3^1_2 30^3_3$ (281.9 nm) was expected to occur (42). The energy levels relevant to this experiment are shown in Figure 1.

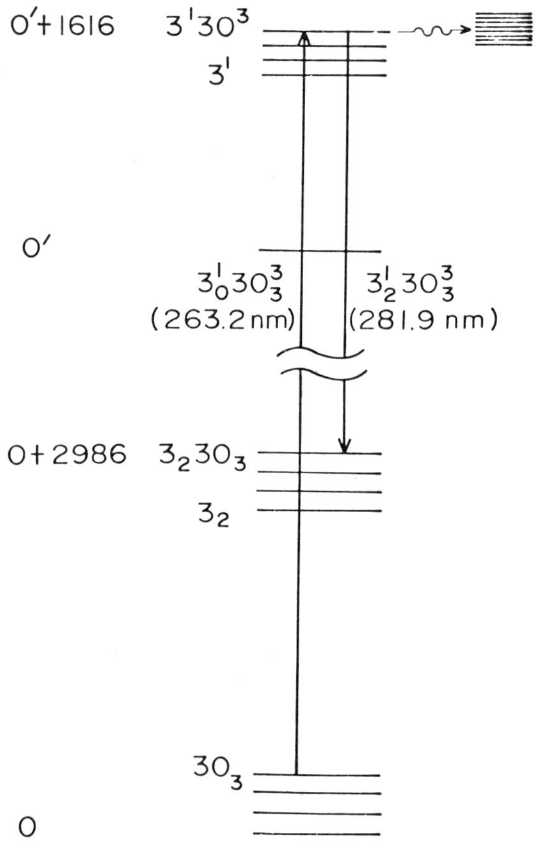

Figure 1. Schematic diagram of the main energy levels of p-difluorobenzene relevant to the picosecond fluorescence gating experiment.

Spectral evolution of the fluorescence from the $3^1 30^3$ level was expected to occur based on the chemical timing experiments of Parmenter et al. (40) and the spectral studies of Kable, Lawrence and Knight (43). Coveleskie, Dolson and Parmenter (40) studied the effect of added oxygen on the fluorescence spectrum of the $3^1 5^1 30^1$ (2191 cm^{-1}) level of p-difluorobenzene vapor and suggested that intramolecular vibrational energy redistribution (IVR) occurs on the 30 picosecond timescale. Kable, et. al., have obtained evidence that the ν_{30} mode may be a promoting mode for IVR in p-difluorobenzene. These experiments suggest that fluorescence originating from a level containing 1600 cm^{-1} excess energy should consist of sharp lines riding on a diffuse background. Furthermore, the intensity ratio of unstructured to structured fluorescence should be ca. 0.3.

The picosecond gating apparatus is based on a passively modelocked Nd^{3+}: glass oscillator/amplifier system which is shown in Figure 2. The entire pulse train (ca. 50 pulses of 8 ps

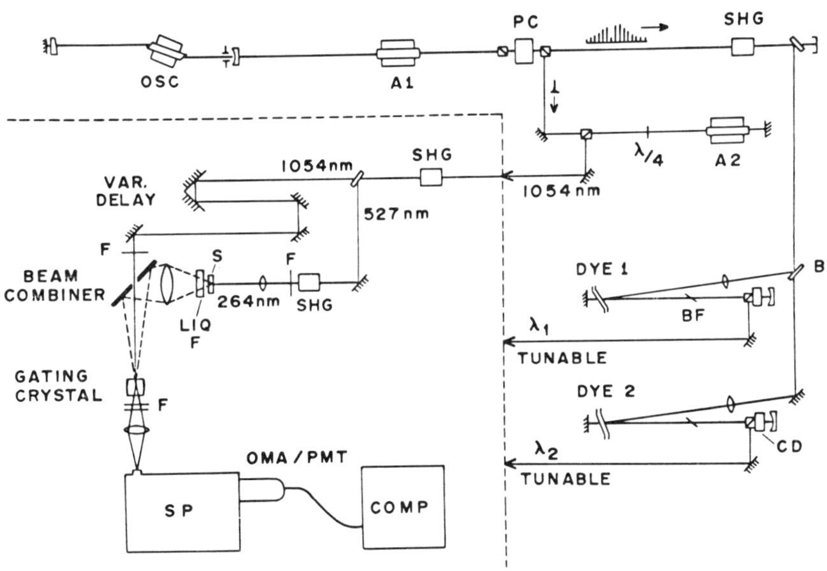

Figure 2. High repetition rate glass laser-dye laser system.

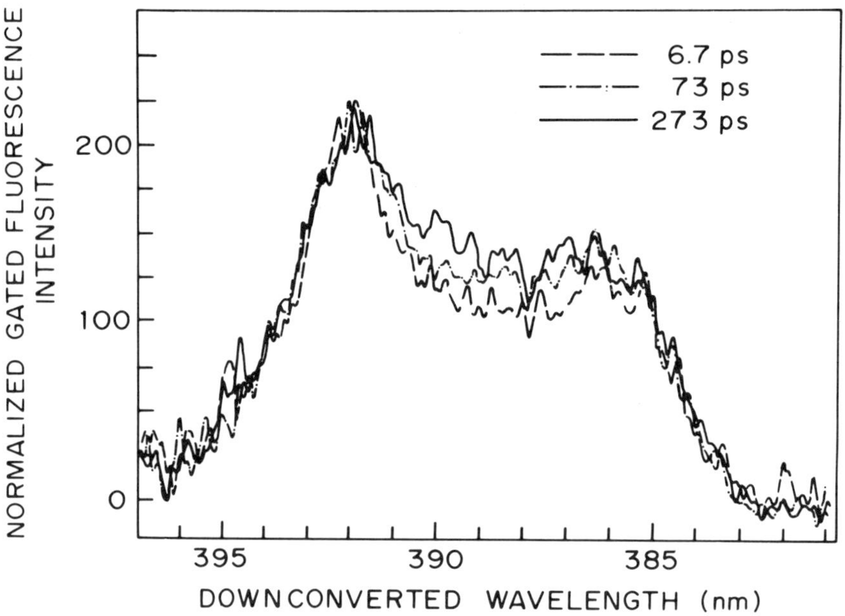

Figure 3. Gated spectra of p-difluorobenzene at 300 K, The three spectra are normalized to the peak signal (after, R. M. Moore, unpublished work).

duration and 100 μJ average energy) was weakly amplified and a single pulse extracted from the train. This pulse was further amplified to about 5 mJ by a double pass amplifier. The second, third and fourth harmonics of the glass laser were then generated with approximate pulse energies of 1 mJ, 500 μJ and 100 μJ, respectively. The weakly amplified pulse train could be doubled or tripled and the resulting harmonics used to synchronously pump two cavity dumped dye lasers. The dye lasers were tuned by birefringent filters to produce 20 μJ pulses of about 2 ps duration.

In the gating experiment fluorescence from the sample was filtered to remove luminescence at the UV detection wavelength, collected by a lens, and directed into a KDP crystal where it was combined with with the fundamental of the laser. The KDP crystal was oriented such that the difference of the fluorescence and laser pulse frequencies is generated. The down converted fluorescence was imaged on the split of a spectrometer and detected with an optical multichannel analyzer or a phototube. By delaying the arrival of the laser pulse, a time scan of the fluorescence spectrum could be obtained. The gate width was measured to be 10 ps.

The downconverted fluorescence spectrum was obtained at a time delay as short as ~ 6 ps. Two of the peaks are shown in Figure 3. Kinetic studies were carried out to ascertain the time dependence of the spectrum. The results are shown in Figure 4.

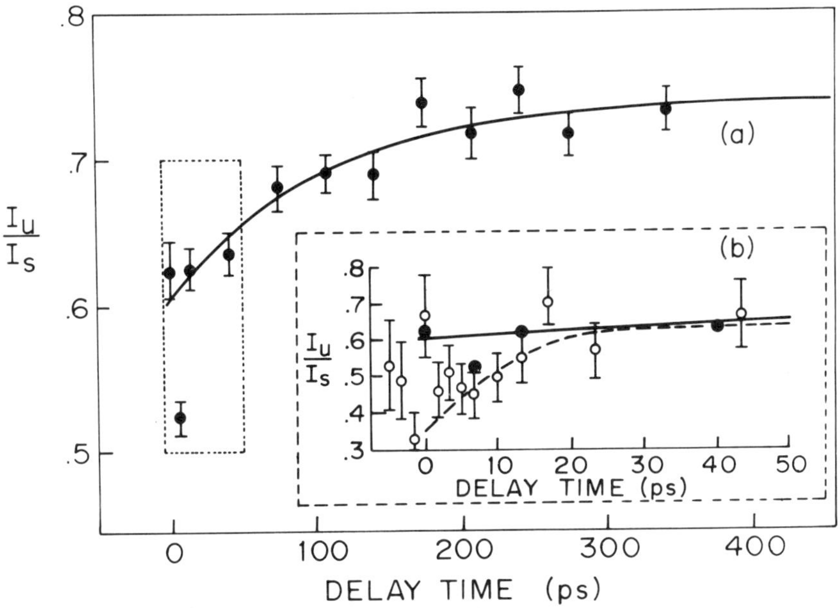

Figure 4. Collected kinetic data for the spectial evolution of the p-difluorobenzene vapor fluorescence (after R. Moore, unpublished work).

These data show that the emission from the unstructured region is delayed in its appearance relative to the emission from the peak region. These preliminary data fit to a model in which the initially pumped $3^1 30^3$ level decays into nearby robivronic levels. The initial fast process is followed by a slower evolution of the background emission. We conjecture that at first there is a rapid change perhaps due to coherent evolution of the pumped levels then over a period of 100 ps, there is an irreversible relaxation.

3. VIBRATIONAL ENERGY RELAXATION OF IONS IN SOLUTION

An important aspect of chemical physics that is currently attracting much attention is the determination of the mechanisms responsible for vibrational relaxation in the ground electronic states of molecules (44-47, 9). Information of this type would help to clarify the way in which energy redistribution contributes to thermally assisted reaction rates. Of primary interest is to uncover vibrational relaxation pathways and rates in solution, where many chemical reactions take place. To this end, the most attention has been paid to room tempterature neat liquids and a few solutions of organic compounds (9). Since these systems exhibit picosecond relaxation times, elaborate techniques involving pulsed tunable IR pumping and picosecond anti-Stokes Raman probing have generally been employed (9, 48, 40).

We have undertaken a course of study to further explore vibrational energy relaxation of ionic solute molecules in aqueous solution at room temperature. This class of systems has been chosen because ionic compounds include the simple diatomic ions, so that we expect to bring our results into relationship with theory. In addition, the solubility of many inorganic salts can be extremely high (up to 10 moles/liter) in water, which makes feasible measurements by current pulsed laser techniques.

T_1 Relaxation of Aqueous CN^- Ion: Our preliminary results of the diatomic anion CN^- in H_2O/D_2O (Na^+ and K^+ salts) are reviewed here (34). Of particular interest to us was the concentration dependence of the vibrational energy relaxation rate ($1/T_1$) and how T_1 measurements related to the CN-stretch Raman linewidths of these same solutions. The elucidation of relationships between the observed relaxation times and Raman spectra were expected to lead to a clear picture of energy relaxation contributions in more complex systems such as the SCN^- ion and higher cyanide containing homologs, and perhaps larger molecular ionic species.

The method employed in this study is an extension of that used by Laubereau and Kaiser (9). The laser system used for our picosecond experiments has been described in Section 2 above.

Briefly, an amplified single pulse from this laser (4 mJ) was used to produce second (527 nm, 0.5 mJ) and third (351 nm, 150 μJ) harmonic pulses (about 8 ps duration). A third pulse from the synchronously pumped Rhodamine-B dye laser (592 nm, 10 μJ) was then combined with the 527 nm pulse and focused colinearly by an achromatic lens into a 5 cm cell containing a saturated solution of NaCN. This cell sufficiently amplifies the dye laser pulse by the SRS process so that the resulting two fields can be used to pump the CN-stretching vibration (2080 cm^{-1}) in a second cell (0.5 to 10 mole/liter solutions). After exciting about 3% of the

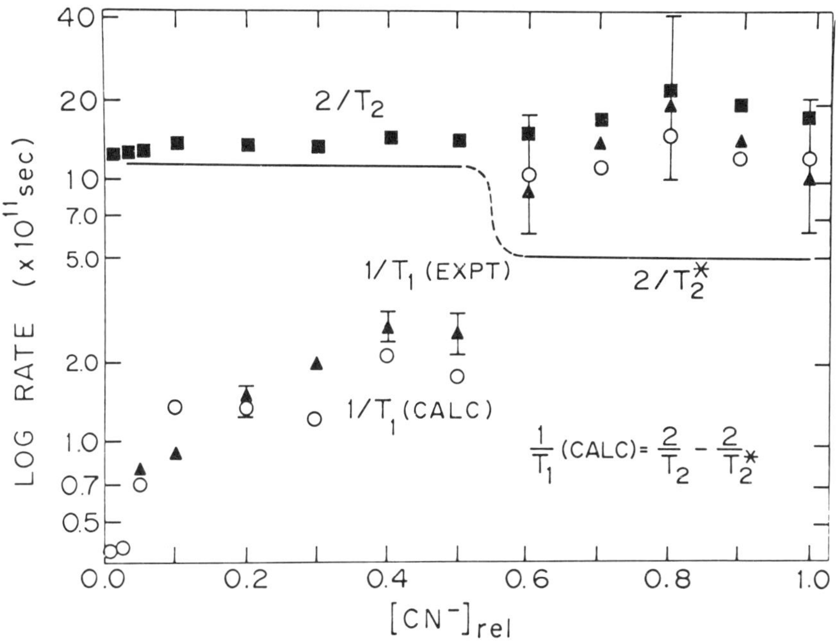

Figure 5. Calculated $1/T_1$ vibrational relaxation rate (0) derived from equation (1) using the experimentally determined dephasing rate $1/T_2$ from isotropic Raman linewidths ( ) and a "pure" dephasing rate $1/T_2*$, indicated by the solid line.

solute species by this stimulated Raman gain process, the instantaneous population in v=1 is probed by the anti-Stokes scattering of the variable-delayed 351 nm pulse. Collection of this anti-Stokes signal (I_{AS}) by standard Raman optics as a function of delay maps out the population relaxation time, T_1.

The time-dependence of the I_{AS} signal for a saturated solution (relative concentration of unity) of NaCN in H_2O or D_2O exhibited a pulse-width limited decay of not greater than 5.0 ps. In more dilute solution $[CN^-]_{rel}$ = 0.2 or 2.0 molar), we observed a lengthening of the T_1 time to about 7.0 $\pm$ 1.0 ps. By monitoring the I_{AS} peak intensity at t = 5 ps as a function of concentration, one can then obtain the $1/T_1$ relaxation rates corrected for excitation and concentration effects and normalized to the $[CN^-]_{rel}$ = 0.2 absolute decay measurement. The log of these $1/T_1$ rates with their respective error limits are shown as the triangles in Figure 5 as a function of $[CN^-]_{rel}$

According to this analysis, it is clear that the T_1 relaxation in concentration dependent. Down to ca. 0.5 of saturation, the value of $1/T_1$ changes only slightly going through a maximum near 0.8 (about six H_2O/CN^- mole ratio). At 0.5 saturation, T_1 is observed to undergo a sudden decrease.

Relation of T_1 Relaxation to Raman Lineshapes

Under particular circumstances, it should be possible to extract relaxation information from frequency domain measurements of vibrational modes. Theoretical efforts have been placed on determining rate contributions from dephasing and energy relaxation mechanisms (50, 51). Overall, the total vibrational dephasing rate, $1/T_2$ (related to the Raman full linewidth at half-maximum $\Delta\nu$ cm^{-1} for a Lorenzian band by $1/T_2 = \pi \Delta\nu c$) is composed of a "pure" dephasing rate, $1/T_2^*$ produced by translational collisional perturbations plus rates due to energy relaxation, $1/T_1$, resonance energy transfer, inhomogeneous broadening, etc. It has been found from a variety of measurements of T_2 in cryogenic liquids (45) and room temperature liquids (9) that the T_1 contribution cannot, a priori, be extracted from the Raman lineshape and that the pure dephasing process dominates the width.

Recent measurements of Raman spectra in aqueous NO_3^- (52) and HgI_4^{-2} (53) solutions have addressed this problem and indicate that temperature variations in HgI_4^{-2} solutions of the 123 cm^{-1} ν_1 band are due to change in T_2^*. The concentration effects observed for the $\nu_1(A_1')$ NO_3^- band in H_2O from 6 to 13 cm^{-1} are attributed to T_2^* inhomogeneous broadening with the Kubo model (50) modulation time (τ_c = 0.3 ps is related to collision times) being nearly constant. In both of these analyses, $1/T_1$ rates were assumed to be negligible. Armed with our measured T_1 times for CN^-, it should be possible to determine the extent that T_1 actually contributes to the Raman linewidth.

A careful analysis of the spontaneous Raman lineshapes for the CN^- solutions shows that there are two highly polarized Lorenzian components which change in magnitude as a function of concentration. The height of the smaller, higher energy peak is most significant at 0.5 concentration. This band is presumably due to ionic aggregation or some type of Na^+ $-H_2O-CN^-$ ion pairing which is also observed in other Na^+ salts (54). All solutions of KCN in H_2O, however, have one Lorenzian peak in the GH-stretch region, and the NaCN/KCN linewidths are the same for very dilute solutions. Deconvolution of the NaCN bands to obtain the linewidth of the much larger "free cyanide" component (populated by Raman gain) shows a distinct correlation between the T_1 dependence obtained by our time-domain measurements and the linewidth.

Neglecting contributions to the total T_2 dephasing time from resonant V-V transfer, we assume a rate relationship of the form:

$$\frac{2}{T_2} = \frac{1}{T_1} + \frac{1}{T_2*} \tag{1}$$

As was found in the nitrate study (52), we also assume that pure dephasing dominates the linewidth. Then, using the observed linewidths ($1/T_2$) corrected for slit width and depicted as squares in Figure 5, one can subtract a chosen $1/T_2*$ rate (the solid line in the figure) to obtain the "best fit" $1/T_1$ (calc) values shown as the circles. These results also suggest that the rate increase at $[CN^-]_{rel} = 0.5$ is indeed a real effect, and that for concentrations $\leq$ 0,05 molar, the dilute τ_1 limit is about 25 ps. It should be noted that the 'dilute' T_2* determined here is 0.4 ps. This time constant might be related to the rotational correlation time or jump model rest time which was determined to be 0.8 - 0.9 ps in 4 molar KCN solutions by means of IR and Raman isotropic and anisotropic profiles (55). The vibrational dephasing mechanism in this case could be mainly determined by the quasi-free rotation time of the CN^- ion.

It appears that for increasingly concentrated solutions at room temperature, the observed CN^- Raman line broadening from the dilute full width of 6.8 cm^{-1} to 10.5 cm^{-1} at saturation is mostly due to an increase in the vibrational population relaxation rate. Further studies are in progress to determine more accurately the dilute T_1 limit, and whether this relaxation is sensitive to local environment (cation type, water structure) or the addition of alternate intramolecular pathways (in SCN^-). It is anticipated that a more complete picture of water (56) and concentrated electrolyte solutions may then lead to a better understanding of vibrational energy relaxation and redistribution in solutions.

4. VIBRATIONAL RELAXATION AND TRAPPING IN MOLECULAR CRYSTALS

The conceptual basis for understanding vibrational dynamics in crystals is the same as that for isolated molecules. A low temperature crystal is an approximate realization of the limit of an infinitely large molecule in a well defined initial state. Since the interactions and densities of crystal states are often known from spectroscopic measurements, crystals represent a useful model for understanding the general principles which govern vibrational relaxation in large systems. In addition, the concepts which arise in crystal relaxation are an important bridge between the pure quantum mechanical description of vibrational relaxation in isolated molecules and the statistical description of relaxation in condensed media at finite temperatures.

Vibrational coherence decay in molecular crystals arises from both quasi-elastic, or dephasing, processes and from population relaxation. Dephasing due to phonon scattering has a pronounced temperature dependence, decreasing as the temperature decreases and vanishing at the absolute zero. In addition, in any real crystal, there are strains and impurities which cause scattering and hence dephasing. Population relaxation can occur through decay to lower vibrational levels of the host material ("self trapping") or through energy transfer to impurities. In order to account properly for all of these processes, it is necessary to know the band structure of the vibration of interest, and for the reason we have chosen to study the 991 cm^{-1} mode of benzene.

The width (14) of the 991^{-1} band is about 1.05 cm^{-1}, with the four k = 0 Davydov components in the relative order A_{1g} (0 cm^{-1}), B_{3g} (.47 cm^{-1}), B_{2g}(.53 cm^{-1}) and B_{1g} (1.05cm^{-1}). By choosing the laser polarizations parallel to the b crystallographic axis, the A_{1g} components may be preferentially excited. The fact that the Raman driven level lies at the band bottom where the density of states is vanishing, provides a great simplification to the interpretation of the observed coherence loss in this state. In addition, only six internal vibrational states lie below 991 cm^{-1} and of these only two, ν_{12} at 976 cm^{-1} and ν_{10} at 855 cm^{-1} lie within the spectrum of one phonon lattice excitation.

We have measured the coherence loss of the benzene 991 cm^{-1} mode (A_{1g} component) by the time resolved CARS technique. In this experiment, two synchronously pumped dye lasers are tuned so that their difference frequency is 991 cm^{-1}. The higher frequency laser is split into two beams of equal intensity. The pulses in one of these beams are made time coincident with the pulses from the second, lower frequency laser to form a "pump pair" which excites the sample by stimulated Raman scattering. The pulses in the other part of the high frequency laser beam can be time delayed with respect to the pump pair. These pulses probe the coherent amplitude remaining at a time delay after the pump pair has excited the crystal by stimulation coherent anti-Stokes emission. The intensity of this emission as a function of time delay is the time resolved CARS signal. For noise burst pulses (57), this signal can be written (29):

$$I_3(\tau) = \int_{-\infty}^{\infty} R^2(t')\, Q(\tau-t')\, dt' + BQ(\tau) \qquad (2)$$

where R(t) is the Raman response function, i.e. the Fourier transform of the Raman lineshape. The "instrument function" Q(t) is the third order correlation function of the pump and probe pulses and is obtained from the CARS signal in liquid benzene, where the dephasing time is only a few picoseconds. The convolution structure obtains in our éxperiment because of the noise burst properties of dye laser pulses. Since the coherence time of the

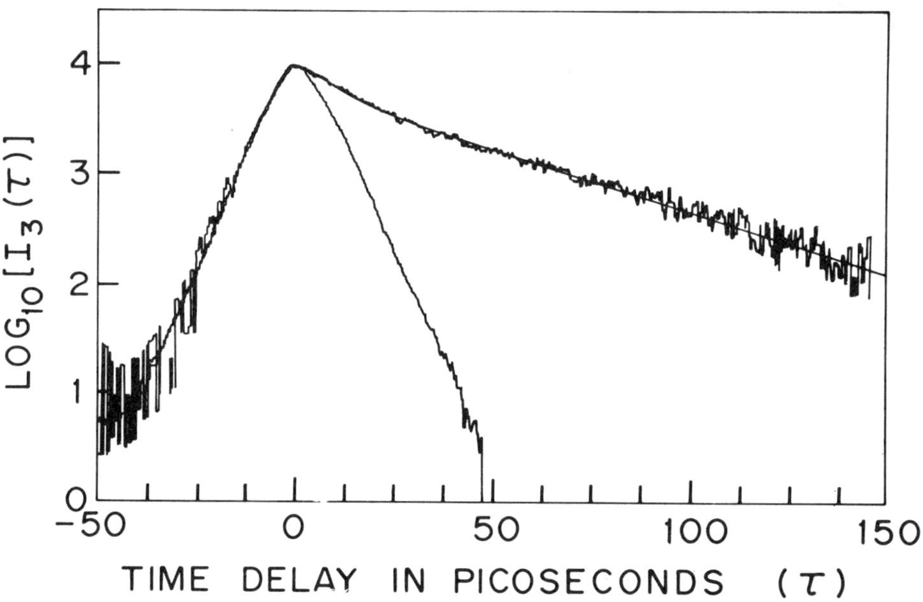

Figure 6. Log of coherent antistokes signal intensity versus delay Time for pure benzene A_g (991 cm^{-1}) at 1.6°K.

noise in our pulses is much less than the time duration of each pulse (2 psec, va. 8 psec) interference among amplitudes at different times during the excitation pulse are averaged out and the signal is approximately the sum of intensities even for coherence times of the order of the pulse duration. The second term in $I_3(\tau)$ is a coherence artifact arising because the high frequency pump and probe pulses are derived from the same dye laser and hence have correlated noise substructure. The use of equation (2) allows us to extract dephasing times which are of the order of the pulse duration (~ 10 psec).

Figure (6) shows the excellent fit to our data obtained using equation (2). In this case, the data is for a pure benzene crystal at 1.6°K. R(t) is taken to be a single exponential with a time constant of 80 psec. We have argued previously (14, 29-31) that this decay represents a population relaxation process. Recently, Califano and coworkers (58) calculated the rate of population decay from this state in the benzene crystal and found a relaxation time of 50 psec. The primary channel into which population relaxes is ν_{10}

We have also studied mixed proto/perdeutero benzene crystals. The addition of 10% perdeutero benzene to the proto host causes a two fold increase in the coherence decay rate. To understand the concentration dependence of the coherence decay it is necessary to consider the effects of impurities in the separated band limit, ie. when the impurity states lie well outside the width of the relevant host vibrational band. In this case frequency matching between host and impurity is poor, so that excitation by the relatively narrow pump pulses doesn't penetrate the impurity centers. Hence the initial state created in our experiment can be written as:

$$|\psi\ (0)> \quad \sum_j \zeta_j\ |j> \tag{3}$$

where ζ_i is a random variable equal to 1 if site j is occupied by a host molecule and 0 is by an impurity. The set $\{|\ >\}$ is the site basis for the host material. In terms of this initial state, the Raman response function is

$$R\ (t) = \ <<\psi \quad |\psi(0)|\psi>>_{conf} \tag{4}$$

The outer brackets in this expression represent a configurational average over impurity sites. In general, R(t) decays because of both impurity scattering and energy trapping. However, it is possible to show (29) that for small impurity concentrations, the coherence decay law is:

$$R(t) \propto e^{-(t/\tau)^{1/2}} \tag{5}$$

for a state like 991 cm^{-1} A_{1g}, which is located near a band point where the density of states is asymptotically vanishing. For this A_{1g} component in particular, detailed calculation shows that

$$\tau = C_B^{-2} A \tag{6}$$

with $A = 2 \times 10^{-11}$ sec. and C expressed as mole fraction. Thus, both the form of the decay law and the characteristic timescale for dephasing make it seem unlikely that impurity scattering is significant in this system, even at impurity concentrations of 15% or more.

The nature of initial state (3) has important consequences for the concentration dependence of energy trapping. The total energy decay rate in a mixed crystal in the separated band limit (29) has the following form:

$$k = k^{(0)} + c_A^2 n^2 k_{AA} + C_A C_B n k_{AA} + C_A C_B n^2 k_{AB} + C_B^2 n k_{AB} \tag{7}$$

Here C_A and C_B are the mole fractions of host and impurity molecules. k_{AA} and k_{AB} are overall site-site energy transfer rate constants forhost-host and host-guest energy transfer, and n is an effective number of nearest neighbors surrounding an excited host site. $K^{(0)}$ represents the single site self-trapping rate constant, and hence is concentration independent. The total rate in equation (7) is the sum of coherent ($\sim n^2$) and incoherent ($\sim n$) energy transfer processes. In the limit of the pure crystal, $C_A = 1$ we have

$$k_o = k^{(0)} + n^2 k_{AA}, \tag{7a}$$

while in the infinite dilution limit $C_B = 1$ we have

$$k_{oo} = k^{(0)} + nk_{AB}. \tag{7b}$$

A striking conclusion of this description is that even if the trapping efficiency of the impurity is equal to the self trapping efficiency ($k_{AB} \sim k_{AA}$) the rate of energy decay will <u>decrease</u> with addition of impurity molecules simply because the coherence of the energy transfer process is beinglost.

Figure (7) shows the "global" concentration dependence of the vibtational energy trapping rate in C_6H_6/C_6D_6 mixed crystals. Included on this graph is a point at 2% proto in perdeutero host (30). Since the infinite dilution rate constant k_{oo} approximately equals the pure crystal rate k_o, equations (7) imply that $n_{AB} \simeq n^2 k_{AA}$ for the 991 cm^{-1} A_{1g} state. This conclusion is consistent with the fact that the number of single quantum decay channels available in C_6D_6 is nearly three times the number in C_6H_6.

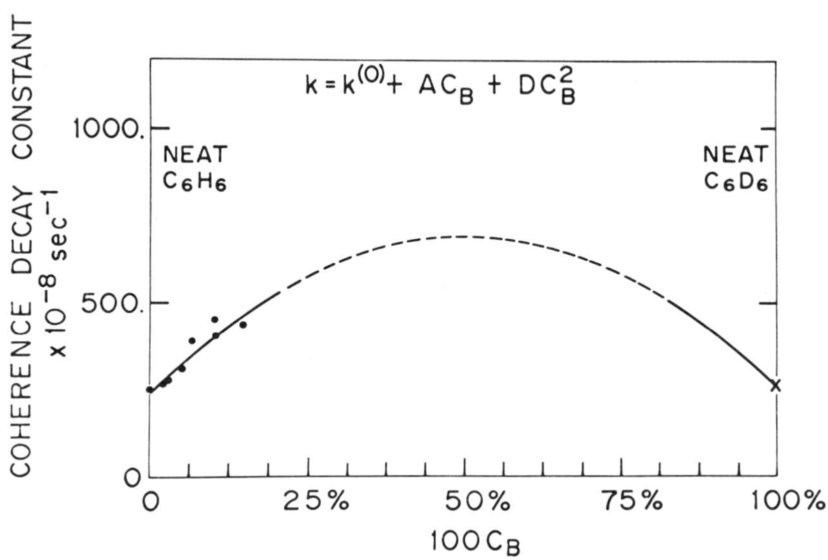

Figure 7. Coherence decay rate of the 991 cm^{-1} mode in mixed proto/perdeutero benzene crystals at 1.6°K as a function of perdeutero concentration (C_B). The line is a fit using equation (7) and the identity $C_A + C_B = 1$. The dashed portion indicates the "percolation" regime where scattering effects may be important. X is the dilute proto in deutero host measurement from reference (30).

REFERENCES

(1) Rentzepis, P. M. 1968, Chem. Phys. Lett. 2, p.117.

(2) Huppert, D., Jortner, J. and Rentzepis, P. M. 1977, Isr. J. Chem. 16, p. 277.

(3) Hochstrasser, R. M. and Weisman, R. B. 1980, in Radiationless, edited by S. H. Lin (Academic, New York).

(4) Hochstrasser, R. M., Lutz, H. and Scott, G. W. 1974, Chem. Phys. Lett. 24, p. 162.

(5) Anderson, R. W., Hochstrasser, R. M., Lutz, H and Scott, G. W. 1974, Chem. Phys. Lett. 28, p. 153.

(6) Anderson, R. W., Hochstrasser, R. M., Lutz, H. and Scott, G. W. 1974, J. Chem. Phys. 61, p.2500.

(7) Anderson, R. W., Hochstrasser, R. M. Lutz, H. and Scott, G. W. 1975, Chem. Phys. Lett. 32, p.204.

(8) Laubereau, A. von der Linde, D and Kaiser, W. 1972, Phys. Rev. Lett. 28, p. 1162.

(9) Laubereau, A. and Kaiser, W. 1978, Rev. Mod. Phys. 50, p. 607.

(10) Alfano, R. R. and Shapiro, S. L. 1970, Phys. Rev. Lett. 24, p.584; Alfano, R. R. and Shapiro, S. L. 1971, 8, p.631.

(11) Greene, B. I. Hochstrasser, R. M. and Weisman, R. B. 1979, J. Chem. Phys. 70, p.1247.

(12) Heritage, J. P. and Jain, R. K. 1978, Appl. Phys. Lett. 32, p.101.

(13) Heritage, J. P. in Picosecond Phenomena II, edited by R. M. Hochstrasser, Kaiser, W. and Shank, C. V. 1980, Verlag, Berlin, Heidelberg, New York.

(14) Ho, F, Tsay, W.-S., Trout, J. and Hochstrasser, R. M. 1981, 83, p. 5.

(15) Duppen, K, Heap, B. M. M and Wiersma, D. A. 1981, Chem. Phys. Lett. 79, p. 399.

(16) Shank, C. V. and Ippen, E. P. 1977, in Dye Lasers, edited by F. P. Schäfer (Springer-Verlag, Berlin, Heidelberg and New York).

(17) Fork, R. L., Greene, B. I. and Shank, C. V. 1981, Appl. Phys. Lett. 38, p.671; Shank, C. V., Fork, R. L. and Yen, R. T. in Picosecond Phenomena III, edited by K. B. Eisenthal, R. M. Hochstrasser, W. Kaiser and A. Laubereau 1982 (Springer-Verlag, Berlin, Heidelberg and New York).

(18) Honig, B. 1978, Ann. Rev. Phys. 29, p.31.

(19) Holten, D. and Windsor, M. W. 1978, Ann. Rev. Biophys. Bioeng. 7, p.189.

(20) Ippen, E. P. Shank, C. V. Lewis, A. and Marins, M. A. 1978, Science 200, p. 1279.

(21) Holten, D., Hoganson, C., Windsor, M. W., Schenck, C. C. Parson, W. W., Migus, A., Fork, R. L. and Shank, C. V, 1980, Biochim. Biophys. Acta 592, p. 461.

(22) Shank, C. V, Ippen, E. P. and Bersohn, R. 1976 Science 193, p. 50.
(23) Greene, B. I., Hochstrasser, R. M. Weisman, R. B. and Eaton, W. A. 1978 Proc. Natl. Acad. Sci. USA 75, p.5255.
(24) Chernoff, D. A. Hochstrasser, R. M. and Steele, A. W. 1980, Proc. Natl. Acad. Sci. USA 77, p. 5606.
(25) Cornelius, P. A., Hochstrasser, R. M. and Steele, A. W. J. Mol. Biol., (in press).
(26) Eisert, W. G., Degenkolb, E. O., Noe, L. J. and Rentzepis, P. M., 1979, Biophys. J. 25, p. 455.
(27) Freed, K. F. and Nitzan, A., 1980 J. Chem. Phys. 73, p.4766.
(28) Moore, R., Doany, F. E., Heilweil, E. J., and Hochstrasser, R. M. Faraday Discuss. Chem. Soc. 75, (in press).
(29) Velsko, and S. P. and Hochstrasser, R. M. "Vibrational Dynamics in Molecular Crystals: The separated band limit", in process of publication.
(30) Bozio, R. DeCola, P. L. and Hochstrasser, R. M. "Vibrational Energy Decay in Solid Benzene by Coherent Raman Spectroscopy", in process of publication.
(31) Ho, F., Trout, J., Tsay, W.-S., Velsko, S. P. and Hochstrastrasser, "Vibrational Energy Trapping in Benzene Crystals", in process of publication.
(32) Abram, I. I. and Hochstrasser, R. M., 1980, J. Chem. Phys. 72, p.3617.
(33) Abram, I. I. and Hochstrasser, R. M., 1981, J. Chem. Phys. 75, p.337.
(34) Heilweil, E. J. Doany, F. E. Moore, R. and Hochstrasser, R. M., 1982, 76, p.5632.
(35) Rothenberger, G. C., Negus, D. K. and Hochstrasser, R. M., "Solvent influence on photoisomerization dynamics", in preparation.
(36) Hochstrasser, R. M. and Prasad, P., in Excited States, 1974 Vol. 1, edited by E. C. Lim (Academic, New York).
(37). Bray, R. G. and Berry, M. J. 1979, J. Chem. Phys. 71, p. 4909.
(38) Hopkins, J. B., Powers, D. E. and Smalley, R. E., 1980, J. Chem. Phys. 72, p. 5039; Hopkins, J. B., Powers, D. E., Mukamel, S. and Smalley, R. E., 1980, J. Chem. Phys. 72, p. 5049; Hopkins, J. B., Powers, D. E. and Smalley, R. E. 1980, J. Chem. Phys. 73, p. 683.
(39) Amirav, A., Even, U. and Jortner, J. 1981, J. Chem. Phys. 75, p. 3770.
(40) Coveleskie, R. A., Dolson, D. A. and Parmenter, C. S. 1980, 72, p. 5774.
(41) Dunn, T. M., private communication.
(42) Coveleskie, R. A. and Parmenter, C. S. 1981, J. Mol. Spectrosc. 86, p.86.
(43) Kable, S. H., Lawrance, W. D. and Knight A. E. W. 1983, J. Phys. Chem. 86, p. 1244.
(44) Flynn, G. and Weitz, E. 1974, Ann Rev. Phys. Chem. 25, p.275. Flynn, G. 1981, Acct. Chem. Res., 14, p. 334.

(45) Renner, G. and Maier, M. 1974, Chem. Phys. Lett. 28, p.614; Ewing, G. 1975, Chem. Phys. Lett. 30, p. 485.
(46) Brueck, S. R. J. and Osgood, R. M., 1976, Chem. Phys. Lett. 39, p. 568.
(47) Knapp, E. W. and Fischer, S. F. 1981, Chem. Phys. 63, p. 203.
(48) Fendt, A., Fischer, S. F. and Kaiser, W. 1981, Chem. Phys. 57, p.55; Chem. Phys. Lett., 1981, 82, p. 350.
(49) Laubereau, A. Fischer, S. F., Spanner, K. and Kaiser, W. 1978, 31, p. 335.
(50) Kubo, R. in "Fluctuations, Relaxation, and Resonance in Magnetic Systems," ed. D. Ter Haar (Plenum Press, New York, 1962).
(51) Fischer, S. F., Laubereau, A. 1975, Chem. Phys. Lett., 35, p. 6.
(52) Kato, T., Tanaka, T. 1979, Chem. Phys. Lett., 62, p. 77.
(53) Baglin, F. G., Wilkes, L. M. 1981, J. Phys. Chem., 85, p. 3643.
(54) Frost, R. L., James, D. W., Appleby, R., Mayes, R. E., 1982, J. Phys. Chem., 86, p.3840.
(55) Lascombe, J., Perrot, M., 1978, J. Chem. Soc. Farad. Disc., 66, p. 216.
(56) Rice, S. A., Sceats, M. G., 1981, J. Phys. Chem. 85, p. 1108; Katoaka, Y. Hamada, H. Nose, S., Yamamoto, T., 1982, 77, p. 5699.
(57) H. A. Pike and M. Hersher, J. Appl. Phys. 41 (1976) 4562.
(58) Califano, J. and coworkers, private communication.

TIME RESOLVED SPECTROSCOPY AND DYNAMICS OF LARGE AMPLITUDE VIBRATIONAL MOTION IN MOLECULAR LIQUIDS AND CRYSTALS

Chr. Flytzanis, G.M. Gale and M.L. Geirnaert

Laboratoire d'Optique Quantique*, Ecole Polytechnique
91128 - Palaiseau, Cedex, France

We present a description of the dynamics of vibrational overtones and two-phonon states in molecular liquids and crystals. Anharmonicity and cooperative excitation are shown to be the two main competing mechanisms that determine the spectral and dynamical features of these states and large amplitude vibrational motion in general. We propose the use of nonlinear optical techniques to study the dynamics of cooperative excitation of pairs of molecules, coherent two-phonon states, large wave vector phonons, phonon localization and soliton states in condensed molecular systems. As a prelude we present the first experimental results of coherent excitation of the first overtone and two-phonon state of the mode ν_2 in liquid and crystalline CS_2 respectively. We also analyze the impact of the photoinduced defects in the time response of the optical nonlinearities in conjugated polymer crystals.

I - INTRODUCTION

The purpose of the present paper is to assess the main spectral and dynamical features of the vibrational overtones in molecular liquids and solids and show that these are related to certain aspects of large amplitude vibrational motion in condensed matter that are totally out of reach by the study of the fundamental vibrational modes alone. Furthermore, we show that nonlinear optical techniques are the most indicated to selectively exploit these new features and bring into light hidden or unsuspected configurations of vibrational motion in dense media and crystals in particular.

Certainly vibrational overtones in molecular liquids and multiphonon states in crystals were unavoidable since the early[1-5]

K. B. Eisenthal (ed.), Applications of Picosecond Spectroscopy to Chemistry, 205–235.

infrared (IR) and Raman (R) spectroscopy and, however undesirable since they destroy the "harmony" of the fundamental modes, they have frequently been studied as by-products and used to complement the spectroscopy of the latter and their basic principles. We recall that these arise from the assumption that the vibrational amplitudes are infinitesimal which implies that the potential energy is a purely quadratic form of the vibrational coordinates. This reduces the problem to that of a set of uncoupled harmonic oscillators, which is elementary ; the total vibrational energy equals the sum of energies of the simple harmonic oscillators or normal modes which ignore each other. Their frequencies, line widths, intensities and polarisations and their relation to the microscopic description is the essential goal of the IR and R spectroscopy.

However, this picture is bound to be inaccurate or even fail when the amplitudes become large and anharmonic terms or interaction among oscillators set in leading to damping and mixing of normal modes and the rise of overtones. In solids in particular multiphonon interaction is the essential phonon breakdown and relaxation mechanism[6-7] yet the two and three- phonon states *per se* have been studied only indirectly and considered as undesirable disturbances from the realm of darkness and unconsciouness of the lattice.

With ever increasing amplitudes, the anharmonicity on one side and the coupling of identical units on the other enter into competition, the later delocalizing the vibrational excitation while the former, acting as an intrinsic lattice trap, localizing and inducing a phonon localization, a perequisite to solid state reactivity or phase transitions. Local modes[8,9] in polyatomic molecules with many identical molecular groups (ex, C-H bonds) or solitons[10-12] in onedimensional conjugated polymers are consequences of such large amplitude vibrational motions in molecular systems which cannot be simply accounted for by straight-forward normal mode analysis.

The competition between the coupling of identical oscillators which essentially introduces a harmonic spectrum and tends to delocalize the excitation over all oscillators and the anharmonicity which destroys it by favoring localized states is an intrinsic feature of the overtones and it is worth while to consider[13,14] the concrete example of the first overtone of a pair of diatomic molecules (ex. HCl). In such a system excitation near the overtone region can take place in two distinct ways : (a) a single molecule may be excited from its ground state v = 0 to v = 2 or (b) the two interacting molecules may both be excited from v = 0 to v = 1. If the anharmonicity of the vibration of the molecule is appreciable, larger than the coupling of the molecules, the state (a) would be appreciably below state (b) and the transitions might be observed independently. The latter should be at almost exactly twice the fundamental frequency; this situation has now been observed in a number of simple molecular liquids[(13-15)] by infrared spectroscopy. In the absence of anharmonicity, the two transitions would be superimposed and spectrally undistinguishable from each other; their dynamics,however, being quite different as will be shown below, they can be studied by time resolved techniques.

In crystals the periodicity strongly enhances the coupling and the degeneracy of the oscillators resulting in the large optical phonon dispersion over the Brillouin zone which in most cases completely masks the impact of the anharmonicity. This is certainly the case in inorganic ionic or covalent crystals where the forces across unit cells are as strong as inside them; in molecular crystals where the intermolecular forces are of the van der Waals type, much weaker than the intramolecular ones, the situation may be drastically different as some isolated cases tend to indicate[(16-17)]; bound two-phonon[(18-24)] states or large amplitude local states may appear which are of relevance to solid state reactivity and phase transitions and the previous simple picture must be modified.

In the present paper, we start by introducing the hamiltonian which incorporates the basic interactions and give a description of its spectrum (section II). In section III we present the main spectral and dynamical features of the first overtone and two-phonon states and we stress the importance of the joint two-phonon density of states. In section IV we outline the principle of the active nonlinear spectroscopy of these states and we single out the time resolved CARS technique in particular. In section V we address some questions related to the coherence of the two-phonon states, their propagation and parametric instabilities. In section VI we present a short discussion of the time response of the optical nonlinearities in one- dimensional conjugated polymer crystals and their relation to the dynamics of photoinduced solitons, the large amplitude intrinsic conjugation defects. We conclude in section VII with some general considerations concerning phonon localization and its relation to solid state phase transitions and topochemical reactivity.

II - ORIGIN AND DESCRIPTION OF VIBRATIONAL OVERTONES IN MOLECULAR LIQUIDS AND SOLIDS

The vibrational overtones in molecular systems are the first stages of large amplitude vibrational motion; they can still be described in terms of the fundamental modes with appropriate perturbative approaches. Here, we shall only concentrate on the basic mechanisms and gross features of their spectrum and leave aside any details connected with specific molecular structures.

The starting point will be a simplified model for the hamiltonian of an assembly of polar and polarisable molecules in interaction with a time varying external electromagnetic (EM) field. Besides the direct interaction of each molecule with the EM field the molecules interact among themselves through the induced dipoles and this introduces an intermolecular term in the total potential energy which in the usual expansion of the dipoles in normal coordinates introduces intermolecular terms in the vibrational potential energy which add to the intramolecular ones. This distinction is irrelevant for the fundamental modes (except in describing the polariton mode in crystals) but essential for the description of the vibrational overtones and large amplitude vibrational motions in condensed matter in general. The competition between these two terms determines the extent to which the vibrational motion will be localized or shared by many identical molecules.

The assembly which later will be specified to represent a sample of either a liquid or a crystal contains N identical molecules located at positions $r_i (i = 1,\ldots N)$; the orientation of the dipole moment μ_i of molecule i will be characterized by a unit vector $\hat{\Omega}_2$. The wavelength of the EM field E will be sufficiently long so that E may be assumed homogeneous over the extension of the molecular sample.

<u>Intramolecular terms</u>

The vibrational motion of a molecule alone in its ground electronic state will be described by a hamiltonian :

$$h = t + v - \bar{\mu}.\bar{E} - \frac{1}{2}\bar{\bar{\alpha}}\bar{E}\bar{E} \qquad (1)$$

only a function of the nuclear configuration $\bar{R}$; t is the cinetic energy operator, v the potential energy, μ the molecular dipole moment and α the molecular polarizability. The total dipole moment of the molecule is :

$$\bar{m} = \bar{\mu} + \bar{\bar{\alpha}}\bar{E} \qquad (2)$$

Within the Born-Oppenheimer approximation the small amplitude motion around the equilibrium configuration $\bar{R}_0$ is obtained by expanding v, $\bar{\mu}$ and $\bar{\bar{\alpha}}$ in Taylor series with respect to $\bar{u} = \bar{R} - \bar{R}_0$ or in terms of normal coordinates q :

$$v = v_o + \frac{1}{2}\sum_\sigma \omega_\sigma^2 q_\sigma^2 + \frac{1}{3}\sum_{\pi\rho\sigma}\beta_{\pi\rho\sigma} q_\pi q_\rho q_\sigma + \frac{1}{4}\sum \gamma q_\pi q_\rho q_\sigma q_\tau \qquad (3)$$

$$\bar{\mu} = \bar{\mu}_o + \sum_\sigma \bar{\mu}_\sigma^{(1)} q_\sigma + \frac{1}{2}\sum_{\sigma\pi}\bar{\mu}_{\sigma\pi}^{(2)} q_\sigma q_\pi + \ldots. \qquad (4)$$

$$\bar{\bar{\alpha}} = \bar{\bar{\alpha}}_o + \sum_\sigma \bar{\bar{\alpha}}_\sigma^{(1)} q_\sigma + \frac{1}{2}\sum_{\sigma\pi}\bar{\bar{\alpha}}_{\sigma\pi}^{(2)} q_\sigma q_\pi + \ldots. \qquad (5)$$

the cubic and higher terms in (3) and the nonlinear terms in (4) and (5) are the physical origin of the overtone spectrum as observed in a very rarefied molecular gas where the isolated molecule situation can most closely be approached in practice. It is evident yet important to stress that the second order term in (4) would vanish in a non deformable electron model as would α as well ; in contrast, the linear term in (4), the one responsible for the infrared oscillator strength of a fundamental mode and all terms in (3) do not need such a mechanism although their magnitude cannot be accounted for without it.

Intermolecular terms

In the presence of the other molecules of the assembly the total dipole moment of molecule i is now :

$$\bar{p}_i = \bar{\mu}_i + \bar{\bar{\alpha}}(\bar{E} - \sum_{j\neq i}\bar{\bar{T}}_{ij}\bar{p}_j) \tag{6}$$

where $-\sum_{j\neq i}\bar{\bar{T}}_{ij}\bar{p}_j$ is the field at molecule i due to moments of all other molecules ; the tensor $\bar{\bar{T}}_{ij}$ is given by :

$$T_{ij} = \nabla_i\nabla_j \frac{1}{r_{ij}} = \frac{1}{r_{ij}^3} \quad (\bar{I} - 3\bar{r}_i\bar{r}_j/r_{ij}^2) \tag{7}$$

where $\bar{I}$ is the unit tensor and $r_{ij} = |\bar{r}_i - \bar{r}_j|$, the distance between molecules i and j. The vector $\bar{p}$ is seen to depend on the magnitude, position and orientation of the moments of all other molecules. The modification of (6) with respect to (2) comes from the local field which on molecule i is :

$$\bar{E}_i = \bar{E} - \sum_{j\neq i}\bar{\bar{T}}_{ij}\bar{p}_j \tag{8}$$

We can rearrange expression (6) to obtain :

$$\bar{p} = \bar{m} + \bar{\bar{\alpha}}\,\bar{E} \tag{9}$$

with

$$\bar{m} = (\bar{\bar{I}} + \bar{\bar{\alpha}}\bar{\bar{T}})^{-1}\,\bar{\mu} \tag{10}$$

$$\bar{\bar{\alpha}} = \bar{\bar{\alpha}}(\bar{\bar{I}} + \bar{\bar{\alpha}}\bar{\bar{T}})^{-1} \tag{11}$$

where $\bar{\bar{T}}$ is now a symmetrical tensor of order 3N x 3N with elements T_{ij} in 3-dimensional subspace for $i\neq j$ and $T_{ij} = 0_{ij}$. I is the unit tensor in the same space and similarly for μ and α. Expressions (10) and (11) are the effective permanent dipole moment and polarizability per molecule respectively.

It can be shown[25,26] that the interaction energy of the molecules with the EM field and the induced interactions is :

$$V_e = \frac{1}{2}\mu T(1 + \alpha T)^{-1}\mu - [(I + \alpha T)^{-1}\mu]E - \frac{1}{2}E\alpha(1+\alpha T)^{-1}E \qquad (12)$$

so that the total potential energy of the assembly of N indentical molecules is :

$$V = V_o + V_e + V_S \qquad (13)$$

where $V_o = \Sigma v_i$ is the non-electrical part of the potential energy which is only a function of R ; in the term V_S we include other short range intermolecular forces whose form is not that simple but play an important role. The first term in (12) contributes to the frequency shifts for infrared active modes and modifications to the potential energy parameters, the second and third terms govern the infrared and Raman transition intensities when one introduces those parts of (4) and (5) which depend on the normal coordinates. To obtain the intensities and positions of the IR and R transitions the spectrum and wave functions of the molecular sample must be determined and at this level we only need to concentrate our attention on that part of (12) that is independent of the electric field or :

$$V = V_o + \frac{1}{2}\mu T(I + \alpha T)^{-1}\mu \qquad (14)$$

and introduce the developments (3), (4) and (5). From this point the distinction between a liquid sample and a crystalline one must be introduced.

<u>Molecular liquids</u>

As usual because of the higher energies involved we treat the vibrational motion quantum mechanically and separated from the other motions (orientation, translation, rotation,...) ; the latter which are very complex in form are of much lower energy and can be treated by classical mechanics and statistical considerations. Along with the anharmonic terms they play a central role in the dynamics and the relaxation of the vibrational motion.[27,28]

For the vibrational motion (a) we make the approximation that the only important anharmonicity will be intramolecular, namely the one contained in (3) and the long range intermolecular term in (14) will be expanded to quadratic terms and (b), we assume that the intermolecular interaction in general may have small influence on the vibrational levels of a single molecule which may be taken to be those of the anharmonic hamiltonian (3). These two approximations constitute the bases of the local-site picture for a liquid, a satisfactory zero-order perturbation. The short range intermolecular forces will introduce anharmonicities which may be considered as perturbations and treated accordingly.

For the vibrational motion then the ground state wave function is :

$$|0> \equiv \Phi_G = \Pi\phi^{\circ}_{\sigma i} \tag{15}$$

where $\phi^{\circ}_{\sigma i}$ is the harmonic oscillator ground wave function of mode σ on molecule i.

The excited states are not so easily constructed. We only state the procedure to follow for the fundamental ($v_\sigma = 1$) and first harmonic overtone states ($v_\sigma = 2$) of mode σ.

For the fundamental we introduce the functions :

$$|1;\sigma_i > \equiv \Phi'_{\sigma i} = \phi'_{\sigma i}\Pi'_{\pi j}\phi^{\circ}_{\pi j} \tag{16}$$

where Π' means that the wave function $\phi^{\circ}_{\sigma i}$ is not included and $\phi'_{\sigma i}$ is the first excited harmonic oscillator wave function (v_σ=1) on molecule i. The states of definitive energy for the N molecules are of the form :

$$\Phi'_1 = \sum_i a_{1i}\ \Phi'_{\sigma i} \tag{17}$$

where the a_{1i} is obtained by diagonalizing the intermolecular potential energy ; $a^*_{1i}a_{1i}$ then measures the extend of localization of the vibration on molecule i.

For the first overtone we introduce two sets of functions :

$$|02;\sigma_i,\sigma_i > \equiv \Phi''_{\sigma i\sigma i} = \phi''_{\sigma i}\Pi_{\pi j}\phi^{\circ}_{\pi j} \tag{18}$$

$$|11;\sigma_i,\sigma_j > \equiv \Phi''_{\sigma i\sigma j} = \phi'_{\sigma i}\phi'_{\sigma j}\Pi'_{\pi k}\phi^{\circ}_{\pi k} \tag{19}$$

where $\phi''_{\sigma i}$ is the wave function of state $v_\sigma = 2$ on molecule i. These states are not states of definite energy but may form such by introducing :

$$\Phi'' = \sum_i a_{ii}\Phi''_{\sigma i,\sigma i} + \sum_{i,j}{}' a_{ij}\Phi''_{\sigma i,\sigma j} \tag{20}$$

and diagonalizing the intermolecular terms where the quantity $a^*_{ii}a_{ii}$ measures the extent that the vibrational excitation is localized on molecule i (local state) and $a^*_{ij}a_{ij}$ shared by two molecules i and j (cooperative state). One may introduce the pair distribution function[14] but it is intuively clear that a_{ij} will be appreciable for nearest neighbors a and b. The problem essentially reduces to the eigenvalues of the 2 x 2 energy matrix

$$\begin{vmatrix} o & W \\ W & \delta \end{vmatrix} \tag{21}$$

where δ is the anharmonicity defect within a molecule and $W = \sqrt{2} < 11|V_{ab}|02>$ is the coupling between zero-order states $|11>$ and $|02>$ through the intermolecular potential of closest neighbors a and b. One of the two eigenvectors of (21) will describe a predominantly local state and the other a predominantly cooperative state and lie higher in energy.

One may similarly proceed to form the first combination overtones and the higher harmonic and combination overtones. The problem becomes rapidly very complex. The previous procedure and its extension is applicable to both liquids and amorphous solids the distinction between the two entering in the way the other motions and their interaction with the vibrational motion is treated.

Crystalline solids

In crystalline solids the periodicity and the closer proximity of identical molecules makes the local-site picture as zero order approximation inconvenient and one must introduce from the outset a non-local approach as zero-order representation. The starting point of a such a representation is naturally provided by the band model or phonon states in the solid which are eigenstates of the harmonic part of the hamiltonian[3,29].

Quite generally, if a lattice contains N primitive cells, with n particles per cell, the 3nN degrees of vibrational freedom are distributed in 3n phonon branches 3 of which are acoustical and the remaining 3n-3 optical. Each phonon state in a branch σ is characterized by its frequency $\omega_{\sigma}(\bar{k})$, wave vector $\bar{k}$ which varies over the entize Brillouin Zone (BZ) and polarisation $\hat{e}_{\sigma}$. In ionic and convalent crystals where the cohesive forces are as strong inside as outside the unit cell this classification is sufficient ; the normal coordinates and wave functions are obtained from the symmetry and the equations of motion and as a rule the widths of the phonon branches are large (appreciable wave-vector dispersion).

In molecular crystals the cohesive forces are of the van der Waals type, much weaker than the intramolecular forces and a further distinction is made between internal and external modes as follows ; in a molecular crystal containing n(nonlinear) molecules per cell z atoms per molecule, 3nzN degrees of freedom are distributed into n(3z -6) internal vibrational modes with weak wave vector dispersion, 3(n-1) translational lattice modes, 3n "librational" modes and 3 acoustical modes. The external modes (translational, librational and acoustical) essentially describe hindered motion of rigid molecules and their energies are much lower than those of the internal ones.

A first approximation to the one phonon states can be obtained by altogether neglecting the anharmonic terms ; the intermolecular term contains the long range dipolar interactions (7) and (14) which introduce drastic modifications of the infrared active modes at a macroscopic level (ex. separation into longitudinal and transverse modes in cubic lattices), but these effects will not be considered here. The inclusion of anharmonic terms has two effects[29,30] (a) renormalization of the one-phonon states and (b) formation of two- and many- phonon states which can be interpreted as superposition of two or more independent phonons for sufficiently weak anharmonicity (continuum of two or many-phonon states) ; thus the combination state of the continuum of two phonon states formed from states of branches σ' and σ'' with wave vectors $\bar{k}'$ and $\bar{k}''$ respectively is characterized by a wave vector :

$$\bar{k} = \bar{k}' + \bar{k}'' \tag{22}$$

and an energy :

$$\omega = \omega_{\sigma'}(\bar{k}') + \omega_{\sigma''}(\bar{k}'') \tag{23}$$

where $\bar{k}'$ and $\bar{k}''$ vary over the entire BZ. If the anharmonic terms become large new states can be formed,[18,20], termed bound two-phonon states which may branch off from this continuum and show a line spectrum, with drastically different features. There have been some model calculations using a unidimensional diatomic chain[22] and more realistic ones in 3-dimensional lattices with triatomic molecules[21,31], that clearly exhibit these features and will be cursively summarized below.

<u>One dimensional diatomic lattice</u>

It is described by a system of coupled anharmonic oscillators with hamiltonian :

$$H = \sum_n \left(\frac{p_n^2}{2m} + \frac{1}{2}m\omega^2 x_n^2 + \beta x_n^3 - \gamma x_n^4\right) + \sum_{n<m} D_{nm} x_n x_m \tag{24}$$

where x_n is the normal vibrational coordinate of a molecule n ; such a form can be derived from (3) and (8). By introducing the local creation and annihilation operators a_n^+ and a_n respectively

$$x_n = (\hbar/2m\omega)^{1/2}\,(a_n^+ + a_n) \tag{25}$$

$$p_n = i(\hbar m\omega/2)^{1/2}(a_n^+ - a_n)$$

within the rotating wave approximation (neglect of phonon non-conserving terms) this can be recast in the form[22,32]

$$H = \sum_n \varepsilon a_n^+ a_n + \sum_{n<m} \Delta_{nm} a_n^+ a_m - \Gamma \sum_n a_n^+ a_n^+ a_n a_n \tag{26}$$

whose spectrum has the following features of interest [22]:

(1) <u>$\Gamma = 0$</u> , harmonic crystal of diatomic molecules and (24) for

nearest neighbor interactions only reduces to

$$H = \sum_q \left(\varepsilon + \frac{1}{2} W \text{ sign } (\Delta) \cos q\right) a_q^+ a_q \qquad (27)$$

where W is the phonon branch width, $W=4\Delta$ and

$$a_q^+ = \frac{1}{\sqrt{N}} \sum_{n=1}^{N} e^{-iqn} a_n^+ \qquad (28)$$

is the creation operator for a delocalized phonon with crystal momentum $\hbar q$; the one phonon and two free phonon states of wave vector k are then :

$$|\psi_k' > = a_k^+ |0>$$

$$|\psi_k'' > = a_{k'}^+ a_{k''}^+ |0>$$

if $k = k' + k''$.

(2) $\Delta_{nm} = 0$, a set of local anharmonic oscillators with energy levels :

$$E_1 = \ell\varepsilon - \Gamma\ell(\ell - 1) \qquad (29)$$

(3) $\Delta, \Gamma \neq 0$, the two-phonon spectrum strongly depends on the ratio Γ/W. The free phonon appearance (27) persists up to $4\Gamma/W = 1$. At this value a two-phonon bound state splits off from the continuum borrowing oscillator strength from the latter and with Γ/W increasing further it emerges into the local anharmonic oscillator first overtone state.

Along with the spectrum a central role is played by the density of states; both can be jointly obtained from the real and imaginary part of the Green's function[33]

$$G(Z) = \frac{1}{Z - H} \qquad (30)$$

which in terms of the free particle Green's function $G^\circ(Z)$, corresponding to $\Gamma = 0$, can be written :

$$G(Z) = \frac{G^\circ(Z)}{1 + 2\,\Gamma G^\circ(Z)} \qquad (31)$$

the energies of the free phonon states $\omega(k') + \omega(k'')$ are the poles of the free particle Green's function while the bound two-phonon state is the pole of (31) or

$$1 + 2\,\Gamma G^\circ(z) = 0 \qquad (32)$$

the density of states $\rho(\varepsilon)$ is obtained from :

$$\rho(Z) = \frac{1}{\pi} \lim_{\delta \to 0} \left\{ \operatorname{Im} G(Z + i\delta) \right\} \tag{33}$$

The problem has a close resemblance to the Koster-Slater model of an impurity [34].

Three-dimensional realistic lattices

There have been detailled calculations of two phonon states and energies for real crystals, ionic [35,36], covalent [37 40] and molecular [21,3]. These calculations require infinite summations over the lattice and to specifically introduce the symmetry.

In inorganic crystals, ionic or covalent the phonon band width is large with respect to the anharmonicity and the free two- phonon model essentially applies. The two-phonon density of states can be obtained in certain cases by means of critical point analysis. These are defined by :

$$\bar{\nabla}_{\bar{k}} \left(\omega_{\sigma'}(\bar{k}') \pm \omega_{\sigma''}(\bar{k}'')\right) = \theta \quad \text{or} \tag{34}$$

$$\bar{\nabla}_{\bar{k}}\omega_{\sigma'}(\bar{k}') = \pm \bar{\nabla}_{\bar{k}}\,\omega_{\sigma''}(\bar{k}'')$$

and their number is larger than that of the one phonon density of states defined by $\bar{\nabla}_{\bar{k}}\omega_{\sigma}(\bar{k}) = 0$. Symmetry and group theoretical methods can be of some help although their use is complicated by the fact that one generally works away from the center of the Brillouin Zone.

In molecular crystals of triatomic molecules [21] like CO_2, NO_2, OCS, CS_2 detailled calculations revealed the possibility for bound two-phonon states to be observed. Similar considerations exists for diatomic molecular lattices [41]. The calculations there bear much [42,43] resemblance to the exciton models and the wave functions are similar to the ones we introduce for liquids. There is however an essential difference between molecular liquids and their crystalline counterparts, namely the normal modes of the molecular crystal are not identical with the ones of its constituant molecules ; this is certainly the case of the modes connected with intermolecular motion within a unit cell.

In solids much attention has been given [44] to the coupling of the two-phonon states with the one-phonon states of a different branch. This so-called Fermi resonance has been extensively studied in ionic crystals and ferroelectrics in particular near their phase transition. In these materials quite often the phase transition is heralded by a specific mode becoming soft and in the process of lowering its frequency it crosses different two-phonon states which since they are formed by pairs of large wave vector one-phonon states, reflect some features of the phase transition.

III - INFRARED AND RAMAN STUDY OF OVERTONES

We discuss here the main features of the overtones and many phonon states as revealed in IR and R spectroscopy of liquids and crystals ; we essentially limit ourselves to the first overtone and two-phonon states.

Refering to the molecular sample of section II its infrared absorption coefficient and Raman differential scattering cross section at frequency ω are given by the fluctuation-dissipation[45] theorem as follows :

$$\alpha(\omega) = \frac{1}{3\hbar V}\frac{1}{2\pi}\int_{-\infty}^{\infty} dt\ e^{-i\omega t} < \bar{M}(0)\ \bar{M}(t)>_{Av} \tag{35}$$

$$S \equiv \frac{d^2\sigma}{d\omega' d\Omega} = \left(\frac{\omega'}{c}\right)^4 \frac{1}{V}\frac{1}{2\pi}\int_{-\infty}^{\infty} dt\ e^{-i\Delta\omega t} < [\hat{\varepsilon}\bar{\bar{A}}(0)\hat{\varepsilon}'][\hat{\varepsilon}\bar{\bar{A}}(t)\hat{\varepsilon}']>_{Av} \tag{36}$$

where $\{\hat{\varepsilon},\omega\}$ and $\{\hat{\varepsilon}', \omega'\}$ are the polarization and frequency of the incident and scattered light respectively, V the volume of the scattering sample and $\bar{M}$ and A its total dipole moment and polarizability ; $< >_{Av}$ implies averaging over all motions that give a time dependance to $\bar{M}$ and $\bar{\bar{A}}$. We may write :

$$\bar{M} = \Sigma\bar{m}_i \qquad \bar{\bar{A}} = \Sigma\bar{\bar{\alpha}}_i \tag{37}$$

where the summations extended over all N molecules of the sample and we stress that $\bar{m}_i$ and $\bar{\bar{\alpha}}_i$ are understood to be given by (10) and (11) respectively. The contributions of the overtones and many phonon states is simply obtained by singling out the corresponding changes in $\bar{M}$ and $\bar{\bar{A}}$. The distinction between liquids and crystals essentially arises from the way the averaging $< >_{Av}$ in (35) and (36) is treated, a consequence of the way the molecules are distributed.

<u>Molecular liquids</u>

In a liquid the quantities m_i and α_i are alteredin time by the vibrational, orientational and translational motion of the molecule itself and by the motion of the other molecules since m_i and α_i depend on the intermolecular distances and orientations.

In the averaging $< >_{Av}$ all these motions must be taken into account appropriately. Over the vibrational motion which has been treated quantum mechanically this amounts to a trace operation with the vibrational matrix ρ_v ; the other types of motions which are complex will be considered as stochastic processes and the corresponding averages over ensembles will be used. To this effect we may write for the effective molecular quantities :

$$\bar{m}(t) = \bar{m}_o(t) + \sum_\sigma \bar{m}_\sigma^{(1)} q_\sigma(t) + \sum_{\sigma\tau} \bar{m}_{\sigma\tau}^{(2)} q_\sigma(t) q_\tau(t) \tag{38}$$

$$\bar{\bar{\alpha}}(t) = \bar{\bar{\alpha}}_o(t) + \sum_\sigma \bar{\bar{\alpha}}_\sigma^{(1)}(t) q_\sigma(t) + \sum_{\sigma\tau} \bar{\bar{\alpha}}_{\sigma\tau}^{(2)} q_\sigma(t) q_\tau(t) \tag{39}$$

where $q_\sigma(t)$ are operators in the Heisenberg representation and their time dependence comes from the vibrational motion while the other quantities in (38) and (39), which are space fixed on the molecular frame obtain theirs from all the other stochastic motions. Introducing (38) and (39) in (37), (35) and (36) we can single out the contributions of the overtones to the IR and R intensities. These can be expressed as Fourier transforms of the correlation functions $< \delta''\bar{M}(0)\ \delta''\bar{M}(t)>$ and $< \delta''\bar{\bar{A}}(0)\ \delta''\bar{\bar{A}}(t) >$ respectively where $\delta''\bar{M}$ and $\delta''\bar{\bar{A}}$ are the parts in (37) that are quadratic in the q_σ's after insertion of (38) and (39) ; the details are given in ref. 46 using the vibrational density matrix in the space of the vibrational states of section II. Here we use a more intuitive approach along the lines of the one introduced in Ref. 14 and concentrate only on the first overtone of mode σ which we assume to be away from any fundamental mode (no Fermi resonance).

The time dependence of the normal coordinate with relaxation can be written :

$$q_\sigma(t) = q_\sigma(0) \exp(i\omega_\sigma t) V_\sigma(t) \tag{40}$$

where $V_q(t)$ incorporates all relaxation processes. Similarly for the product $<q_\sigma(t)q_\sigma(t)>$ we may write

$$<q_\sigma(t)q_\sigma(t)> = Q_{\sigma\sigma}(t) = V_L(t)Q_L(0)e^{i\omega_L t} + V_C(t)Q_C(0)e^{i\omega_C t} \tag{41}$$

where L and C refer to the predominantly local and cooperative states respectively of hamiltonian (21) and ω_L, ω_C are its eigenvalues ; in (41) we anticipated that the relaxation processes are different for the two states. We now see that this leads to two bands in the IR or R overtone spectrum centered at frequencies ω_L and ω_C with band shapes given by the Fourier transform of certain correlation functions ; the details are given in Ref. 46 where simplified models of the different types of stochastic motions are also discussed. We only point out here that if states L and C are almost degenerate[14] the corresponding relaxation processes being different their shapes will be different ; this will be difficult to infer by frequency resolved spectroscopy while the time resolved techniques proposed in section IV make it easier.

Experimental IR and R spectroscopy of overtones in molecular liquids has essentially been concerned with the positions of the overtones[4] but only very exceptionally[14] with their shapes.

Overtones have also been studied in some mixtures where the distinction between local and cooperative states may be more clear. All these effects have their analogues in the collision induced processes in gases[15] where more detailled spectral studies exist.

Crystals

In crystals the periodicity allows all motions, vibrational, translational, orientational, to be treated quantum mechanically in terms of the delocalized phonon modes ; for ionic and covalent inorganic crystals these are optical or acoustical while for molecular crystals they are optical, translational, librational and acoustical.

When the spectrum contains no bound two-phonon states and the total dipole moment $\bar{M}$ and polarisability $\bar{\bar{A}}$ of the crystalline sample are expressed in power series of the free phonon coordinates

$$\bar{M} = \bar{M}_o + \sum_{\sigma k} \bar{M}^{(1)}_{\sigma\bar{k}} Q_{\sigma\bar{k}} + \sum_{\sigma\bar{k},\sigma\bar{k}'} \bar{M}^{(2)}_{\sigma\bar{k},\sigma\bar{k}'} Q_{\sigma\bar{k}} Q_{\sigma'\bar{k}'} \qquad (42)$$

$$\bar{\bar{A}} = \bar{\bar{A}}_o + \sum_{\sigma\bar{k}} \bar{\bar{A}}^{(1)}_{\sigma\bar{k}} Q_{\sigma\bar{k}} + \sum_{\sigma\bar{k},\sigma\bar{k}'} \bar{\bar{A}}^{(2)}_{\sigma\bar{k},\sigma'\bar{k}'} Q_{\sigma\bar{k}} Q_{\sigma'\bar{k}'} \qquad (43)$$

and introduced in (35) and (36) respectively one obtains for the two phonon contributions

$$\alpha''(\omega) = \frac{1}{3hV}\,\frac{1}{2\pi}\int_{\infty}^{\infty} dt\; e^{-i\omega t}\,\mathrm{Tr}\,[\rho\,\delta''_o\bar{M}(0)\;\delta''_o\bar{M}(t)] \qquad (44)$$

$$S'' = \left(\frac{\omega}{c}\right)^4 \frac{1}{V}\,\frac{1}{2\pi}\int_{-\infty}^{\infty} dt\; e^{-i\Delta\omega t}\,\mathrm{Tr}\rho\,[\hat{\varepsilon}\delta''_o\bar{\bar{A}}(0)\hat{\varepsilon}']\;[\hat{\varepsilon}\delta''_o\bar{\bar{A}}(t)\hat{\varepsilon}'] \qquad (45)$$

where ρ is the density matrix in the phonon state space, $\delta''\bar{M}$ and $\delta''_o\bar{\bar{A}}$ that part of (42) and (43) that is quadratic in phonon coordinates with total wave vector equal to that of the light which can be assumed essentially zero. In view of (22) this means that only two-phonon states with $\bar{k} + \bar{k}' \simeq 0$ will be included ; these lie on opposite points of the BZ, but otherwise combinations to all types of phonons are in principle allowed[47].

When two-phonon bound states are present some care must be used in deriving[21] the expressions of $\delta''\bar{M}$ and $\delta''\bar{\bar{A}}$ which will not be explicited here.

There have been detailled calculations for some real crystalline lattices[21,36]. The group theoretical methods[39,40,48] have been extensively used to derive the allowed combinations and two-phonon density of states have been calculated. It should be pointed out that $\bar{M}^{(2)}_{\sigma k,\sigma' k'}$ and $\bar{\bar{A}}^{(2)}_{\sigma k,\sigma' k'}$ vary over the Brillouin Zone which introduces complications in relating the observed spectrum with the critical points of the two-phonon density of states ; this is

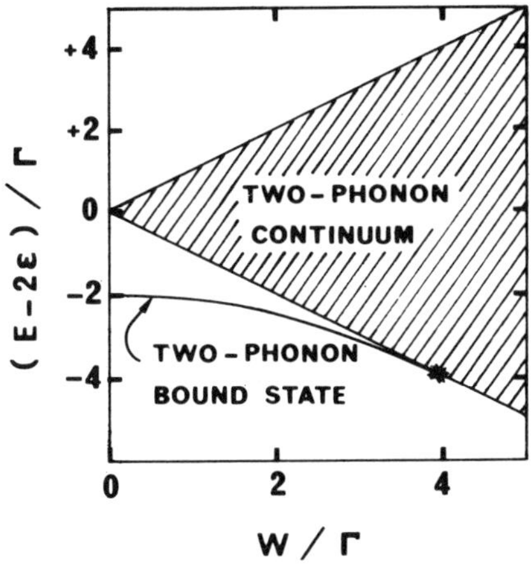

A. Energy levels of the two-phonon spectrum of the one-dimensional diatomic chain (from Ref.22) exhibiting the branching off of the bound two-phonon state.

B. Energy levels and density of states of the two-phonon spectrum of a realistic three-dimensionallattice (from Ref.21)

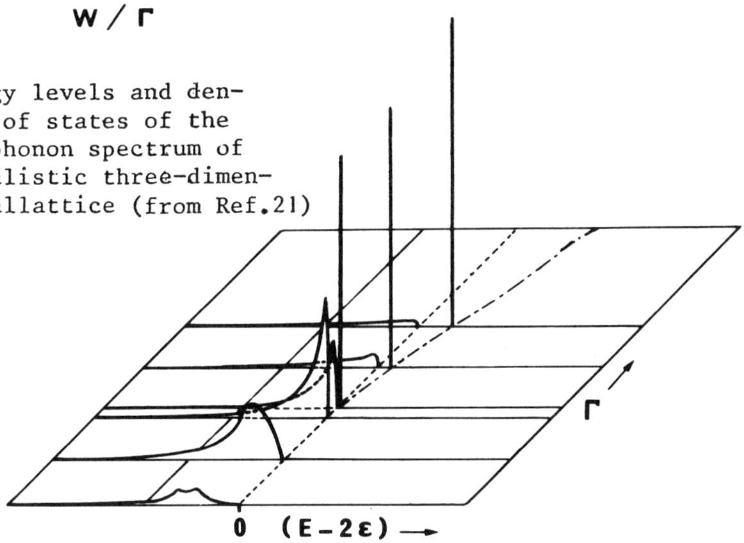

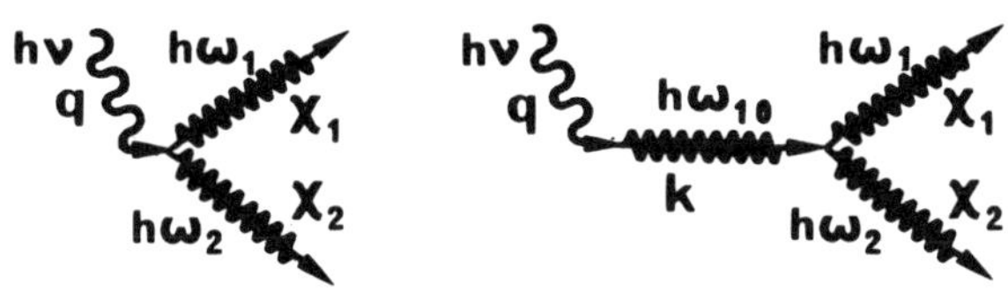

Figure 2

mainly true for ionic and covalent inorganic crystals; in molecular crystals such a complication seldom arises on the other hand bound two-phonon states may exist here and substantially alter the appearance of the overall two-phonon spectrum (see Figure 1).

It has been customary both in IR-and R-spectroscopy of two-phonon states to distinguish two mechanisms[31,40] of optical excitation of such states (see Figure 2) (a) direct interaction of the light radiation with the lattice to produce two phonons, (b) an intermediate (virtual) state is produced and rapidly decays into two phonons. The degree of coherence of the two phonons is an essential parameter which distinguishes the two processes but its determination is very difficult; it plays a crucial role in the coherent excitation of these two-phonon states by nonlinear techniques as will be discussed below. Incidentally the second process (b) is closely related to the phonon breakdown .

There have been many experimental studies of two-phonon states by IR- and R-spectroscopy and the reader may consult references (22,36,40). Evidence for the existence of bound two-phonon states in ionic and covalent crystals has been controversial, ex. Diamond[26,27]; on the other hand a mounting evidence[13,16,49,50] of such states in molecular crystals exists. The bound two-phonon state because of its coupling and overlap with the free two-phonon states has a finite linewidth but its shape may be difficult to determine. As a rule its linewidth in crystals like CO_2, NO_2, CS_2 or OCS ranges to a few cm^{-1} or less which indicates long living states (few pcs or more) and thus appropriate for coherent excitation.

The above discussion was based on the Born description of lattice dynamics. A quite differnt approach in interpreting the two-phonon Raman spectrum was proposed by Raman itself[48] by doubling the size of the unit cell and assuming that only $\bar{k}=0$ vibrations are involved. This does not alter the fundamental phonon spectrum but introduces a different point of view concerning the two-phonon spectrum which now consists of a limited number of lines instead of discrete lines and a continuum as in Born's theory. This interpretation introduces some simplifications but also features that have not been confirmed yet and gained only limited acceptance up till now.

As stated previouly in conventional R and IR spectroscopy there is no restriction on the individual phonon wave vectors other than their sum should equal the total photon wave vector which is negligible compared to the Brillouin Zone. The situation is quite different in static electric field induced IR and R spectra[51,52]. Here one may have two-phonon states again but both phonons have wave vectors in the center of the BZ; furthermore one of them is resonantly driven.

IV - TIME RESOLVED STUDY OF OVERTONES AND MANY-PHONON STATES

In the conventional IR and R spectroscopical study of overtones one encounters the same difficulties as in the corresponding study of the fundamental modes [4] : limited instrumental resolution, overlapping with neighboring bands, Fermi resonances, contribution of rotational motion, etc. In the case of the fundamental modes some of these difficulties can be overcome with the use of nonlinear optical techniques and time-resolved ones with picosecond laser pulses in particular [53].

We propose and discuss below their extension to the study of the overtones and many phonon states and present the first [54] experimental investigation. These techniques contain [53] two stages :

a) an excitation stage where the molecules are selectively prepared in the desired state by stimulated Raman process (or coherent excitation with frequency difference equal to the transition frequency of the state) or IR resonant radiation.

b) a probing stage where a second weak pulse properly delayed with respect to the one used at the excitation stage monitors the instantaneous state of the excited system.

The essential advantages of the time resolved techniques over the conventional spectroscopy are : a) direct access to the correlation functions $<\delta\bar{M}(o)\delta\bar{M}(t)>$ and $<\delta\bar{\bar{A}}(o)\delta\bar{\bar{A}}(t)>$ without going through an inverse Fourier transform, b) one can drive the vibrational state selectively into specific non equilibrium situations with high intensity picosecond pulses while the conventional spectroscopy only allows studies close to the equilibrium, c) there are some additional advantages when the Raman or coherent excitation process is used (coherent excitation of the molecules in the bulk, the frequencies can be tuned into the transparency region, the influence of the rotational motion is reduced) furthermore the selection rules can be quite different.

The two fundamental questions to be settled in order to apply these techniques and extract the desired information are a) set up the polarisation in the medium that has its origin in the overtone excitation and b) solve the nonlinear propagation equation of the fields coupled through the nonlinear polarisation set up by the overtone mode.

Excitation stage

We assume that the dynamics of the overtone are introduced through the vibrational overtone coordinate $Q_{\sigma\sigma}(t)=q_\sigma(t)q_\sigma(t)$ whose expectation value $<Q_{\sigma\sigma}>=\mathrm{Tr}\rho_V\, Q_{\sigma\sigma}$ sets up a change in the polarisation

$$\delta P = \frac{N}{2V} \left(m^{(2)} <Q_{\sigma\sigma}> + \alpha^{(2)} <Q_{\sigma\sigma}> E\right) \tag{46}$$

where ρ_v is the density matrix in subspace of the overtone states introduced in section II and its evolution is governed by

$$\frac{d\rho_v}{dt} = \frac{1}{i\hbar} \left[\hbar''_e, \rho_v\right] + \frac{1}{i\hbar} \frac{d\rho_v}{dt}\Big|_{Rel} \tag{47}$$

where $h''_e = h_o + h''_{\sigma\sigma}$, $h''_{\sigma\sigma}$ is the effective interaction hamiltonian

$$h''_{\sigma\sigma} = \frac{1}{2} m^{(2)} Q_{\sigma\sigma} E - \frac{1}{2} \alpha^{(2)} Q_{\sigma\sigma} E^2 \tag{48}$$

and h_o is the unperturbed hamiltonian derived from (1), (3) and (14) with the anharmonic terms included.

This set of equations completely describes the excitation stage. The linear and quadratic terms in the electric field are related to the IR resonant and Raman (or coherent) excitations respectively. They are differently implicated: in the IR excitation $E = E_o \cos\omega t$ while in the Raman (or coherent) excitation $E = E_1 \cos\omega_1 t + E_2 \cos\omega_2 t$, where $\omega_1 - \omega_2 \simeq \omega_o$, the overtone frequency. The crucial question is the nature of the relaxation mechanisms that occur in (47) and their implications on the coherence time of the overtone (or bound two-phonon) state and this will be discussed now.

Relaxation mechanisms

The mechanisms that are responsible for the relaxation of the fundamental modes are also involved in the relaxation of the overtones and two-phonon states but the competition of the anharmonicity and cooperativity there endows these states with new features that have drastic implications on their dynamics and relaxation. In particular the coupling and overlap of local (or bound) and delocalized states within the overtone (or two-phonon) subspace plays a key role and we shall cursively discuss this point separately for liquids and crystals.

Liquids. The overtone spectrum of simple molecular liquids was discussed in section II. The relevant subspace consists of the ground state $|0>$ and the local and cooperative states $|L>$ and $|C>$, which are predominantlv $|02>$ and $|11>$ respectively. Within this subspace of the overtone coordinate $<Q_{\sigma\sigma}>$ takes the form (41) and can be viewed as a two-component vector $[Q_L, Q_C]$ with $Q_{L,C} = Q_{L,C}(0) V_{L,C}(t)$; the relaxation functions V_L and V_C have to be determined through (40). The solution in time domain is quite involved and particular attention must be paid (14,46) in solving the density matrix equation since states L and C are very close and coupled to each other through the term W in (21). This problem is discussed in Ref.46; one expects that state L is more stable than state C and V_L will be weakly damped and approximately of the form $\exp(-t/T_L)$ while $V_C(t)$ is heavily damped and in some cases can be considered as a (virtual) quasi-reservoir. Thus in general T_L will not be simply related to the coherence time T_2 of the fundamental modes. Amajor complication in liquids arises from the interplay and simultaneously presence of quantum and classical (stochastic) motions.

Crystals . The situation is quite different here since all motions can in principle be treated quantum mechanically but additional complications arise because of polariton (retardation) effects, strong anisotropy and selection rules which drastically influence the relaxation processes. We shalldisregard such macroscopic effects, treat all quantities like sca and single out the anharmonic coupling as the intrinsicly microscopic relaxation mechanism within the two-phonon subspace. As discussed in section II the relevant states here are the zero phonon state $|0\rangle$, states of the continuum of free two-phonon states $|\psi''_F\rangle$ and the bound two-phonon state $|\psi''_B\rangle$. Within the Born lattice theory only to the later we may ascribe a finite coherence lifetime limited by its coupling and decay to the free two-phonon states (within the Raman theory[48] the situation may be quite different). The determination of this lifetime can be most conveniently done with the use of the Green's function formalism outlined in section II but even so the calculation is in general quite involved. We shall illustrate the method with an intuitively simpler approach which borrows much from the one used by Koster and Slater[34] in their treatment of the localized impurity electron states in a crystal or the one used to calculate the impurity effects in lattice dynamics[29,30].

We assume that the coupling responsible for the loss of coherence of the bound two-phonon state by decav to free two-phonon states is the anharmonic terms in (24) or (26). We introduce creation and annihilation operators A_B^+ and A_B rescpectively, for the bound two-phonon state so that the relevant part in (26) responsible for this decay can be singled out in the form

$$h_{BF} = \sum_F \Gamma_{BF} a^+_{q'} a^+_{q''} A_B \tag{49}$$

where the summation actually implies appropriate integrations over the Brillouin zone Γ_{BF} is the appropriate anharmonicity strength and F labels all combinations (q',q"). We also have $|\psi''_B\rangle = A^+|0\rangle$ and $|\psi''_F\rangle = a^+_{q'} a^+_{q''}|0\rangle$, where a_q is given by (28), which are eigenstates of an effective hamiltonian h_o.

The eigenstates of the total hamiltonian $h_o + h_{BF} = h$ will be of the form

$$|\psi''\rangle = c_B|\psi''_B\rangle + \sum_F c_F|\psi''_F\rangle \tag{50}$$

and their eigenvalues are given by

$$\varepsilon - E_B = 2\sum_F \frac{\Gamma_{BF}}{\varepsilon - E_F} = g(\varepsilon) \simeq \Gamma \int \frac{\nu(z)dz}{\varepsilon - z} \qquad (51)$$

where $E_F = \omega(q') + \omega(q'')$, $\nu(z)$ is the joint two-phonon density of states. This equation is equivalent to the set of equations (30) through (33) and has as many roots as there are terms in the summation (the total number of free two-phonon states) plus one which comes from the bound two-phonon state.

Let us now assume that at time t=0 only the bound two-phonon state was excited so that

$$|\psi''(0)\rangle = |\psi''_B\rangle \qquad (52)$$

For later times this state will evolve under the influence of h_{BF} given by (49) and will get mixed with the $|\psi''_F\rangle$'s so that at time t we may write in terms of the eigenstates (50)

$$|\psi''(t)\rangle = \Sigma_\kappa c_\kappa e^{-i\varepsilon_\kappa t} |\psi''_\kappa\rangle \qquad (53)$$

The time evolution of the initial state (52) then is simply obtained from

$$V_B(t) = \langle\psi''(0)|\psi''(t)\rangle = \frac{1}{2\pi i}\int_C e^{-izt} \frac{dz}{z - E_B - g(z)} \qquad (54)$$

where the contour C can be expanded to an infinite circle. Setting

$$g(z+i\delta)_{\delta\to 0} = \pi(h(z) - i\rho(z)) \qquad (55)$$

we obtain under certain conditions on the behaviour on $\nu(z)$ which are in general satisfied in practice

$$V_B(t) = \langle\psi''(0)|\psi''(t)\rangle = e^{-i\varepsilon_o t} e^{-t/T_B} \qquad (56)$$

where ε_o is determined from $\varepsilon_o - E_B - h(\varepsilon_o) = 0$ and the bound two-phonon lifetime (coherence time) is given by

$$1/T_B = \nu(\varepsilon_o) = \rho(\varepsilon_o) \qquad (57)$$

We thus obtain an exponential behaviour and the lifetime is related to the two-phonon density of states calculated at the energy ε_o.

It is intuitively plausible from the previous considerations that the lifetime of an overtone in a liquid will be shorter than the lifetime of the corresponding bound two-phonon state in the molecular crystal since many degrees of freedom are frozen in the later case. Furthermore in the later case we expect T_B to be very sensitive to structural phase transitions since then the joint two-phonon density of states $\rho(\varepsilon_o)$ changes .

It is also implicit from the above considerations that in practice one can have access to a large density of two-phonon states by coherently driving the bound two-phonon state which then transfers its coherence to the free two-phonon states.

Nonlinear Propagation. Probing stage

In the probing process the electromagnetic field of frequency ω_p is scattered off the vibrational overtone amplitude and produces fields shifted in frequency by the overtone frequency, $\omega_S=\omega_p-\omega_o$, and $\omega_A=\omega_p+\omega_o$. The equation of the probe fields starts from

$$\Delta E' - \frac{1}{c^2}\frac{\partial^2}{\partial t^2}(\varepsilon E') = \frac{4\pi}{c^2}\frac{\partial^2 P'_{NL}}{\partial t} \qquad (57)$$

where E' is the total (incident and scattered) probing field, $E'=E_p\cos\omega_p t + E_S\cos\omega_S t + E_A\cos\omega_A t$, and P'_{NL} the nonlinear polarisation which in the case of coherent excitation simply takes the form

$$P'_{NL} = \frac{1}{2}\frac{N}{V}\alpha^{(2)} \langle Q_{\sigma\sigma}\rangle E' \qquad (58)$$

and $\langle Q_{\sigma\sigma}\rangle$ is the solution of (47) after the excitation. Equation (57) can be linearized with the usual assumptions and the subsequent considerations follow the lines of the treatment[(53)] for the fundamental modes with the additional complication that $\langle Q_{\sigma\sigma}\rangle$ is a two component amplitude with different frequencies and dynamics.

As in the case of fundamental modes one may distinguish between coherent and incoherent scattering, the former measuring the evolution of $Q_{\sigma\sigma}$ and the second the evolution of the populations of of the overtones

Since the overtone is expected to be most closely related to its fundamental modes one may also probe the amplitude and population of the later and gain important information abour coherence and energy pathways in liquids and solids. In particular one can get a better understanding of the phonon breakdown mechanisms whose study has been hampered by many difficulties[(53)].

Experimental Investigations

The first demonstration[(54)] of time resolved picosecond coherent antistokes Raman Scattering on overtones of CAHORS (Coherent Antistokes Higher Order Raman Scattering) was done on the $2\nu_2$ overtone (795 cm^{-1}) in liquid CS_2 and in the corresponding two-phonon states (801 cm^{-1}) in crystalline CS_2. The experimental results confirm the previous considerations and open the way to systematic studies of the dynamics of overtones and two-phonon states in large class of molecular systems.

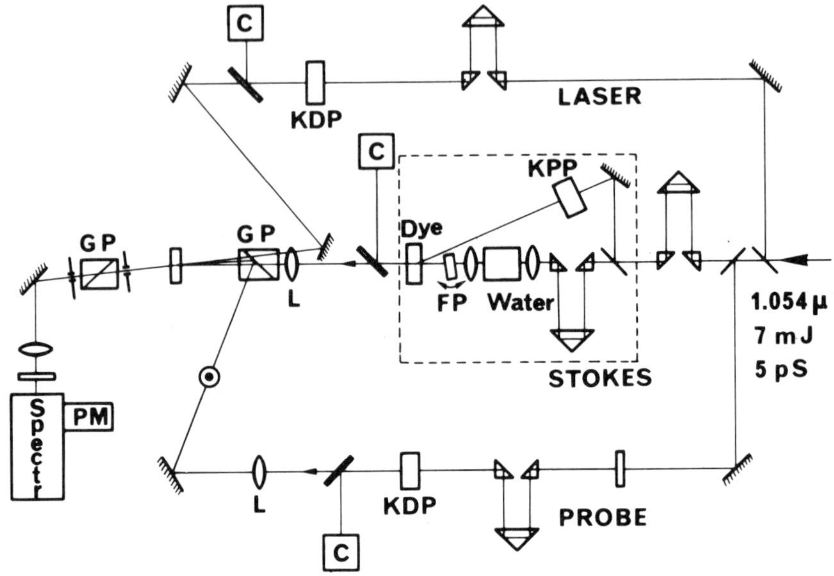

Figure 3

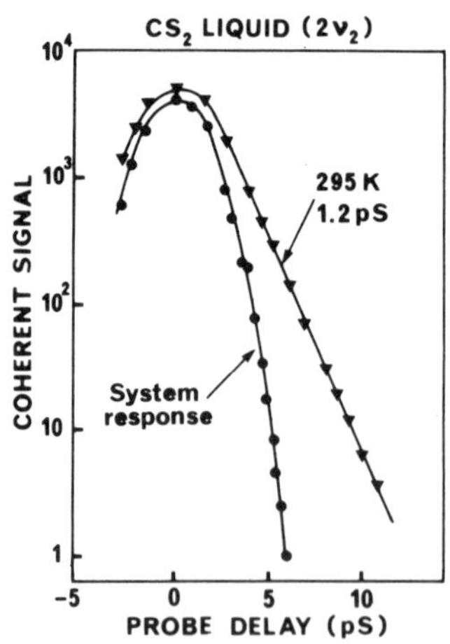

Figure 4

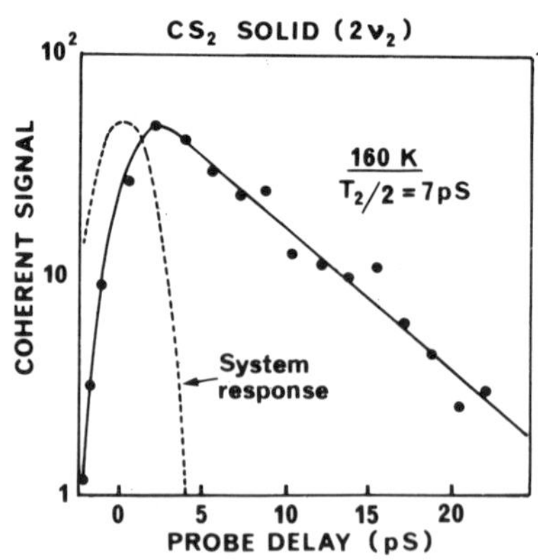

Figure 5

The experimental system is schematically shown in Fig. 3. Single pulses of 5ps duration are generated by a passively mode-locked laser system consisting of a Nd/ phosphate glass laser oscillator an electrooptic pulse selector and a double pass double stage amplifier chain. Pulse energy is 7 mJ at 1.054 μm with a repetition rate of 0.5 HZ.

A first beam splitter removes 30 % of the initial infrared intensity which is doubled in frequency, after an optical delay line, to serve as a pump pulse transient stimulated Raman excitation of the vibration. 20 % of the remainder is doubled for use as a probing pulse and the rest of the infrared beam is directed into a Stokes generator to provide a frequency shifted pulse for Raman excitation.A frequency tunable low divergence frequency shifted pulse is generated by amplification in a dye amplifier (pumped by an intense second harmonic pulse) of a small frequency band of the self phase modulation continuum obtained by focusing an IR beam into a water cell. Tunability is obtained by rotation of the Fabry-Perot frequency selector. Coherent excitation at a frequency ω_o is obtained by focusing the two synchronized pulses ω_L and ω_S (with $\omega_L-\omega_S \simeq \omega_o$) into the sample with parallel polarisations. Probing of the coherent amplitude is obtained by anti-Stokes scattering of a suitably delayed probe pulse ω_p (here $\omega_p=\omega_L$) with crossed polarisation. The small divergence of the three optical pulses in the sample allow a specific geometry for non-collinear k-matching to be adjusted which permits excellent angular discrimation, by the use of small apertures, against anti-Stokes production during the excitation process. Addition noise rejection is provided by a crossed polariser placed behind the sample. Off-axis anti-Stokes scattering of the probe pulse is measured in a small solid angle of acceptance by a photomultiplier after frequency selection by suitable filters and a monochromator. The 2mm thick sample of CS_2 is placed in a cryostat for the study of the temperature variation of T_2 in the liquid and the solid (T_f=162°K).

The temporal response function of the system is obtained by non-resonant time resolved CARS in a 2mm sample of CH_3I, which provides the cross correlation curves of the three input pulses by the almost instantaneous response of the driven molecules.

The fundamental mode and overtone spectrum of CS_2 both in liquid and crystalline phase are known(55-57) where as can be noticed the anharmonic shift of the overtone $2\nu_2$ is very small indicating that local and cooperative states are almost degenerate.

The experimental values are shown in Figs. 4 and 5 for the overtone $2\nu_2$ for liquid (T=300°K) and crystalline (T≃160°K) phases. In the liquid an exponential decay with $T_2 \simeq 2,2$ psec was obtained ; this value of T_2 is close to the one measured in orientational relaxation time measurements (58) for this molecule as can be expected for this mode. An exponential decay is also obtained in the crystalline phase. A dramatic increase of the coherent decay time on crystallization is observed which implies a phase transition of the relaxation mechanism (54).

V - NONLINEAR SPECTROSCOPY OF OVERTONES. PARAMETRIC INTERACTIONS AND INSTABILITIES

The time-resolved techniques outlined in section V have their analogue in frequency domain ; this allows extension of nonlinear spectroscopical techniques (59) like CARS (60), RIKES (61) to the overtones and two-phonon states. Furthermore coherent excitation of two-phonon states under certain conditions may lead to parametric interactions, instabilities and generation of coherent large wave vector phonons. We wish to briefly discuss these points here and for this purpose we need to derive the corresponding nonlinear susceptibilities.

Nonlinear Spectroscopy

Equations (39) (40) and (41) can also be resolved in frequency domain in explicit form and when introduced in (43) one obtains an expression of the nonlinear polarisation when also the nonresonant electronic contribution is added. Thus for the polarisation at frequency ω_A one has

$$\delta P = \chi^{(3)}(\omega_1,-\omega_2,\omega_p)\ E_1E_2{}^*E_p \tag{44}$$

with $\omega_1-\omega_2 \simeq \omega_o$ and $\chi^{(3)}=\chi_E^{(3)}+\chi_Q^{(3)}$ where $\chi_E^{(3)}$ is the nonresonant electronic contribution (the same one as in CARS experiments) and $\chi_Q^{(3)}$ is the resonant part related to the overtone resonance. It can be derived classically as follows.

To the overtone two-component amplitude $Q=[Q_L,Q_C]$ from (41), corresponds classically a conjugated force $F_Q=-\nabla_Q h''_{\sigma\sigma}$ that drives this amplitude according to equation

$$\ddot{Q} + \frac{1}{T}\dot{Q} + \Omega_o\ Q = \frac{1}{2}\ \alpha^{(2)}E^2 \tag{45}$$

where $E=E_1\ \cos\omega_1 t + E_2\ \cos\omega_2 t$; all quantities are two component vectors and Ω_o the 2x2 matrix (21). With $\omega_1-\omega_2 \simeq \omega_o$ only the component $\omega_1-\omega_2$ in the right hand member will appreciably drive the amplitude. If we momentarily assume a single component the solution of (45) is simple and one obtains from (43) and (44)

$$\chi_Q^{(3)} = \frac{N}{4}\ \frac{[\alpha^{(2)}]^2}{\omega_o^2-(\omega_1-\omega_2)^2+i(\omega_1-\omega_2)T^{-1}} \tag{46}$$

which together with χ_E introduces an interference effect when $\omega_1-\omega_2$ is tuned around ω_o similar to the one seen in CARS experiments[60] on fundamental modes. The general solution of the two-component matrix equation (45) is given in ref (59) using the coupled anharmonic mode scheme. The whole problem can actually be treated with the Breit-Wigner formalism[62,44] when one of the components has a very broad spectrum that it can be considered as a continuum.

Similarly one can treat other effects enhanced by overtone resonances for instance higher order Raman induced Kerr effect scattering (HORIKES). There have been some CARS-experiments of overtone in benzene[63] and in diamond[64] but none in molecular crystals where the lines are much narrower. In the case of liquid benzene[63] the overtones borrows oscillator strength from the totally symmetric CH-streching mode.

Two-phonon parametric interactions and instabilities

The coherent excitation with two intense fields allows one to generate coherent bound two-phonon states with wave vector $k \simeq 0$ and large amplitude. Their amplitude can be viewed as a field amplitude satisfying an inhomogeneous wave equation. Since the overlap $<\psi''_F/\psi''_B>$ of the bound and free two-phonon states can be substancial[22] the bound two-phonon can break down with finite probability into two free phonons coherently propagating with wave vecrors k' and k" such that $\bar{k}'+\bar{k}'' = \bar{k} \simeq 0$ and amplitudes q'_σ and q''_σ that satisfy the one phonon propagation equation with a source term. The source term is related to $<\psi''_F/\psi''_B>$ whose square also measures the probability of transfering the coherence from the bound to the free two-phonon amplitudes with total wave vector conserved or $\bar{k}'+\bar{k}'' = \bar{k}$.

A similarity with the parametric amplification and oscillation[65] in three wave interactions in nonlinear optics can be drawn as follows : the overlap $<\psi''_F/\psi_B>$ corresponds to the second order susceptibility $\chi^{(2)}(\omega_p;\omega_s;\omega_i)$ and Q, q'_σ and q''_σ correspond to the field amplitudes of the pump (p), signal (s) and idler (i) respectively. The phonon propagation equations can be linearized within the envelop approximation similar to the one used in nonlinear optics[66]. This striking analogy can be the starting point of nonlinear phonon optics.

With the coherent excitation of the bound two-phonon mode and its coupling with free two-phonon states one has the possibility to parametrically generate large amplitude one-phonon states throughout the Brillouin zone and study their propagation, coherence and dynamics.

VI - TIME RESOLVED STUDY OF INTRINSIC DEFECTS IN CONJUGATED POLYMER CRYSTALS

In simple molecular crystals where the molecules retain their individuality inside the crystal, large amplitude vibrational motion is essentially revealed in the spectrum and dynamics of their overtones as discussed above. In organic crystals with large electron conjugation in one direction like in the polyacetylenes or the polydiacetylenes, because of the strong electron-lattice coupling built in through the conjugation, large amplitude vibrational motion may lead to a new type of states like the solitons[10-12] with electron and lattice motion closely interwoven. Since these defects are intrinsic to conjugated chains and can be photoinduced, their dynamics will drastically affect the time response of the optical properties of the material. Time resolved study of the optical Kerr effect in particular, which in these materials has very high values[67-70] because of the electron delocalization and is highly anisotropic (many orders of magnitude larger along the chain direction than perpendicular to it) can be very sensitive to probe the dynamics of these defects. We briefly discuss below the basic aspects of these defects and point out a new class of "delocalized defects" arising from chain pairing.

<u>Dynamics of photoinduced defects</u>

The most striking implication of electron lattice coupling in one-dimensional conjugated chains like polyacetylenes and polydiacetylenes

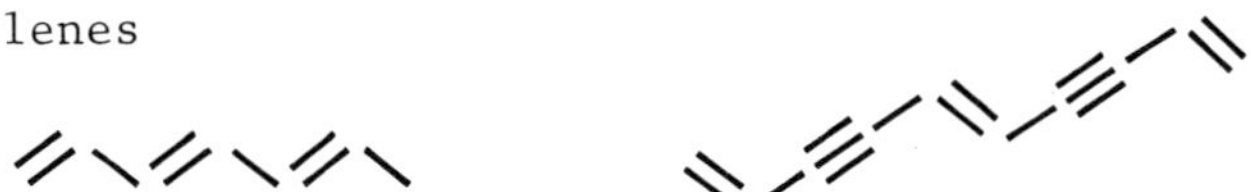

A

is the appearance of the semiconducting state[6] : the equal bond one-dimensional lattice (metallic state) is unstable with respect to a lattice distorsion and this so called static Peierls instability is the origin of the opening of the instrinsic band gap at the edge of the Brillouin zone with an infinite density of states there and the presence of bond alternation.

Clearly equally probable are the forms

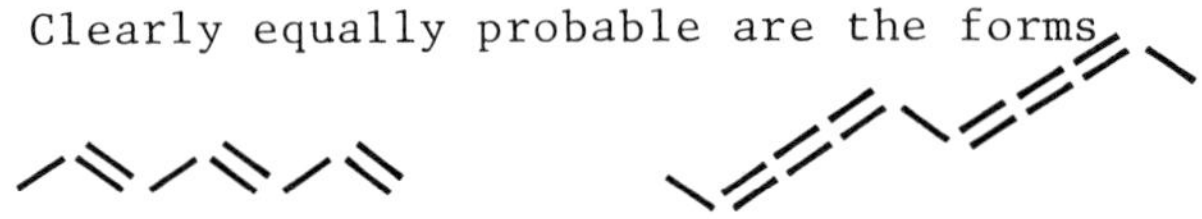

B

obtained from forms A by interchanging single (and triple in the case of polydiacetylenes) and double bonds. As a consequence intrinsic conjugation defects may appear along a chain separating the two equivalent structures. These "walls" or Pople-Wamsley[10] defects can be visualized as polarons or soliton states[11,12] where a non-bonding p-orbital with an unpaired electron is dressed with a amplitude vibrational motion and propagates with the velocity of sound.

In polyacetylenes, these defects are characterized by the appearance of new states[10-12] at mid gap with oscillator strength borrowed from the band states ; the problem bears much resemblance to the Koster-Slater problem[34] and can be most conveniently treated with the Green's function formalism[33]. These defects can be created by photoexcitation of frequency ω_1 near the absorption peak but also below it and will diffuse then with the sound velocity. This introduces a time scale of the order of 1-5 ps (the time it takes an acoustic distorsion to more over a repeat unit of the chain) right after the excitation during which the appearance of the absorption peak and the optical properties of the material below it will be markedly different from those observed with pulses longer than 10-100 ps. This behaviour can be monitored with a second weak pulse appropriately delayed with respect to the first one of frequency ω_2 sweeping the whole transparency and main absorption peak region ; this essentially measures the correlation function of the polarisation $<\delta P(t)\delta P(0)>$ which is related to the Optical Kerr Coefficient $n_2=2\pi\chi^{(3)}(\omega_1,-\omega_1,\omega_2)/n_o$ through the fluctuation dissipation theorem.

The experimental investigations[71-73] in particular the ones performed in polyacetylenes, using an excitation and probe technique to measure the changes in absorption, show a transient absorption behaviour which may be traced to a soliton type motion ; however some noticeable modifications must be introduced in the single soliton model[10-12] in order to account for the observed spectrum.

The conjecture was made in (70,74) that the defects are photoexcited in pairs or

which essentially recombine and annihilate sqeezing away the islands of the in-between form. This in particular implies that the states do not appear in midgap but higher up and their transient behaviour will depend on the recombination dynamics. This is certainly the case in the polydiacetylenes where the two forms A and B are not degenerate as in polyacetylenes.

Chain pairing defects

In conjugated polymer crystals, in addition to the localized defects or solitons discussed above, one may have a new class of delocalized defects arising through chain pairing of adjacent chains. Because of the periodicity and the juxtaposition of identical chains even small overlap of interactions between molecular sidegroups may have drastic changes on the main absorption peak ; splitting or broadening of the latter may result depending on the coupling strength between two neighboring chains.

A chain pairing model was postulated[75] to describe the low temperature splitting of the main absorption peak in PTS but such a chain pairing persists also at room temperature since there are two chains per unit cell and is certainly responsible for part of the broadening of the main absorption peak.

In the presence of an intense laser beam at fréquency ω_1 this chain pairing may be modified for a short time (few picoseconds) either by the photoinduced solitons or the direct absorption of photons by the sidegroups along the chains and this will lead to additional modifications of the main absorption peak. Furthermore changes in the chain pairing may interfere with the soliton motion in different ways and in particular cause separtation of the two partners of a soliton pair (one of the partners may be derailed on the neighboring chain). Some of these changes are inversion symmetry preserving some not and one may achieve topologically quite different situations.

The polydiacetylenes because of their large values of $\chi^{(3)}$ are strong candidates for nonlinear optical devices in different applications; however the time response of these nonlinearities may be limited by the intrinsic defects discussed above and caused by the competition of resonant chemical structures (forms A and B, chain pairs). A possible remedy is to control this competition without affecting the delocalization; this requires to control the polymerization process in these materials.

VII - PHONON LOCALISATION AND TOPOCHEMICAL REACTIVITY. CONCLUSION

With the last section we touch a fundamental aspect of large amplitude vibrational motion namely its relation to structural changes in condensed matter and more specifically to structural phase transitions and topochemical reactivity. Phonon localization caused by critical changes in anharmonic phonon interactions as the temperature changes is an essential perequisite to these phenomena. Depending on the type of the phonon localization and its dynamics one may subsequently have reversible (phase transitions, ex. ferroelectricity) or irreversible (topochemical reactivity, ex. polymerization) changes of the material system. Such a point of view is being adopted in few cases but no unified approach as yet has been put forward for this type of phenomena and even a description of their complexity lies outside the scope of the present work.

We only wish to point out here two novel aspects which arise in connection with the possibility offered by coheerent excitation with laser beams of a large amplitude vibrational motion connected with a structural change of the material: (a) one may induce these changes selectively in confined regions of the material essentially the dimensions of the laser beam cross section and study their stability and dynamics, (b) more importantly one may force these changes by being "off resonance" so that no real heating up of the material occurs, the temperature does not change as the system is forced to the new state.

The behaviour of interacting anharmonic oscillators with external forces present contains[76] a wealth of new phenomena which only now are being understood and their intensive study will allow a better understanding of the dynamics of structural changes in condensed matter.

ACKNOWLEDGEMENTS

The authors acknowledge generous financial assistance throughout this work from Direction des Recherches, Etudes and Techniques (D.R.E.T. Contract No 81-239) of the Ministry of Defense.

REFERENCES

1. E. Fermi and F. Rasetti, Z. Physik, 71, 689 (1931)
2. R.S. Krishnan, Proc. Indian Acad. Sc., 24, 25 (1946)
3. M. Born and K. Huang, Dynamical Theory of Crystal Lattices, Clarendon Press, Oxford (1954)
4. G. Herzberg, Infrared and Raman Spectra of Polyatomic Molecules, D. van Nostrand, New York (1945)
5. M. Lax and E. Burstein, Phys. Rev. 97, 39 (1955)
6. R.E. Peierls, Quantum Theory of Solids, Clarendon Press, Oxford (1955)
7. B. Szigeti, Proc. Roy. Soc. (London) A258, 377 (1960)
8. B.R. Henry and W. Siebrand, J. Chem. Phys., 49, 5369 (1968)
9. I. Abram, A. de Martino and R. Frey, J. Chem. Phys., 66, 5727 (1981)
10. J.A. Pople and S.H. Wamsley, Mol. Phys., 5, 15 (1962)
11. S.A. Brazovskii, JETP. Lett., 28, 606 (1978)
12. W.Su, J.R. Schrietter and A.J. Heeger, Phys. Rev. B22, 2099 (1980)
13. A. Ron and D.F. Hornig, J. Chem. Phys., 39, 1129 (1963)
14. J. Chesnoy, D. Ricard and C. Flytzanis, Chem. Phys., 42, 337 (1979)
15. see for references the article of G. Birnbaum in Vibrational Spectroscopy of Molecular Liquids and Solids, Plenum Press(1980)
16. M.P. Gush, W.F. Hare, E.J. Allin and J.C. Welsh, Phys. Rev. 106, 1101 (1957)
17. V.S. Gorelik, G.C. Mitin and M.M. Sushchinkii, Soviet. Phys. Sol. St., 16, 1019 (1974)
18. M.H. Cohen and J. Ruvalds, Phys. Rev. Lett., 23, 1378 (1969)
19. V.M. Agranovich and I.I. Lalov, Sov. Phys. JETP 34, 350 (1972)
20. J. Ruvalds and Zawadowski, Phys. Rev., B2, 1172 (1970)
21. F. Bogani, J. Phys., C11, 1283 (1978) ; ibid, C11, 1297 (1978)
22. J.C. Kimball, C.Y. Fong and Y.R. Shen, Phys. Rev., B23, 4946 (1981)
23. T. Holstein, R.O. Orbach and S. Alexander, Phys. Rev., B26, 4721 (1982)
24. J. Jortner and S. Rice, Phys. Rev., B26, 4727 (1982)

25. M. Mandel, P. Mazur, Physica 24, 116 (1958)
26. J.C. Decius, J. Chem. Phys., 49, 1387 (1968)
27. see articles by S. Bratos, W.A. Steele and D.W. Oxtoby in Vibrational Spectroscopy of Molecular Liquids and Solids, Plenum Press (1980)
28. H. Morawitz and K.B. Eisenthal, J. Chem. Phys., 55, 887 (1971)
29. A.A. Maradudin, E.W. Montroll and G.H. Weiss, Solid State Phys. Suppl. 2, ed. F. Seitz and A. Turnbull, Academic Press, New York, 1963
30. see for instance J.J. Eliott, in Lattice Dynamics and Intermolecular Forces, ed. S. Califano, Academic Press, New York (1975)
31. see for instance the articles of V. Schettino and D.A. Dows in Lattice Dynamics and Intermolecular Forces, ed. S. Califano Academic Press, New York, 1975
32. J. Hubbard, Proc. Roy. Soc. (London) A276, 238 (1964)
33. D.N. Zubarev, Sov. Phys. Uspekhi, 3, 320 (1961)
34. G.F. Koster and J.C. Slater, Phys. Rev., 95, 1167 (1953)
35. M. Born and M. Bradburn, Proc. Roy. Soc., A188, 161 (1947)
36. see for instance, J.R. Hardy and A.M. Karo, The Lattice Dynamics and Statics of Alkali Halide Crystals, John Wiley, New York (1980)
37. R. Turbino and J. Birman, Phys. Rev. Lett., 35, 670
38. C.H. Wu and J.L. Birman, J. Phys. Chem. Solids, 36, 305 (1975)
39. J.L. Birman, Phys. Rev., 127, 1093 (1962)
40. see for instance R. Loudon, Adv. Phys., 13, 423 (1964)
41. J. Klafter and J. Jortner, preprint
42. see for instance J. Jortner and S.A. Rice, J. Chem. Phys., 44, 1966
43. A.S. Davydov, Theory of Molecular Excitons, McGraw-Hill Book co, New York (1962)
44. J.F. Scott, Review of Mod. Phys., 46, 83 (1974)
45. R.G. Gordon, J. Chem. Phys., 43, 1307 (1965)
46. M.L. Geirnaert, G.M. Gale and C. Flytzanis, to be published
47. S.S. Mitra, J. Chem. Phys., 39, 3031 (1963)
48. C.V. Raman, Proc. Ond. Ac. Sci., A43, 327 (1956)
49. S. Aurilier, S.S. Mitra and H. Wu, J. Chem. Phys., 64, 2202 (1976)
50. F. Legay and Legay Sommaire, Chem. Phys., 65, 49 (1982)
51. F. Anastrassakis, S. Iwasa and F. Burstein, Phys. Rev. Lett., 17, 1051 (1966)
52. E. Burstein, in Dynamical Processes in Solid State Optics, ed. R. Kubo and H. Kamimura, Syobako (Tokyo) and W.A. Benjamin Inc. (New York) p. 34 (1966)
53. A. Laubereau and W. Kaiser, Rev. Mod. Phys., 50, 608 (1978)
54. M.L. Geirnaert, G.M. Gale and C. Flytzanis, to be published
55. A. Anderson, P.J. Grout, J.W. Leech and T.S. Sun, Chem. Phys. Lett., 21, 9 (1973)
56. A.V. Iogansen, Opt. Spect., 45, 36 (1979)
57. A.V. Iogansen, B.V. Rassadin, G.I. Romantsova and N.M. Grushina, Opt. Spectr., 44, 645 (1979)

58. see for instance J.M. Halbout and C.L. Tang, Journal de Physique (Paris), Suppl. C2-1983, 135
59. C. Flytzanis, Theory of Nonlinear Optical Suceptibilities in Quantum Electronics, a Treative vol IA, edited by H. Rabin and C.L. Tang, Academic Press, N.Y., p. 9 (1975)
60. M. Levenson, C. Flytzanis and N. Bloembergen, Phys. Rev., B6, 3692 (1972)
61. D. Heiman, R.W. Hellwarth, M.D. Levenson and G. Martin, Phys. Rev. Lett., 36, 189 (1976)
62. G. Breit and E. Wigner, Phys. Rev., 49, 519 (1936)
63. H. Gruhl, N. Periasamy, Chem. Phys. Lett., 95, 352 (1982)
64. G.L. Eesley and M. Levenson, Opt. Lett., 3, 178 (1978)
65. J.A. Giordmaine, Lectures on Parametric Optics in Quantum Optics, ed. R.J. Glauber Int. School of Physics Enrico Fermi, Academic Press, N.Y. (1967)
66. J. Armstrong, N. Bloembergen, J.Ducuing and P. Pershan, Phys. Rev., 127, 1918 (1962)
67. C. Sauteret, J.P. Hermann, R. Frey, F. Pradère, J. Ducuing, R.R. Chance and R.H. Baughman, Phys. Rev. Lett., 36, 956 (1976)
68. G. Agrawal and C. Flytzanis, Chem. Phys. Lett., 44, 366 (1976)
69. G. Agrawal, C. Cojan and C. Flytzanis, Phys. Rev. B17,776
70. C. Flytzanis, in Proceedings of the American Chemical Society Symposium on Nonlinear Optics Piezoelectric and Pyroelectric Polymers (Kansas City 1982) Ed. D. Williams
71. M. Lequime and J.P. Hermann, Chem. Phys., 24, 431 (1977)
72. C.V. Shank, R. Yen, R.L. Fork, J.Orenstein and G.L. Baker, Phys. Rev. Lett., 49, 1660 (1982)
73. Z. Vardeny, J. Strait, D. Moses, T.C. Chung and A.J. Heeger, Phys. Rev. Lett., 49, 1957
74. C. Flytzanis, to appear
75. G.P. Agrawal, C. Cojan and C. Flytzanis, Phys. Rev. Lett., 38, 711 (1977)
76. B.V. Chirikov, Phys. Reports 52, 263 (1979)

---

* Laboratoire propre du C.N.R.S.

FLUORESCENCE STUDIES OF VERY SHORT-LIVED EXCITED STATES OF AROMATIC MOLECULES

Bruce P. Boczar, Elisa A. Mangle, Steven A. Schwartz
and Michael R. Topp
Department of Chemistry, University of Pennsylvania
Philadelphia, Pennsylvania 19104 USA

INTRODUCTION

When a molecule absorbs light, a chain of events is set in motion the outcome of which may have considerable importance in nature, such as in photosynthesis or the visual process, and in modern technology in such diverse areas as photocatalysis and information processing and storage. A detailed knowledge of the fundamental processes in a molecule following excitation by light also leads to valuable information about molecular structural fluctuations as well as the interactions between a molecule and its immediate environment. One of the most basic issues of interest at present is to achieve a fundamental understanding of of internal energy transfer within a molecule, a sequence of processes spanning roughly seven decades of time in the submicrosecond region. At the focus of this is the region below 100 ps, where intramolecular vibrational energy redistribution and electronic energy transfer within a given spin manifold take place.

For the most part, processes on a subnanosecond time-scale require laser techniques. Although picosecond laser technology is required for much of this time-region, pure spectroscopy (i.e., in the frequency-domain) has been used to make measurments on phenomena occurring on a femtosecond time-scale.

Whether or not direct fluorescence spectroscopy is feasible depends on the possible presence of stronger signals in the same spectral region. For highly excited singlet states of aromatic hydrocarbons, there is a sufficient spacing in the energies of the various potentially fluorescent states that direct spectros-

K. B. Eisenthal (ed.), Applications of Picosecond Spectroscopy to Chemistry, 237–258.

copy is readily possible. On the other hand, subtle changes in the spectrum from a single electronic state, such as involving vibrational energy relaxation, require some gating device to isolate the short-duration components. Time-resolution also has the advantage, for processes of slightly longer duration, of removing interference coming from events synchronous with the irradiation pulse, primarily scattered laser light.

As our theme, we shall examine the phenomenon of vibrational energy redistribution within a molecule. This will cover several different experimental approaches. As a starting point, it is useful to examine the very shortest extremes of time to determine where the phenomenon begins to be observed.

UPPER STATE FLUORESCENCE OF AROMATIC MOLECULES

In recent years,[1] we have developed a method for observing the fluorescence spectra of highly excited electronic states of aromatic molecules, which uses two-photon excitation (Fig. 1). The excitation process uses two consecutive allowed transitions

Fig. 1

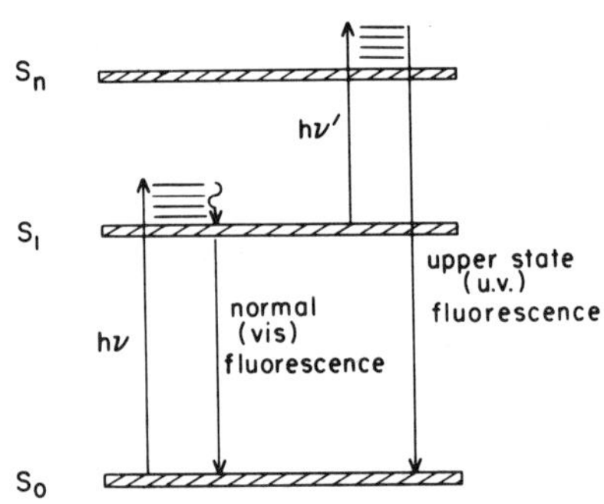

and, depending on the experimental configuration and the laser sources used one may control the amount of molecular relaxation between the two excitation events. If a nanosecond source is used and the sample is in fluid solution at room temperature, one may assume complete thermalization of the molecule within the first excited state.

The two-photon approach allows the elimination of any scattered light from the spectral region of the molecular fluorescence, which is an important condition in view of the very low quantum-yield of the emission (10^{-4}-10^{-6}).

For experiments on the molecule 3,4,9,10-dibenzpyrene in the first instance, two nitrogen-pumped dye lasers were used.[1] One, with a fixed frequency near 380 nm, served to excite the molecule into the S_2 state, following which rapid electronic and vibrational relaxation assured an S_1 population, essentially thermalized at the ambient temperature. The second laser was tunable over a wide range (400-920 nm) with typical results as shown in Fig. 2. The

Fig. 2

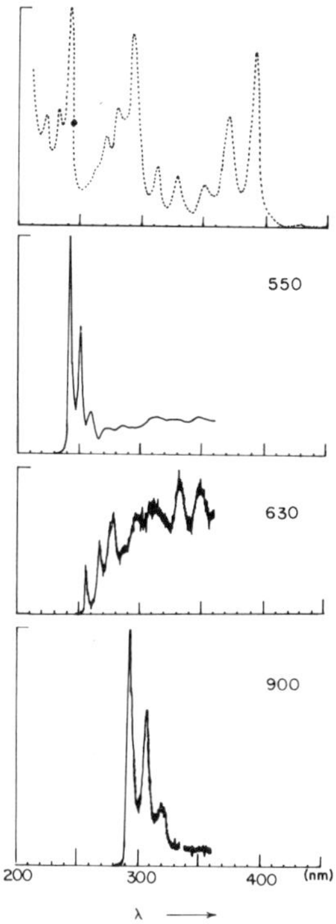

normal absorption spectrum is shown in the upper frame and three fluorescence spectra appear below it. In all cases, the emission spectrum started at the high-frequency edge with a sharp line, the energy of which exactly coincided with that of a thermalized S_1 molecule plus the energy of the second photon (indicated on the figure).

If we assume an initial population according to the scheme:

$$S_0(t) \xrightarrow{h\nu} S_m^{\#} \rightsquigarrow S_1(t) \xrightarrow{h\nu'} S_n^{\#}$$

then the emission feature at the highest frequency corresponds to $S_n^{\#} \rightarrow S_0(F)$. Here: (t) represents a thermalized distribution within a given electronic state, (#) represents some vibronic state directly reached in the irradiation process, (F) is a distribution of states within S_0, selected by the Franck-Condon profile from the emitting state. Thus, the three highest frequency lines in all three fluorescence spectra shown correspond predominantly to $S_n^{\#} \rightarrow S_0(F)$ transitions. The region to longer wavelengths in each spectrum exhibits transitions of the type: $S^*_{m<n} \rightarrow S_0(F)$. In all cases, we label n as the state originally reached by the excitation event, while all the m* are states reached by radiationless internal conversion. For excitation at wavelengths 550 nm and 900 nm (n = 6 and 4 respectively), the transitions $S^{\#}_n \rightarrow S_0(F)$ are very strong and dominate the spectrum. However, as the $S_5 \rightarrow S_0$ transition is weak, transitions of the type $S^*_{m<n} \rightarrow S_0$ are stronger. In fact, contributions can clearly be seen from S_4 (diffuse) and S_3 (structured) when the molecule is irradiated at 630 nm.

The appearance of "sharp" lines in the spectrum is a consequence of the localization of the excess vibrational energy of the molecule. Furthermore, the implication is that the vibrational energy does not randomize on the time-scale of the lifetime of the upper states, since the fluorescence spectra are dependent on the excitation wavelength. This is readily seen from Fig. 3, where a series of spectra is shown for different excitation wavelengths across the profile of the S_6 spectrum. The energies range over about 3400 cm^{-1} and we observe a steady evolution of the spectrum.

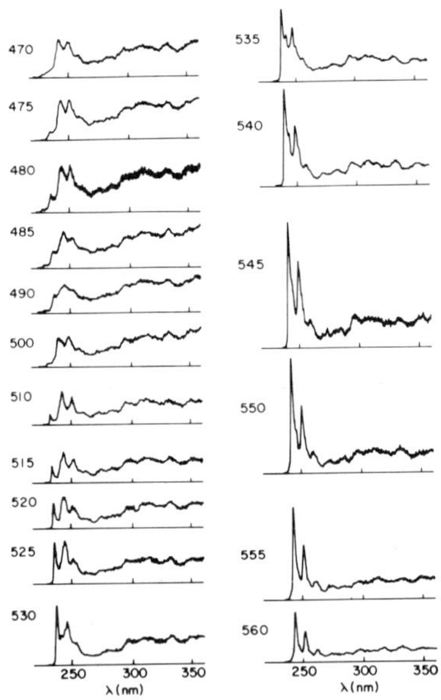

Fig. 3

Although it is not readily evident from the spectrum in Fig. 2 excitation of the molecule in the vicinity of 550 nm generates a progression of lines in fluorescence which are considerably narrower than either the S_0 or S_1 absorption transitions to the S_6 state. This is interpreted in terms of an analogy to fluorescence line narrowing due to molecular site selection, such as is normally observed in cryogenic media. However, the scale of everything is quite different. In fluid solution at 300°K, the minimum linewidth of transitions involving the S_6 state of this particular molecule is about 150 cm^{-1}. This is equivalent to a lifetime near 35 fsec, which has been confirmed independently with an uncertainty of 30% by quantum-yield measurements.

Some questions about vibrational relaxation can also be answered using experiments of this type. These address problems of fundamental interest to the field of picosecond spectroscopy.

Use of Picosecond Pulses

The first example is laid out by the energy level diagram shown in Fig. 4. Using nanosecond excitation sources, it can

Fig. 4

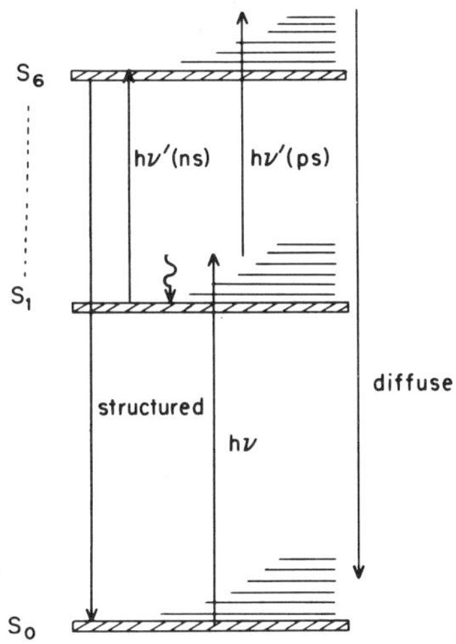

be guaranteed that the average S_1 molecule sampled by the second photon is thermalized. However, if all of the S_1 molecules are sampled on a picosecond timescale, then the effects of incomplete vibrational relaxation would be evident from the upper state fluorescence spectrum. An early experimental result on 3,4,9,10-dibenzpyrene using picosecond pulses is shown in Fig. 5.[2] The lower frame, which involved a delay of a few nsec between absorption of two pulses (355 nm and 532 nm) showed a very similar result to that obtained with a nitrogen laser. However, superimposition of the of the pulses, which guaranteed an average delay of less than 25 psec, showed a quite different result. Not only was the structure in the short-wavelength band mostly absent, but the intensity ratio between different electronic bands was quite different, showing a substantial dependence of the internal conversion process on the internal energy content of the radiatively prepared state.

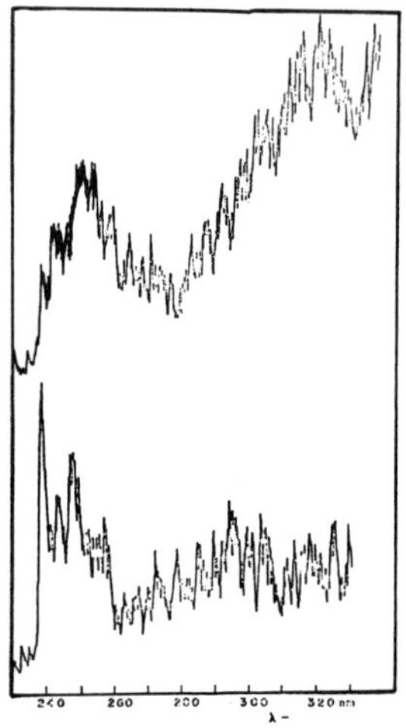

Fig. 5

Centrosymmetric Molecules

A second example is as follows. In the molecule 3,4,9,10 dibenzpyrene, which forms the subject of the above experiments, the symmetry is C_{2v}, which guarantees for all transitions a non-vanishing electronic transition dipole moment. The consequence of this is that, independently of the excitation wavelength, there is always a component in the upper state fluorescence spectrum resulting from the excitation level. This can be seen from Fig. 2 for dibenzpyrene, even for the weak transition excited by 630 nm radiation. However in a centrosymmetric molecule, the level reached in a two-photon excitation process, such as demonstrated in Fig. 6, has formally zero electronic transition moment to the ground state. Although in the real case there is vibronic induction, there still remains a quite large separation according to the electronic parity labelling. The difference between the one- and two-photon absorption spectra of 1,2,5,6-dibenzanthracene shown in Fig. 7, shows this to be the case.[3]

A principal aim, then, of upper state fluorescence experiments involving centrosymmetric molecules is to effectively eliminate fluorescence to the ground state from a state excited according to:

$$S_{0g}(t) \xrightarrow{h\nu} S^{\#}_{1u} \rightsquigarrow S_{1u}(t) \xrightarrow{h\nu'} S^{\#}_{ng}.$$

Thus, we can impose the condition that upper state fluorescence may <u>only</u> be observed from states which have been populated by a radiationless relaxation process. Furthermore, since these upper states fluoresce without undergoing significant vibrational energy

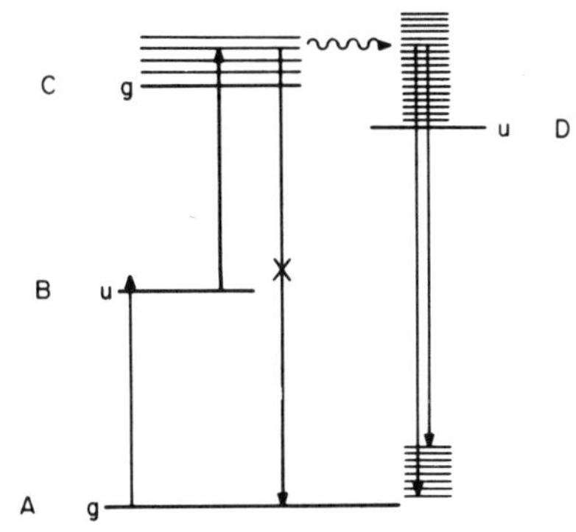

Where C → A is forbidden, we can monitor the details of D → A

Fig. 6

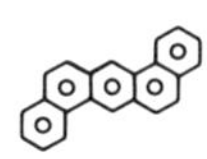

Fig. 7

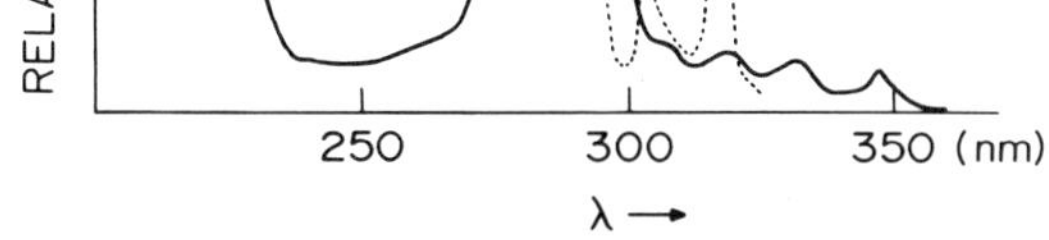

relaxation, the vibrational energy distribution resulting from the initial radiationless process can be deduced from the fluorescence spectrum. Thus, Fig. 8 shows data obtained by irradiation of a hexane solution of 1,2,5,6-dibenzanthracene.[3] Our assumption of separation of states according to electronic parity labelling is supported by the absence of any significant emission at the wavelength corresponding to the excitation energy. Thus, while the points indicated by arrows in Fig. 8 would always have a sharp fluorescence line in a non-centrosymmetric molecule, such a feature has not been observed in the spectrum of a centrosymmetric molecule.

Fig. 8

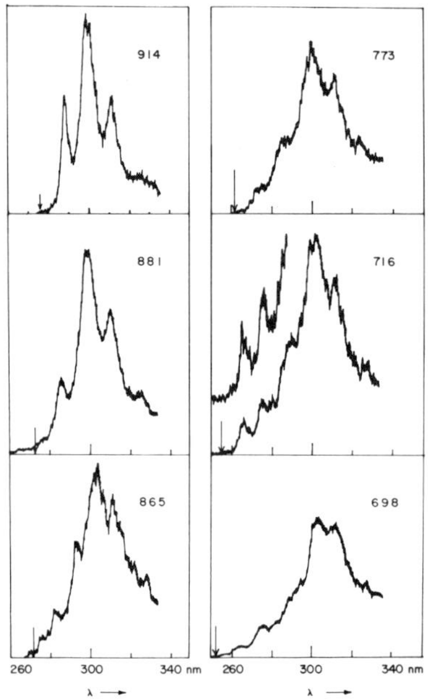

Nevertheless, there are other well-resolved features in the spectrum which are quite sensitive to the excitation wavelength. Furthermore, these features appear to be the most clearly defined when the excitation energy is near 2900 cm^{-1} and again near 5700 cm^{-1} above the origin of the fluorescent "u" state. This probably implies some specific coupling in the internal conversion process involving "C-H" modes of the lower electronic state.

Conclusions from Two-Photon Upper State Fluorescence Studies

We may summarize a few points relevant to the forthcoming discussion, as follows:

1. Intramolecular vibrational energy redistribution is slower than 100 fs.

2. Relaxation of the solvent around an aromatic hydrocarbon in hexane at 300°K is slower than 50 fs, but probably faster than 200 fs.

3. Some evidence has been obtained for specific state-to-state coupling in radiationless internal conversion.

(a) Competition between coupling to different electronic states has been seen to depend on the symmetry and vibrational energy content of the prepared levels.

(b) A memory for the initial excitation energy may be carried through the radiationless conversion process, indicating that an "intermediate" radiationless coupling case may be found for large molecules at quite high excess energies.

The next part of this review will describe approaches to probe aspects of electronic and vibrational relaxation in real time using 10^{-11} s resolution. For the most part, this involves a technique, where the signal of interest is isolated by a fast optical gate. The procedure is strictly analogous to an electronic boxcar technique, except that the necessary time-resolution is achieved using the laser pulses themselves for the essential gating operations.

PICOSECOND TIME-RESOLVED FLUORESCENCE SPECTROSCOPY

Quite recently, a new development in laser technology was reported by us, which appears to solve a number of the principal problems in picosecond fluorescence spectroscopy. Although it had been evident that the spectral bandwidth of the 1064.2 nm component of the $^4F_{3/2} \longrightarrow {}^4I_{11/2}$ transition of Nd^{3+} in YAG was wide enough to support sub-10 ps pulses, this had not been achieved experimentally in a simple manner compatible with the requirements of fluorescence excitation spectroscopy. A preliminary solution was achieved by the use of two separate resonant cavities. One laser generated what amounted to a noise burst anywhere from 30 ps to more than 100 ps in duration. The second cavity was injected with this pulse, which served to bleach an intracavity saturable absorber. The important condition for this second cavity, known as a regenerative amplifier, is that internally generated quantum noise is insufficient to promote cavity oscillation at the level of saturable absorber concentration used. Our design for the regenerative amplifier,[4] patterned loosely after a preliminary report by Murray and Lowdermilk,[5] was used successfully as a spectroscopic source.

However, a very important recent development[6] was to incorporate the essential functions of the two-cavity approach into a single, branched cavity. This contributed extra stability and facilitated the operation of the instrument as a whole by relaxing alignment criteria and by eliminating an electro-optic switch and one laser amplifier rod. Furthermore, the output pulse duration was, at 8 ps, essentially equivalent to that produced by an optimized two-cavity system.

The instrument operates as follows (see Fig. 9).[6] A cavity is constructed which contains two branches. The rear mirror and

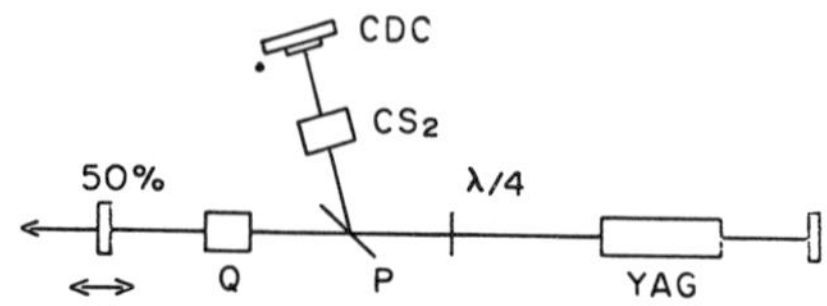

Fig. 9

laser element are common to both subunits. One subunit (the main cavity) is straight and contains a Q-switch and the (50%) output coupler. This cavity may only oscillate on horizontal polarization. The other subunit (the oscillator) is terminated by a side arm containing a 100% reflector and a contacted dye cell. Additional loss for the oscillator is generated by polarization mismatch to the Brewster-angled laser rod, and by a cell of CS_2. The side arm is coupled to the main cavity by a thin film polarizer. Thus the two cavities are formally orthogonal.

The oscillator Q-switch is adjusted for conventional passively mode-locked operation, while the main cavity is blocked by a Q-switch optical density roughly four times higher. The two cavities are weakly coupled together by the use of a λ/4 plate in the common arm set to within 10° of the "null" position. Operationally, this plate is adjusted over a range of about 2° to achieve optimum stability and pulse duration.

The remaining condition is the optical synchronization of the two sub-units, which requires less than about 300 microns difference between the lengths.

Figure 10 shows a schematic representation of the rest of the apparatus for generating tunable picosecond pulses. The output

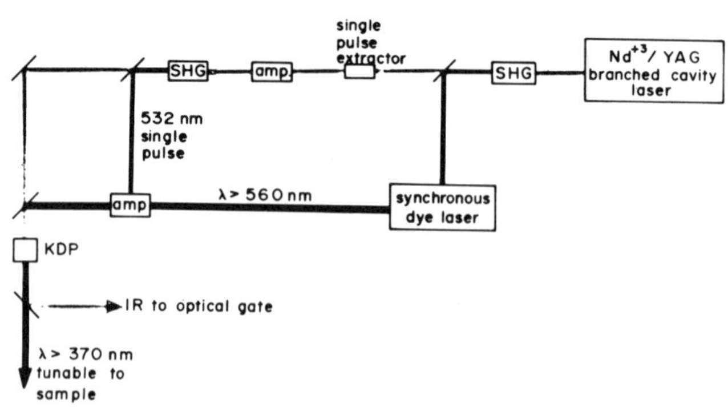

Fig. 10

from the branched cavity laser is frequency-doubled and the second harmonic pulse-train used to pump a dye laser. The dye laser, which is optically synchronous with the branched cavity laser, is prism-tuned to give a bandwidth of 2 cm^{-1}. A single pulse from the residual fundamental train is amplified and second harmonic radiation from this is used to pump a dye amplifier. The amplified dye laser radiation is then combined with the residual amplified fundamental pulse to yield a sum-frequency ultraviolet pulse 8 ps wide, completely tunable and with up to 100 μJ energy content.

The signal-detection end of the apparatus is more standard (see Fig. 11). Briefly, the infrared fundamental pulse, after

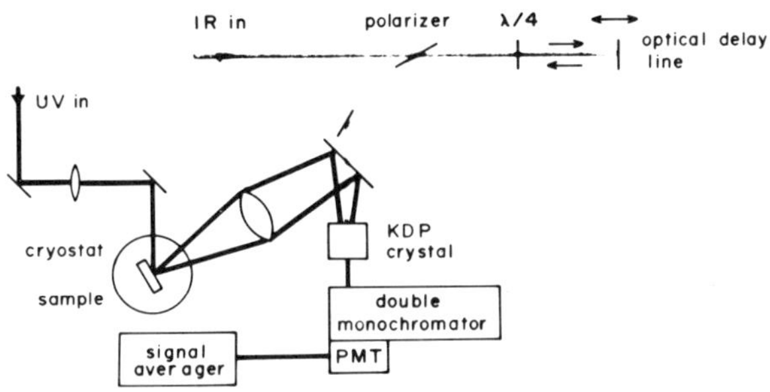

Fig. 11

travelling through a scannable delay line is combined with fluorescence excited by the tunable pulse in a KDP crystal.[7] Here the fluorescence is sampled by generation of the sum frequency, so that the duration and spectral bandwidth of the detected portion of the fluorescence signal are limited only by the gating pulse.

FLUORESCENCE FROM THE SECOND EXCITED SINGLET STATES OF AROMATIC MOLECULES

The second excited singlet state of many aromatic molecules lies close enough in energy to the first excited state that often thermal activation gives rise to significant S_2 to S_0 fluorescence signals. This has been known for many years. However, it is the observation of S_2 signals as subpicosecond transient phenomena that is of most interest to us as important information is revealed about competing processes on this time-scale.

Normally it is quite hard to see S_2 signals distinctly from S_1 fluorescence, and a good example of the contribution of the picosecond time-resolved fluorescence spectrometer to solving this problem is given for diphenyloctatetraene in Fig. 12. While a "conventional" fluorescence spectrometer integrates all signals for at least the few nanoseconds of the normal fluorescence decay time, our laser spectrometer can be set to detect

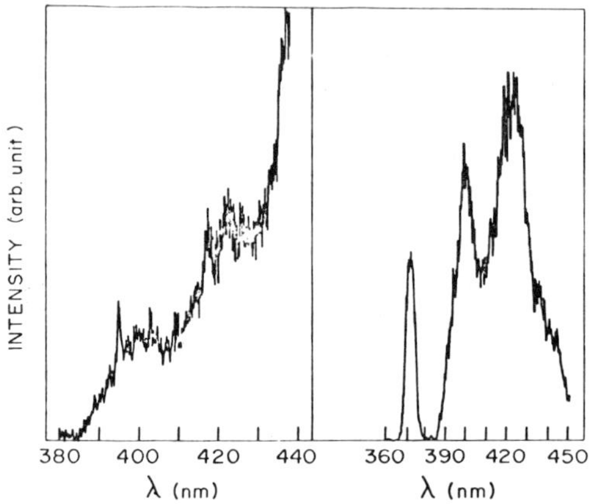

Fig. 12

only events within 5 ps of the irradiation pulse.[8] Thus, the vast majority of the S_1 to S_0 fluorescence is rejected and the S_2 to S_0 signal increases in relative prominence by perhaps two orders of magnitude. The signal may be unambiguously assigned as due to the S_2 state from its excitation spectrum and from the fact that it can only be detected synchronously with the laser pulse. Using relative intensity measurements we have found S_2 lifetimes for several molecules to be in the region of a few tenths of a picosecond.

Figure 13 shows a sequence of fluorescence spectra for 3,4,9,10-dibenzpyrene excited at different wavelengths corresponding to different regions of the S_0 to S_2 absorption. First, it is clear that no substantial amount of vibrational energy migration is occurring on this time-scale, and yet the spectra do not show the narrow line features which characterized the fluorescence spectra from higher energy states. This implies that the lifetime of the S_2 is sufficiently long for the molecular site memory to be erased. The effects of sequence congestion are also quite evident from the spectra. The feature at 432 nm is due to the S_1 to S_0 fluorescence electronic origin. Under these circumstances, the relative intensity of the S_2 and S_1 fluorescence signals depends directly on the instrumental time-resolution. This is a major reason for having optimally short pulses; although the S_2 signal cannot be directly time-resolved, its spectrum is very clearly observed using 10 ps time-resolution.

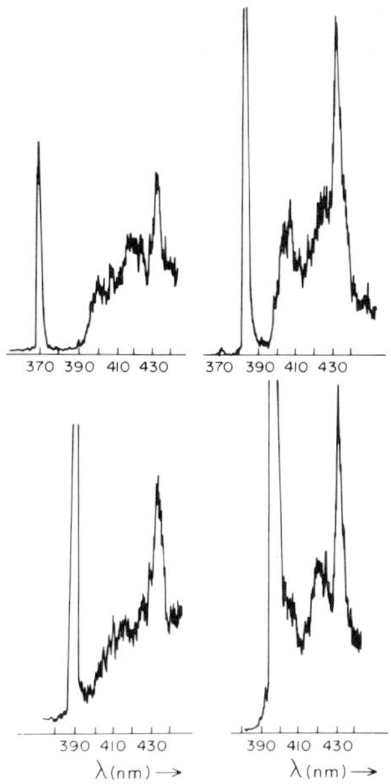

Fig. 13

STUDIES AT HIGHER SPECTRAL RESOLUTION - INTERNAL RELAXATION OF EXCESS VIBRATIONAL ENERGY IN PERYLENE

One of the most fascinating current problems in chemical physics is the study of the internal transfer of energy following the preparation of some vibronic state of a molecule. In aromatic hydrocarbons, this process has been suspected of occurring on a time-scale anywhere from 10^{-13} to 10^{-10} s in fluid solution, while it is known to be considerably slower in isolated small molecules. For the most part, we are interested in the study of vibrational energy relaxation in the electronically excited molecules. Although many parallels may be drawn with the equivalent ground state process, it is more natural experimentally to use direct fluorescence detection. The preceding discussion has shown that vibrational energy relaxation does not occur significantly on the timescale of the lifetime of higher electronic states than S_1.

The new apparatus for picosecond time-resolved fluorescence spectroscopy has proved to be very useful for this type of study, principally because the tunability of the excitation wavelength allows a much more detailed analysis.

Although it is harder to obtain useful quantitative information, the presence of vibrationally unrelaxed molecules can readily be detected in fluid solutions at room temperature. Figure 14 shows, for example two scans of the same region of the fluorescence spectrum of perylene in isopropanol-ether(5:2) at 300°K.

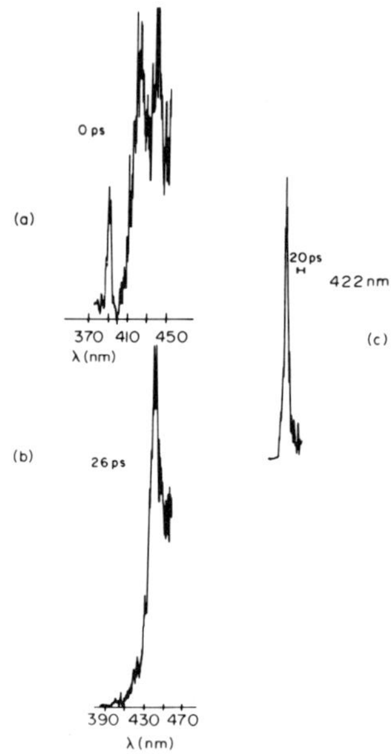

Fig. 14

The scan at 0 ps, (i.e., coincident with the maximum of the irradiating pulse) shows the exciting line, as well as two broad fluorescence bands. While a band near 445 nm (the position of the electronic origin) persists at 26 psec, the 420 nm peak relaxes only slightly more slowly than the excitation pulse profile. This broad band near 420 nm may quite confidently be assigned to vibrationally unrelaxed molecules, but further assignment is difficult at this level of spectral resolution, since inhomogeneous broadening scrambles the radiatively prepared states.

TIME-RESOLVED FLUORESCENCE SPECTROSCOPY OF PERYLENE IN A SHPOL'SKII MATRIX AT 4°K

In n-heptane solution at 4.2°K, the spectrum of perylene becomes very well resolved with linewidths of a few wavenumbers being easy to obtain. Furthermore, there is only weak guest-host coupling so that many weak lines are resolvable in the spectrum. Since the spectrum obtained under conventional conditions is so very well resolved, such deviations as may result from the presence of vibrationally unrelaxed molecules are straightforward to detect. Finally, under these cryogenic conditions, fluctuations in guest-host interactions are close to a minimum, and therefore one might expect the perturbations by the host on internal relaxation of the guest to be at a minimum. In fact, an important aspect of these studies is to maintain a comparison between the Shpol'skii matrix and the behaviour of the (isolated, cold) molecule in a supersonic jet.

A version of the emission spectrum of perylene recorded on our instrument with 30 cm^{-1} resolution is shown in Fig. 15[9] The spectrum, which was not gated, shows all of the usual features with one exception. The concentration used was 5 x 10^{-6} M and, under these conditions, the electronic origin was observed to be inverted. That is, strong reabsorption of light exactly on the resonance leads to a local minimum in emission while the wings are artificially strengthened. This concentration was about optimal for our study since, although this band was deformed, the sample absorption was maintained sufficiently high for useful signals to be observed on weaker transitions. Reabsorption artifacts to the high-frequency side of the electronic origin would also be expected to distort some features of the fluorescence spectra from vibrationally unrelaxed molecules. However, the small Franck-Condon factors of virtually all of the transitions reduce the scale of the problem.

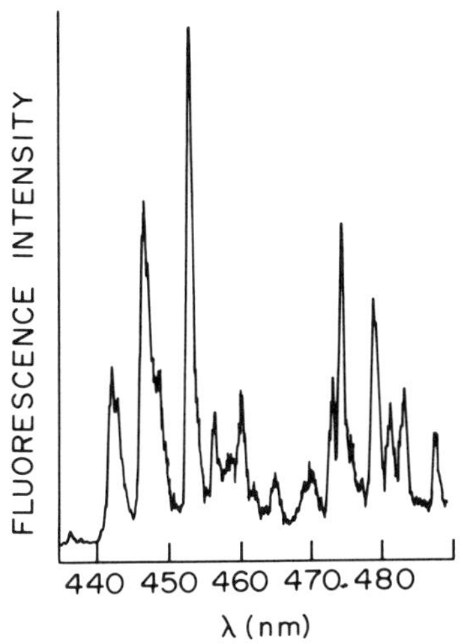

Fig. 15

One factor to take into account upon the adoption of a high-resolution fluorescence technique is the possible contribution of stimulated emission. A pulsed laser uses high fluxes and the linearity of the system response has to be assured. The reason for this consideration is that, although the linewidth of the

S_1 to S_0 transition can in principle approach the subwavenumber region, this is not true of the competing upward transitions. (The limitation on linewidth is imposed by the short lifetime of the upper states, the minimum being on the order of 150 cm^{-1}.) This implies a very large difference in cross-section on resonance between downward and upward transitions from S_1 (assuming the same oscillator strength), and virtually guarantees that stimulated emission will occur. This behaviour is to be contrasted with the room temperature situation. Here, all transitions are grossly broadened and usually the presence of strong sequence-congested transitions in absorption from S_1 is sufficient to preclude stimulated emission.

Figure 16 shows the result of irradiating perylene in a Shpol'skii matrix at full power with our tunable laser. A greatly enhanced line at 453 nm is due to the occurrence of stimulated emission on the 358^0_1 fluorescence transition. This emission intensity appears to rise over about 100 psec, and persists well into the nanosecond region. As may also be seen from Fig. 16, attenuation of the irradiation pulse (5x) eliminated this effect.

Fig. 16

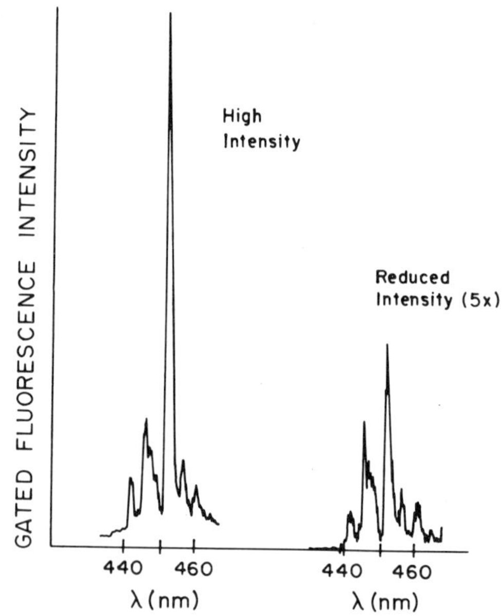

Figure 17 shows the result of an analogous experiment to that depicted in Fig. 14. The excitation wavelength was tuned to a vibronic resonance equivalent to the first overtone of the 1290 cm^{-1} mode (i.e., 1290^2_0). Thus, the excitation wavelength was 399.5 nm, with a linewidth of about 4 cm^{-1} FWHM. What is seen is a progression having an origin approximately 1290 cm^{-1}

above the electronic origin, having periodicity approximately 360 cm^{-1}. This is assigned as $1290^2_1 358^0_n$ where n = 0, 1, 2. A scan of the time-profile of the principal signal at 422 nm showed a rapid rise followed by a relaxation over about 60 psec. Thus, on the basis of these preliminary experiments, a tentative assignment of the lifetime of the 1290^2 state is given as 60 psec. The lifetimes we are measuring for this and similar transitions are much longer than recently reported in the literature.[11] We attribute this to our better control of reabsorption artifacts and of the excitation conditions.

Fig. 17

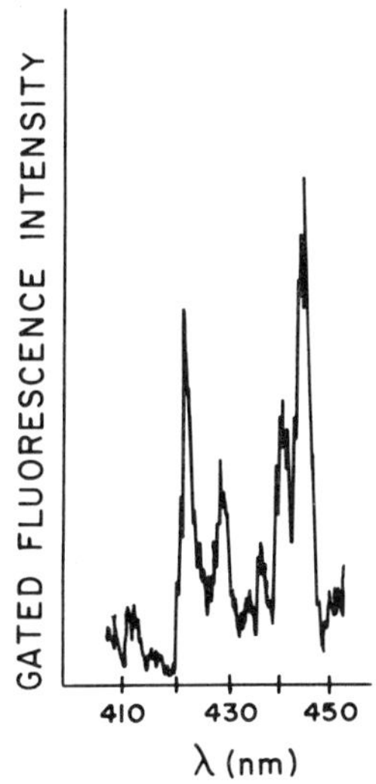

Another feature of interest here is the ability to use the excitation spectrum to identify various modes independently of an absolute calibration. Thus, as Fig. 18 shows, in the excitation range 396 nm to 400 nm there are three principal maxima in absorption, which correspond exactly with those originally obtained by Shpol'skii and Personov.[9] All of these resonances appear in the excitation spectrum for gated fluorescence near 445 nm, since all possible modes of excitation must have a strong component of emission here (i.e., $\Delta v = 0$). However, the excitation spectrum for emission at 422 nm shows only two resonances in this region ($1290^1_0 1400^1_0$ and 1290^2_0). We have assigned the 422 nm transition therefore, for the two cases as $1290^1_1 1400^1_1$ and 1290^2_1 respectively. The third resonance (actually there are two close together, $1400^1_0 548^2_0$ (minor) and $1602^1_0 358^1_0 548^1_0$ (major)) does not contain the 1290 cm^{-1} mode.

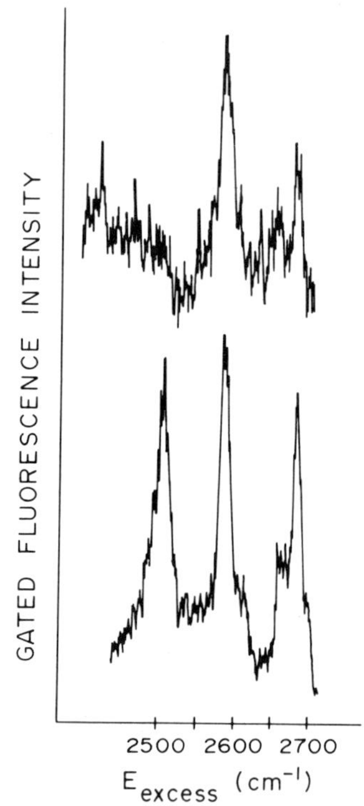

Fig. 18

SUPERSONIC JET EXPERIMENTS

One of the most important questions surrounding the study of molecular vibrational energy redistribution is the possible role of the solvent medium. One logical approach, therefore, to this problem is to carry out complementary experiments in which the molecules are studied in an isolated condition. Furthermore, the internal temperature of the molecules should be as close as possible to the conditions found in the solid matrix. Supersonic jet spectroscopy is well suited to this application, since rotational temperatures of only a few degrees Kelvin are readily attained. Although such low vibrational temperatures are not usually attained in a supersonic jet, sufficiently low temperatures are generally reached that the molecules may be considered to be effectively in a well-defined state when excited.[12] An excitation spectrum for the fluorescence of perylene under supersonic jet conditions is shown in Fig. 19,[13] uncorrected for the spectral response of instrumental sensitivity and for the dye laser output spectrum. The exciting source was a conventional nitrogen-pumped dye laser.

This spectrum is predictable to within a few wavenumbers from literature data on crystals. The major point is that the vibronic excitations are very well resolvable and easily comparable to the low temperature crystal case. As forthcoming work will show, a detailed understanding of band assignments is vital for the formulation of relaxation mechanisms, especially in the range 800-1150 cm^{-1} excess vibrational energy.

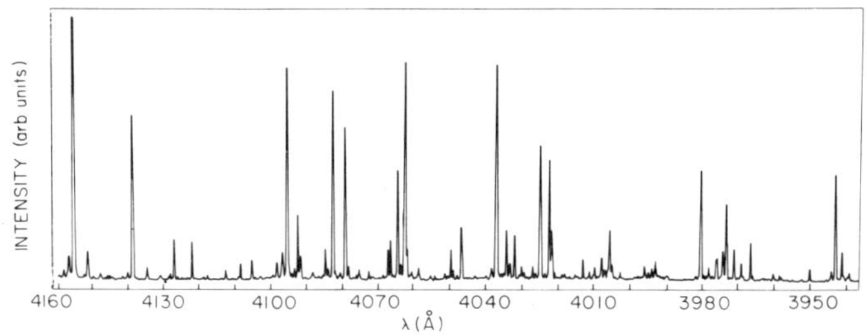

Fig. 19

A few points are to be brought out from this study. First, a fluorescence spectrum of perylene in a jet, as for several other systems noted in the literature, depends on the excitation wavelength.[14,15] Figure 20 shows some samples. Excitation into the electronic origin of the molecule gives rise to the prototype fluorescence spectrum where all lines exhibit predictable Franck-Condon factors. Apart from a slight complication associated with a low frequency bending mode, the absorption and fluorescence spectra are virtually mirror images, as far as resolution permits.

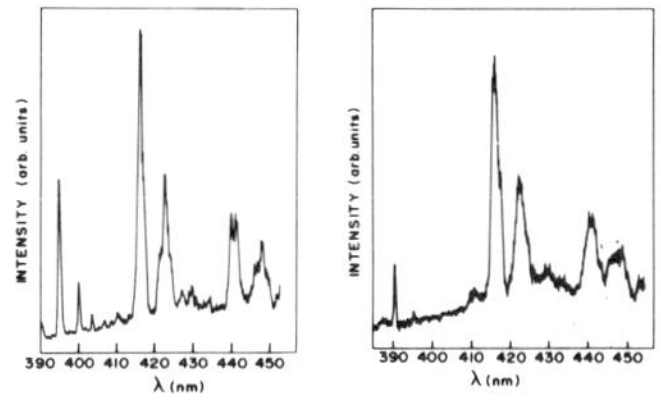

Fig. 20

Excitation according to 353_0^2, exciting the first overtone of a prominent optically active mode gives rise to a spectrum exhibiting features characteristic of an unrelaxed level higher in energy than the origin, showing features to both the blue and red of the electronic origin, all lines being of about the same width and with a markedly different Franck-Condon profile from the 0_0^0 excitation. Excitation according to 1295_0^1 gives two spectra, which were not separated in this nanosecond study. The weakened blue-shifted spectrum was still identified with a now shorter-lived initial vibronic level, while broad features having the Franck-Condon profile roughly of the 0_0^0 excitation indicated also the presence of relaxed molecules.

One immediate conclusion one may gather from this analysis even without picosecond time-resolution is that, once the excitation energy exceeds about 1200 cm^{-1}, the lifetime of the initially excited vibrational levels within S_1 steadily and rapidly drops into the picosecond region. As the fourth frame in Fig. 20 shows, for 1600 cm^{-1} excess energy, the unrelaxed fluorescence is extremely weak, and virtually all of the observed fluorescence comes from states which have been generated by radiationless redistribution of energy from the initially excited level. Even in an isolated molecule, this process is very fast, our best estimates at present indicating that the relaxation rate is about the same at these excess energies in either the isolated molecule or matrix-isolated case.

Another feature of the data obtained from the supersonic jet data relates to the outcome of the radiationless relaxation process and its dependence on the exact nature of the radiatively prepared state. In the region where the redistribution process is relatively slow and where the number of receptor states is still tractable ("intermediate coupling" case),[16] it is probable that specific couplings can play a major role in determining the rate and selectivity of the radiationless redistribution process. That is, symmetry, energy gap and Franck-Condon factor requirements rather than simply the total density of proximate states can determine the mechanism of the process. Important diagnostic factors here are the dependence of the fluorescence spectrum, the intensity of the unrelaxed portion and the linewidth of the broader bands of the relaxed portion of the spectrum on the excitation energy and more specifically on the exact nature of the initial state.

Recent results on isolated perylene molecules are beginning to reveal some general patterns of behaviour:

1. High-frequency fundamental vibrations (>1300 cm^{-1}) and their combinations relax on a subnanosecond time-scale, as seen above.

2. However, combinations containing only low-frequency-modes (<800 cm^{-1}) relax only slowly over a wide range of energies. In some cases this has been followed above 2000 cm^{-1} excess energy.

One important piece of information, which has come from this work, is that the rate of radiationless relaxation depends critically on the radiatively prepared state. Contrary to a recent assertion in the literature,[16] this is not erratic behaviour. Instead, it forms a part of a very elegant pattern which we are beginning to understand through the use of ultra high-speed spectroscopy.

It will be important in the near future to follow up these high-resolution studies with experiments incorporating picosecond fluorescence technology.

ACKNOWLEDGEMENT

This work was supported by the National Science Foundation directly (CHE-82-07889) and through the Materials Science Program (DMR-79-23647).

We are grateful to Kee-Ju Choi, Leslie Hallidy (Diverdi) and Horn-Bond Lin whose work forms the basis of the first part of this paper.

REFERENCES

1. Topp, M. R., Lin, H. B. and Choi, K. J. 1981, Chem. Phys. 60, pp. 47-59.

2. Choi, K. J., Boczar, B. P. and Topp, M. R. 1981, Chem. Phys. 57, pp. 415-423.

3. Lin, H. B. and Topp, M. R. 1979, Chem. Phys. Lett. 67, pp. 273-278.

4. Choi, K. J. and Topp, M. R. 1981, J. Opt. Soc. Am. 71, pp. 520-524.

5. Murray, J. E. and Lowdermilk, W. H. 1978, in Picosecond Phenomena (Shank, C. V., Ippen, E. P. and Shapiro, S. L. eds.), Springer Series in Chemical Physics, Vol 4, pp. 281-284.

6. Boczar, B. P. and Topp, M. R. 1983, Appl. Opt. 22, pp. 1611-1613.

7. Hallidy, L. A. and Topp, M. R. 1977, Chem. Phys. Lett. 46, pp. 8-14.

8. Felder, T. C., Choi, K. J. and Topp, M. R. 1982, Chem. Phys. 64, pp. 175-182.

9. Shpol'skii, E. V. and Personov, R. I. 1960, Opt. & Spectr. 8, pp. 172-176.

10. Hochstrasser, R. M. and Nyi, C. A. 1980, J. Chem. Phys. 72, pp. 2591-2600.

11. Aaviksoo, J., Anijalg, A., Freiberg, A., Lepik, M., Saari, P., Tamm, T. and Timpmann, K. 1982, in Picosecond Phenomena III (Eisenthal, K. B., Hochstrasser, R. M., Kaiser, W. and Laubereau, A. eds.), Springer Series in Chemical Physics, Vol 23, pp. 192-195

12. Levy, D. H., Wharton, L and Smalley, R. E. 1977, in Chemical & Biochemical Applications of Lasers (Moore, C. B. ed., Academic Press) Vol 2 pp. 1-41.

13. Schwartz, S. A. and Topp, M. R. (unpublished work).

14. Lambert, W. R., Felker, P. M. & Zewail, A. H. 1981, J. Chem. Phys. 75, pp. 5958-5960.

15. Amirav, A., Even, U and Jortner, J. 1981, J. Chem. Phys. 75, pp. 3770-3793.

16. Bouzou, C., Jouvet, C., Leblond, J. B., Millie, Ph. and Tramer, A., 1983, Chem. Phys. Lett. 97, pp. 161-166.

PART IV ISOLATED MOLECULES

SINGLE VIBRATIONAL LEVEL DEPENDENCE OF PICOSECOND FLUORESCENCE IN THE CHANNEL 3 REGION OF BENZENE

K. Yoshihara, D.V. O'Connor, M. Sumitani, Y. Takagi and N. Nakashima

Institute for Molecular Science
Myodaiji, Okazaki 444, Japan

An investigation of fluorescence spectra, quantum yields and decay times in collision-free benzene vapour excited to high excess energies in S_1 has been carried out. The laser excitation source provides wavelength tunable picosecond pulses in the range 225-250 nm. At longer wavelengths quantum yields and spectra are normal and lifetimes are single-exponential. In the next region, corresponding to ca. 2800-3800 cm^{-1} excess energy, the quantum yields decrease sharply and decay curves are non-exponential. Most of the structure in spectra is washed out, only the strong progressions in ν_1 being visible. In the third region, above 4000 cm^{-1}, the quantum yields decrease but at a slower rate while the observed decay is again exponential.

INTRODUCTION

Benzene has long served as a prototype molecule in the investigation of the photophysics of aromatic hydrocarbons. Single vibronic level fluorescence in S_1 benzene has proved to be a powerful tool in elucidating the mechanisms of radiative and non-radiative energy dissipation (1-3). At vibrational energies in excess of ca. 2800 cm^{-1}, however, the fluorescence efficiency of benzene decreases so drastically that a study of the high excess energy region by fluorescence techniques has usually been regarded as virtually impossible. This sharp decrease in quantum yield has led to the suggestion that at higher excess energies a new non-radiative channel, the so-called channel 3, comes into play. The dynamic properties falls into picosecond timescale and the nature of this channel is not yet fully investigated.

K. B. Eisenthal (ed.), Applications of Picosecond Spectroscopy to Chemistry, 261–272.

In the absence of detectable fluorescence the channel 3 anomaly has been studied principally by measurements of absorption linewidths (4,5), from which it was inferred that most levels with excess energies in the region of, for example, 5000 cm^{-1} lose energy(non-radiatively) at a rate greater than $2x10^{11}$ s^{-1}. However, since factors other than population relaxation may contribute to line broadening, the reliability of these estimates is questionable. The need for direct fluorescence measurements is therefore severe. Here we describe fluorescence quantum yeilds following one-photon excitation of single absorption bands under collision-free conditions (6).

Spectrally resolved fluorescence resulting from excitation of collision-free single vibronic levels has been extremely useful in elucidating the spectroscopy of the benzene $^1B_{2u} \leftarrow {}^1A_{1g}$ electronic transition (7). We have been engaged in a spectroscopic study aimed at determining the channels through which excess vibrational energy is dissipated from highly excited S_1 vibronic levels.

It has been predicted theoretically (8,9) that internal conversion may be the dominant decay pathway in benzene at high excess energies. In a recent publication (10) two of us reported the results of a flash photolysis study which confirms this prediction experimentally. A formation of the vibrationally highly excited S_0 state ("hot benzene") was observed at excitation into the S_1 and S_2 manifolds (10,11).

EXPERIMENTAL

Picosecond Tunable UV Light by Mixing of Raman and Optical Parametric Generation.

The experimental arrangement is shown in Figure 1 (12). A single pulse with a Gaussian beam profile was selected from a very stable mode-locked oscillator. After passage through three amplifiers the beam was split into two lines and amplified again. The first beam was frequency doubled and then quadrupled in KDP crystals. The fourth-harmonic at 266 nm with an energy of 7 mJ was condensed by a telescope and sent to a 1 m Raman cell containing high-pressure hydrogen or nitrogen for generation of stimulated Raman light with conversion efficiencies of 30-40 % in both cases.

The beam in the second line was frequency doubled and the second-harmonic with an energy of about 10 mJ was tightly focused onto the second crystal in the optical parametric generation(OPG) by a cylindrical lens with a 50 cm focal length. The OPG consists of two 4 cm long crystals of 99 %-deuterated KDP (KD*P)

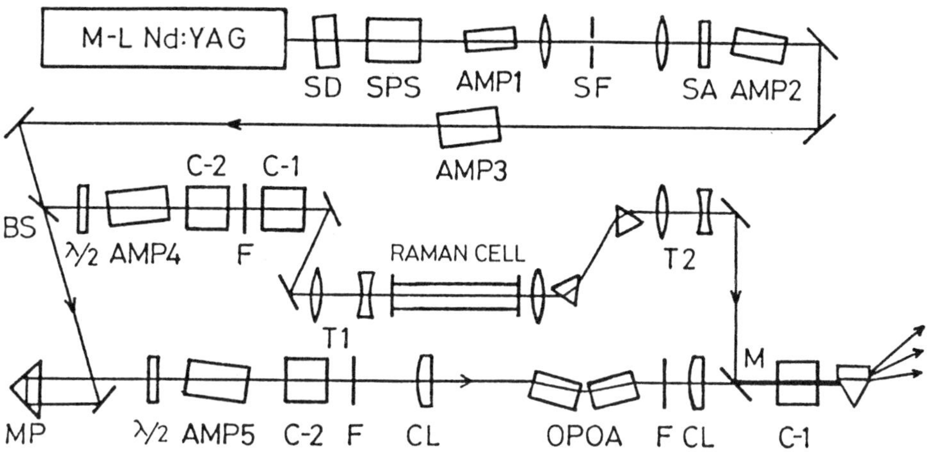

Fig. 1. Optical arrangement for generation of tunable UV light. M-L Nd:YAG, Mode-locked Nd:YAG laser, SD; Saturable dye, SPS; Single pulse selector, SF; Spatial filter, SA; Soft aperture, BS; Beam splitter, λ/2; Half-wave plate, C-1,2; Nonlinear crystal (KDP, type-1,2), F; Filter, T1; 5:2 telescope, T2; 5:3 telescope, MP; Movable prism, CL; Cylindrical lens, OPG; Optical parametric generator, M; Half mirror for beam mixing.

with 61° cut. The energy conversion efficiency for OPG reached 40 % at the degenerate wavelength (1064 nm). Two components of this beam with different wavelengths and mutually orthogonal polarizations cover almost the same spectral region between 900 and 1400 nm.

Upconversion was performed by mixing the optical parametric light with the fourth-harmonic at 266 nm, the N_2 Raman beam at 283 nm, and the H_2 Raman beam at 299 nm. In this way three different spectral ranges between 215 and 245 nm were covered. The energy exceeded 100 μJ over the entire tuning range. The spectral band width was 0.5 Å and the pulse duration of the output was about 10 ps.

Fluorescence Quantum Yields, Lifetimes, and Spectra.

Decay curves following single shot laser excitation of individual absorption bands were measured with a fast channel plate photomultiplier and a 1 GHz oscilloscope. Each decay curve was accumulated over a number of laser shots by means of a high-speed transient digitiser system (13) and the integrated area was then normalised to the intensity of the exciting light, which was monitored at each shot. Fluorescence lifetimes were measured with a calibrated streak camera for the picosecond (14) and with a channel plate photomultiplier for the nanosecond region.

Fluorescence spectra were measured with a monochromator and a solid-state detector array. Collision-free requirements in spectral measurements are, however, much more severe than lifetime measurements, mainly because the integrated intensity of the long lived(relaxed) fluorescence is relatively large. This was therefore avoided by collecting only early time emission by gating the detector with a fast pulser.

RESULTS

Fluorescence Quantum Yields and Lifetimes.

In Table 1 are listed the measured quantum yields and decay times for a number of transitions covering an excitation wavelength range of some 20 nm and extending to 5650 cm^{-1} excess vibrational energy in S_1. Although the spectral bandwidth of the excitation source was narrow enough to pump many individual bands reasonably cleanly, at some excitation wavelengths more than one transition was excited and the designations in Table 1 refer to the most intense transition at the specified excitation wavelength.

Inspection of the data in Table 1 reveals both a specific level and an excitation wavelength effect on the quantum yields, the latter perhaps becoming dominant at shorter excitation wavelengths. Thus in the two series $7_0^1 1_0^n$ and $6_0^1 1^{n+3}$ the $7_0^1 1_0^n$ member lies only 210 cm^{-1} to lower energy yet the yield is clearly higher even at some 5000 cm^{-1} excess energy ($7_0^1 1_0^2$ and $6_0^1 1_0^5$). However the yield from the transition 7_0^1 is 6 times larger than that from $6_0^1 1_0^3$ whereas that from $7_0^1 1_0^2$ is only about twice as large as that from $6_0^1 1_0^5$. The level effect is less marked in other series as might be expected from a consideration of the anomalously narrow linewidths of the $7_0^1 1_0^n$ absorption bands (4). For example, if the yields for the hot band transitions $6_1^2 1_0^n$ are compared to those for the $6_0^1 1_0^n$ transitions at slightly shorter excitation wavelengths it will be seen that at lower excess energies the $6_1^2 1_0^n$ yields are indeed smaller but that for $6_1^2 1_0^4$ and $6_0^1 1_0^4$ at 236.6 nm and for $6_1^2 1_0^5$ and $6_0^1 1_0^5$ at 231.6 nm they are almost equal (16).

In order to illustrate graphically the channel 3 threshold we have plotted in Figure 2 the present quantum yield results for three series of transitions, $6_0^1 1_0^n$, $6_1^2 1_0^n$ and $7_0^1 1_0^n$. Although the simple equation, $\phi_f = k_r/(k_r + k_{nr})$, may be inapplicable in

Table 1 Quantum yields and decay times of benzene fluorescence

Transition	$\Delta E(cm^{-1})$	λ(nm)	ϕ_f	τ_1(ns)	τ_2(ns)
$6_0^1 1_0^2$	2367	247.1	1.8×10^{-1}	47	
$6_0^1 11_0^2 1_0^1$	2470	246.5	1.4×10^{-1}	33	
$6_0^1 16_1^1 1_0^2$	2610	248.1	1.0×10^{-1} reference	30	
$6_0^1 10_0^2 1_0^1$	2614	245.7	1.0×10^{-1}	27	
$6_0^1 16_2^2 1_0^2$	2853	249.1	6.3×10^{-2}	—	
$6_0^1 17_0^2 1_0^1$	2856	244.0	1.4×10^{-2}	2.8	13
$6_1^2 1_0^2$	2888	247.7	4.6×10^{-2}	—	
7_0^1	3080	242.8	1.6×10^{-2}	4.0	18
$6_0^1 1_0^3$	3290	241.7	2.5×10^{-3}	0.7	2.9
$6_0^1 16_1^1 1_0^3$	3533	242.6	1.4×10^{-3}	0.8	3.4
$6_0^1 10_0^2 1_0^2$	3537	240.2	1.3×10^{-3}	0.15	1.0
$6_0^1 2_0^1$	3651	239.6	9.7×10^{-4}	0.075	0.5
$6_1^2 1_0^3$	3811	242.2	1.8×10^{-3}	0.7	3.7
$7_0^1 1_0^1$	4003	237.5	9.0×10^{-4}	0.13	
$6_0^1 1_0^4$	4213	236.4	5.7×10^{-4}	0.090	
$6_1^2 1_0^4$	4734	236.9	6.1×10^{-4}	0.057	
$7_0^1 1_0^2$	4926	232.4	4.7×10^{-4}	0.080	
$6_0^1 1_0^5$	5136	231.4	2.8×10^{-4}	0.023	
$6_1^2 1_0^5$	5657	231.9	2.5×10^{-4}	0.026	

certain regions of excitation a plot of $\log(1/\phi_f - 1)$ ($= \log(k_{nr}/k_r)$) gives a good idea of the variation of k_{nr} with excess energy. From Figure 2 (assuming k_r varies only slightly within a series (1)) a rapid increase in k_{nr} in the 2800 - 3800 cm^{-1} excess energy region is evident. Nearly the same increase in k_{nr} was observed C_6D_6 in the 2500 - 3500 cm^{-1} excess energy region (15).

On the basis of the tabulation in Table 1 the wavelength range studied can be divided roughly into three regions. At longer wavelengths quantum yields are normal while lifetimes are

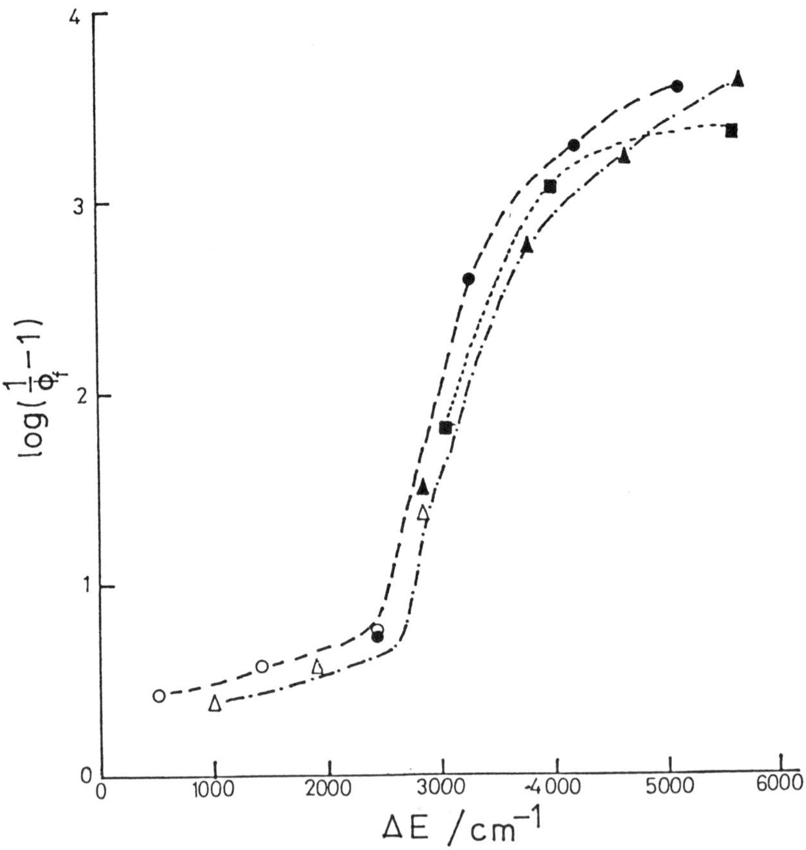

Fig. 2. Plots of $\log(1/\phi_f - 1)$ vs ΔE for the transitions $6_0^1 1_0^n$ (circles), $6_1^2 1_0^n$ (triangles) and $7_0^1 1_0^n$ (squares). Unfilled symbols are data taken from Spears and Rice (ref. (2)). The lines delineate members of the same series and have no further significance.

single-exponential and relatively long (Region I). The fluorescence in this region has been studied in great detail previously (2,3). In the next region (Region II), corresponding to ca. 2800 - 3800 cm^{-1} excess energy, the quantum yields decrease sharply and decay curves are non-exponential. The onset of non-exponential decay seems to occur at the channel 3 threshold. In the third region (Region III), above 4000 cm^{-1}, the quantum yields decrease but at a slower rate while the observed decay is again exponential (14).

Typical examples of the decays in Region I, II, and III are shown in Figures 3 and 4. Figure 3(a) and (b) are the decays of the $6_0^1 1_0^2$ level (ΔE = 2367 cm^{-1}, Region I) and of the 7_0^1 level (ΔE = 3080 cm^{-1}, Region II), respectively. Figure 4 shows the decay of the $7_0^1 1_0^1$ level (ΔE = 4003 cm^{-1}, Region III).

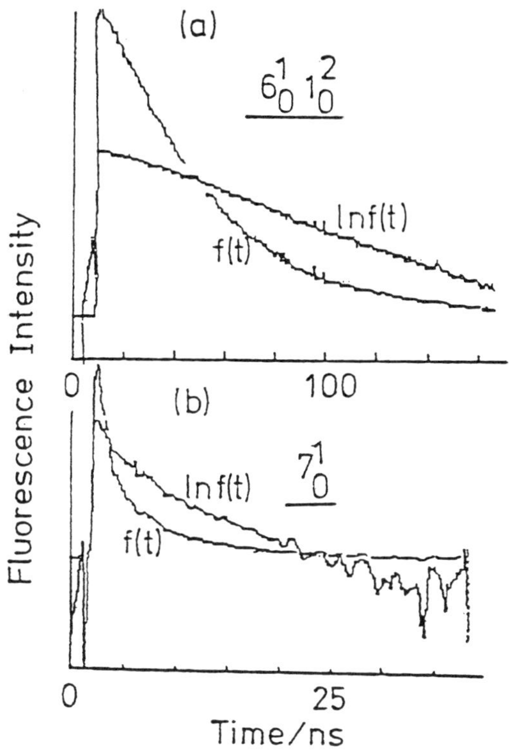

Fig. 3. Fluorescence decay of benzene. (a) Excitation at 247.1 nm ($6_0^1 1_0^2$), excess energy (ΔE): 2367 cm^{-1}. Single exponential decay of 47 ns. (b) Excitation at 242.8 nm (7_0^1), ΔE = 3080 cm^{-1}. Non-exponential decay. Simulation with two exponents gives 4.0 and 18 ns. Channel-plate type photomultiplier detection.

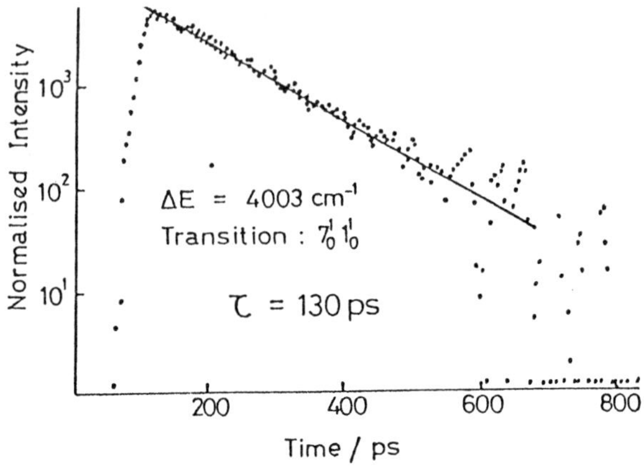

Fig. 4. Fluorescence decay of benzene excited at 237.5 nm ($7^1_0 1^1_0$), ΔE = 4003 cm^{-1}. Streak camera detection. Single exponential fit to observed data τ = 130 ps.

Fluorescence Spectra.

Spectra measured after excitation of the four transitions $6^1_0 1^2_0$, $6^1_0 1^3_0$, $6^1_0 1^4_0$ and $6^1_0 1^5_0$ are presented in Figure 5(a). It can be seen that the spectrum following $6^1_0 1^3_0$ excitation reveals a wealth of vibrational bands, most of which can be assigned, in accordance with the Franck-Condon intensity distribution, to fluorescence from $6^1 1^3$ (16).

Fluorescence from the levels $6^1 1^4$ at ΔE = 4213 cm^{-1} and $6^1 1^5$ at ΔE = 5136 cm^{-1} yields spectra that are almost featureless. Nevertheless the spectral profiles have some features that are indicative of fluorescence from the levels excited. For instance from $6^1 1^4$ there is an indication of a peak at $\nu_1'' = 5$ (275 nm) in agreement with the calculated intensity for $\nu_1' = 4 \rightarrow \nu_1'' = 5$ (ref. 16). Similarly from $6^1 1^5$there is a shoulder at 275 nm, corresponding to the transition $6^1 1^5 \rightarrow 6_2 1_6$ which is expected to be relatively strong.

In Figure 5(b) are plotted the spectra that result from excitation of the four transitions 7^1_0, $7^1_0 1^1_0$, $7^1_0 1^2_0$ and $7^1_0 1^3_0$. The spectrum from the level 7^1 has been observed and analysed previously (17). Spectra from upper members of the $7^1 1^n$ series lie atop an increasing background of congestion. But, although the bands are also drastically broadened, the spectra retain features of the levels initially excited. For example, in the $7^1 1^1$ spectrum the expected lack of intensity is the $\nu_1' = 1 \rightarrow \nu_1'' = 1$ transition at 268 nm is obvious. Also noteworthy are the two

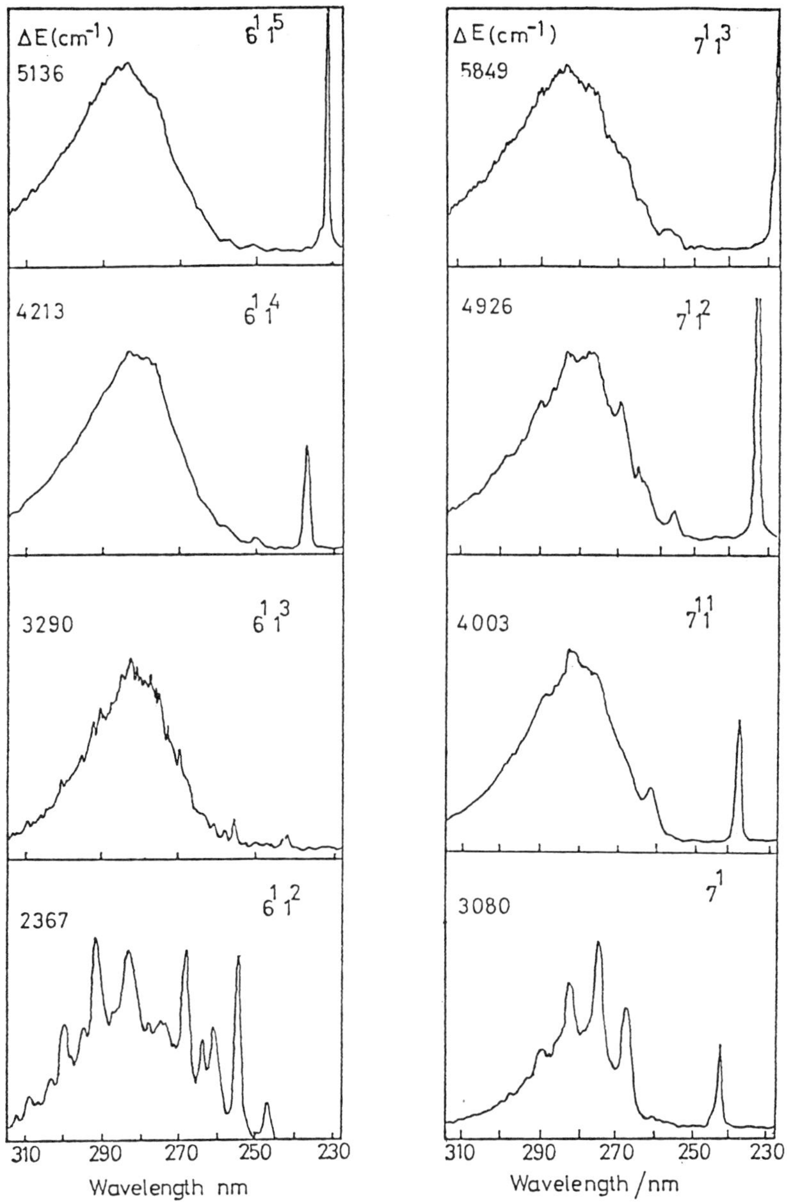

Fig. 5. Spectra resulting from excitation of $6^1 1^{n+2}$(a) and of $7^1 1^n$(b) for n = 0, 1, 2, and 3. Gated spectra except $6^1 1^2$ and 7^1. Experimental details: see ref. 16.

minima in the $6_1 7_1 1_n$ progression from the level $7^1 1^2$ corresponding to transitions to $\nu_1'' = 1$ (262 nm) and $\nu_1'' = 3$ (276 nm) (ref. 16).

DISCUSSION

Fluorescence Quantum Yields and Lifetimes.

A rapid increase in k_{nr} in the 2800 - 3800 cm^{-1} excess energy region is clear in Figure 2. This rapid increase is not maintained at higher excess energies, in agreement with observations in multiphoton ionization measurements of the benzene $S_1 \leftarrow S_0$ two-photon transition (18). The curves in Figure 2 correspond closely with a similar curve in the report of Wunsch et al. (19) obtained for the $14_0^1 1_0^n$ progression obtained by two-photon excitation. Different selection rules govern one-photon and two-photon absorption so that the levels populated by each are different. From our yields, however, k_{nr} values in good agreement with their data for similar excess energies can be derived. On the basis of intensity measurements they calculated non-radiative rate constants which, at 5000 cm^{-1} excess energy, are of the order of 10^{10} s^{-1}, in some disagreement with the results based on line broadening of the order of 2 x 10^{11} s^{-1}. This has been attributed to the fact that line broadening results from processes other than population relaxation (20).

The non-exponential decay of the intermediate excess energy region (Region II) can be explained at least in two ways. The absorption spectrum of region is sequence congested. It gives the broad back ground and well-characterized sharp band. Thus it is difficult to avoid from exciting some hot bands, which eventually cause non-exponential decay. The other possibility comes from modes of coupling between the prepared state and back-ground interacting levels. The intermediate coupling may give non-exponential decays. In order to clarify these points experiments with jet cooled molecules are required.

Fluorescence Spectra and Channel 3.

It is now well established that the ultinate fate of most benzene molecules that are excited to levels above the channel 3 threshold is formation of ground state molecules. Two of us have shown (10) that the quantum yield for formation of vibrationally hot S_0 benzene following excitation at 248 nm is about 60 %. More recently Otis et al. (21), in an examination of the energy decay following excitation of the $6_0^1 1_0^3$ transition, observed a quantum yield of internal conversion of 75 %. In neither investigation was the time resolution sufficient to follow the decay through possible intermediate steps. In another report (22) it has been suggested that the rapid increase in k_{nr} could result from a complete redistribution of the initial excess energy into levels having excitation in out of plane modes. The spectra presented in this paper which display, at all excess energies, some features characteristic of the level initially excited appear to rule out this proposal. They are compatible however with a mechanism involving partial energy redistribution such as that proposed by Riedle et al. (20). It should be emphasised, however, that the general features of the spectra could be due to other effects, notably excitation of overlapping transitions and congested fluorescence from levels in which ν_1 is highly excited.

The clear distinction in linewidth between the absorption transitions $6_0^1 1_0^n$ and $7_0^1 1_0^n$ is not observed in emission. All levels in the channel 3 region, except 7^1, exhibit grossly broadened spectra. While it is true that the vibrational features in the $7^1 1^3$ spectrum, for instance, can be easily distinguished in comparison to those in the spectrum from $6^1 1^5$ at lower excess energy, this effect is easily explained by the reduced number of quanta excited in the progression forming mode ν_1.

Once again it appears that conclusive information must come from experiments with jet cooled molecules. These experiments, which are at present in progress in our laboratory, offer the possibility not only of reducing hot band excitation but also, and this is now of particular importance, of investigating the role played by rotations in channel 3 decay.

ACKNOWLEDGEMENT

The authors are grateful to Drs. Y. Udagawa and K. Kamogawa for their contribution in measuring fluorescence spectra.

REFERENCES

1. For a review see: C.S. Parmenter, Adv. Chem. Phys., 22, 365 (1972).
2. K.G. Spears and S.A. Rice, J. Chem. Phys., 55, 5561 (1971).
3. W.R. Ware, A.M. Garcia, C.S. Parmenter, M.D. Schuh, and K.Y. Tang, Chem. Phys., 17, 377 (1976).
4. J.H. Callomon, J.E. Parkin, and R. Lopez-Delgado, Chem. Phys. Lett., 13, 125 (1972).
5. K. Aron, C. Otis, R.E. Demaray, and P. Johnson, J. Chem. Phys., 73, 4167 (1980).
6. M. Sumitani, D. O'Connor, Y. Takagi, N. Nakashima, K. Kamogawa, Y. Udagawa, and K. Yoshihara, Chem. Phys. Lett., 97, 508 (1983).

7. A.E.W. Knight, C.S. Parmenter, and M.W. Schuyler, J. Am. Chem. Soc., 97, 1993 (1975).
8. M.G. Prais, D.F. Heller, and K.F. Freed, Chem. Phys., 6, 331 (1974).
9. H. Hornburger and J. Brand, Chem. Phys. Lett., 88, 153 (1982).
10. N. Nakashima and K. Yoshihara, J. Chem. Phys., 77, 6040 (1982).
11. N. Nakashima and K. Yoshihara, J. Chem. Phys., 78 (1983), in press.
12. Y. Takagi, M. Sumitani, N. Nakashima, D.V. O'Connor, and K. Yoshihara, Appl. Phys. Lett., 42, 489 (1983).
13. Y. Takagi, Rev. Sci. Instrum., 53, 1677 (1982).
14. Y. Takagi, M. Sumitani, N. Nakashima, D.V. O'Connor, and K. Yoshihara, J. Chem. Phys., 77, 6337 (1982).
15. D. O'Connor, M. Sumitani, Y. Takagi, N. Nakashima, K. Kamogawa, Y. Udagawa, and K. Yoshihara, in preparation.
16. D. O'Connor, M. Sumitani, Y. Takagi, N. Nakashima, K. Kamogawa, Y. Udagawa, and K. Yoshihara, J. Phys. Chem., 87 (1983), in press.
17. A.E.W. Knight, C.S. Parmenter, and M.W. Schuyler, J. Am. Chem. Soc., 97, 2005 (1975).
18. K. Aron, C. Otis, R.E. Demaray, and P. Johnson, J. Chem. Phys., 73, 4167 (1980).
19. L. Wunsch, H.J. Neusser, and E.W. Schlag, Zeit. f. Naturforsch., 36A, 1340 (1981).
20. E. Riedle, H.J. Neusser, and E.W. Schlag, J. Phys. Chem., 86, 4847 (1982).
21. C.E. Otis, J.L. Knee, and P.M. Johnson, J. Chem. Phys., 78, 2091 (1983).
22. H. Homburger and J. Brand, Chem. Phys. Lett., 88, 153 (1982).

PICOSECOND PHOTO-CHEMISTRY AND SPECTROSCOPY IN SUPERSONIC BEAMS

Peter M. Felker[a] and Ahmed H. Zewail[b]

Arthur Amos Noyes Laboratory of Chemical Physics,[c]
California Institute of Technology
Pasadena, California 91125 USA

In this article we describe the development and the applications of the picosecond-jet technique, which utilizes a picosecond laser and a supersonic jet beam of large molecules. The applications include studies of coherence (quantum beats), photodissociation, isomerization, intramolecular hydrogen bonding, and partial solvation.

I. INTRODUCTION

The dynamics of intramolecular vibrational-energy redistribution (IVR), bond breakage, and bond formation in large isolated molecules is currently a very important and challenging problem. A fruitful approach to this problem involves the study, via direct measurements in the time domain, of the decay parameters of energetically excited molecules. Using such an approach one can determine unimolecular rate constants as well as study any quantum mechanical coherence phenomena that may be involved in a decay process. Furthermore, in cold beams these rates and coherence effects can be studied as a function of the energy and character of individual vibrational modes in a molecule. The results of such studies are crucial to understanding the nature of energy flow within molecules and to assessing the possibility of vibrational mode-selective laser chemistry.[1]

Over the past four years we have been examining vibrational level-selective rate processes and coherence effects using the technique of picosecond-jet spectroscopy.[2,3] In this technique a picosecond laser selectively excites molecular vibronic levels and allows for time domain studies of rates and coherence effects.

K. B. Eisenthal (ed.), Applications of Picosecond Spectroscopy to Chemistry, 273–291.

The measurements are made on molecules that have been cooled vibrationally and rotationally in a supersonic jet expansion. Such cooling greatly reduces the spectral congestion that arises from thermal effects and allows for the single vibronic level excitation of even very large molecules. With the picosecond-jet technique we have studied a number of photochemical and photophysical processes, which will be reviewed here. These include:

1) Coherence and energy redistribution in isolated large molecules (anthracene, stilbene).[2,4-6]

2) Quantum beats and Zeeman effect probing of radiationless processes in pyrazine.[7]

3) Intramolecular trans-cis isomerization of t-stilbene,[6] 1, 4-diphenylbutadiene,[8] and styrene.[9]

4) Intramolecular proton transfer in methyl salicylate.[10]

5) Chromophore selective excitation and photochemistry of intramolecular exiplex formation in anthracene-$(CH_2)_3$-N,N-dimethyl aniline.[11]

6) Bond breakage and IVR of partially-solvated molecules in the jet.[12]

7) Bond rotation dynamics and IVR in bianthryl.[13]

Here, we review some of the work that has been done in this laboratory. We divide the review into three areas. First, we describe the picosecond-jet technique and discuss the time resolution obtainable with various detection schemes. Next, results obtained on intramolecular coherence effects (quantum beats) and IVR in large molecules are presented. Finally, results pertaining to novel photochemical processes such as isomerization, dissociation, and intramolecular exiplex formation in isolated molecular systems are reviewed.

II. THE PICOSECOND LASER-FREE JET APPARATUS

The arrangement we have used to interface the picosecond laser excitation source with the molecular beam (free-jet) is schematically shown in Fig. 1. The lasers used are synchronously pumped dye laser systems, whose coherence width, coherence time, and pulse duration have been characterized[14] by the second harmonic generation (SHG) autocorrelation technique. The pulse width of these lasers is typically 1-2 ps, or, when a cavity dumper is used, 15 ps. The bandwidths of the lasers can be varied using intracavity filters and etalons. Typical bandwidths of the frequency doubled laser output are between 2 and 0.4Å.

The detection of molecular decays can be accomplished using one of three techniques (See Fig. 1):

1) Time-correlated single photon counting (detecting dispersed or total fluorescence) using a fast photomultiplier (or microchannel plate). This method can achieve a resolution of 100-200 ps without deconvolution. With deconvolution one can obtain the time

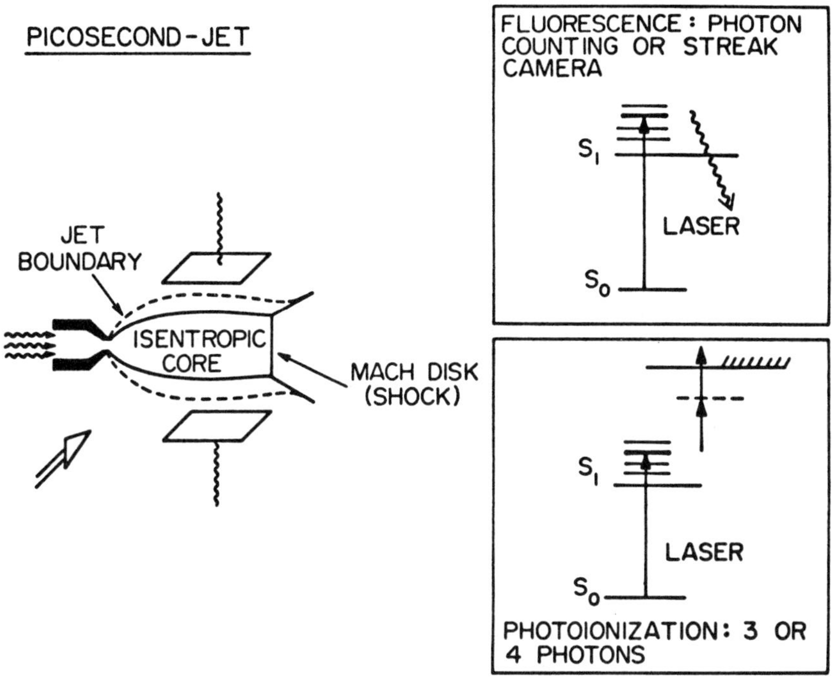

Figure 1. Schematic diagrams representing the experimental configuration and detection schemes used in the picosecond-jet technique. The picosecond laser crosses the free jet expansion at right angles to the axis of expansion. Either fluorescence is collected at right angles to both laser and jet, or photoion current is measured by collection plates mounted in the vacuum chamber. Fluorescence detection of decays requires one excitation event followed by detection and processing by fast electronic devices. The measurement of decays by photoionization requires a pump pulse and a probe pulse. Time resolution depends essentially only on the temporal characteristics of these pulses. The "3 or 4 photons" in the figure refers to the case of t-stilbene.[15]

constants of single exponential decays having lifetimes of 30-50 ps.

2) Streak camera detection of fluorescence. Synchroscan streak cameras can give a resolution of better than 10 ps with a repetition rate of up to 200 MHz.

3) Pump-probe multiphoton ionization detection. This technique has been developed to measure short-time transient behavior.[3,15] Resolution is limited by the pulse width of the laser (∿ 2 ps). An illustration of the capabilities of this detection technique appears in Fig. 2, in which decay results are shown for t-stilbene in a bulb (25°C) excited to $S_1 + \sim 200\ cm^{-1}$ and

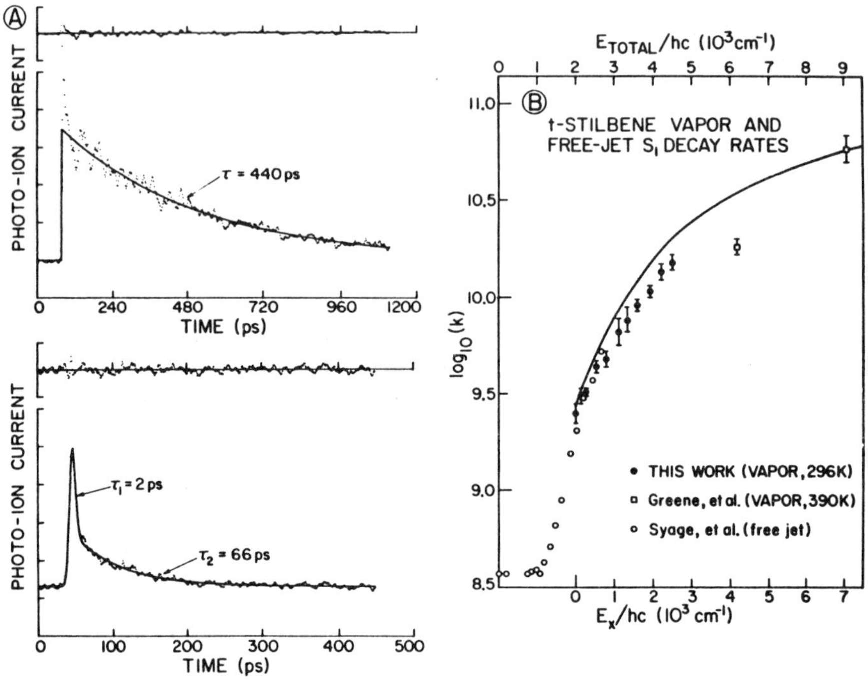

Figure 2. A) Representative multiphoton ionization-detected decays of trans-stilbene vapor[15] ($T = 296K$, $P = 5 \times 10^{-5}$ Torr). The upper plot shows the decay for $\lambda_{pump} = 3079.5\text{Å}$ and $\lambda_{probe} = 6195\text{Å}$; $E_x = 200\ cm^{-1}$. The lower plot is for $\lambda_{pump} = 2876\text{Å}$ and $\lambda_{probe} = 5752\text{Å}$; $E_x = 2500\ cm^{-1}$. Note the different timescales for the two plots. The solid lines are the best fit single exponentials for the upper plot, and the best fit convolution of a Gaussian pulse and a biexponential decay for the lower plot. Also shown are the weighted deviations. B) Excess energy dependence of trans-stilbene S_1 decay rates. Results from several experimental studies (points) and calculated[15] thermally averaged rates (line) are plotted. The lower abscissa is the excess energy for room temperature (296K). The free-jet data should not be referred to this axis. The upper abscissa is the total energy and may be used as a common axis for all the experimental data. For further details see Ref. 15.

$S_1 + \sim 2500\ cm^{-1}$.

In the experiments reviewed herein the vibrational and rotational temperatures were typically < 40K and 1-5K, respectively, depending upon the particular free-jet expansion parameters (carrier gas, carrier gas pressure, etc.). For a given molecule, we typically obtain excitation spectra prior to any picosecond experiments. This is done to characterize the free-jet sample and to reveal the excited state vibronic structure. Excitation spectra are obtained using a frequency doubled YAG-pumped dye laser, a feedback arrangement that allows the wavelength of the laser to be scanned while maintaining UV power and alignment, and A/B boxcar detection that allows normalization of the spectra.[11]

III. INTRAMOLECULAR QUANTUM BEATS AND ENERGY REDISTRIBUTION

The study of quantum beat modulated decays in large molecules is useful for at least two reasons. Firstly, quantum beat spectroscopy is a Doppler-free technique which can provide sub-MHz resolution without requiring a highly monochromatic laser source. Such features are very attractive in the study of isolated large molecules, which have a large number of closely spaced eigenstates. Secondly, and perhaps more importantly, molecular quantum beats are manifestations of excited state dynamical behavior. Thus, in certain cases beats can provide information concerning the parameters important to radiationless transitions (intersystem crossing or internal conversion) or to vibrational energy redistribution in isolated molecules. Before we discuss the results of molecular quantum beat spectroscopy from this laboratory, it is useful to sketch some of the characteristics of the quantum beat phenomenon. Therefore, we first consider simple two-level beats. Then, we show how the form of beat-modulated decays can reveal the nature of excited state interactions in a molecule. Finally, we review the experimental results.

1. Simple Two-Level Beats

Suppose that two molecular eigenstates $|e_1\rangle$ and $|e_2\rangle$ are excited coherently by a laser source. Coherent excitation is possible if the coherence time (τ_c) of the laser is such that $\tau_c^{-1} >> |E_1-E_2|/\hbar$, where E_1 and E_2 are the energies of $|e_1\rangle$ $|e_2\rangle$. The parameters associated with laser pulse coherence (τ_c, pulse duration-τ_p, and pumping time-T) have been discussed for our lasers elsewhere.[14] Typical values of τ_c^{-1} ($\sim 10^3$GHz) allow coherent excitation of eigenstates separated by up to several cm^{-1}. This coherent preparation of molecular eigenstates does not depend crucially on the overall coherence of the laser pulse, in contrast to experiments that rely on the generation of a macroscopic coherence in a sample[16]-e.g., photon echos. The fact that our laser pulses are not transform limited-i.e., $\tau_c < \tau_p$ - does not prevent the coherent excitation of eigenstates.

Upon coherent excitation of $|e_1\rangle$ and $|e_2\rangle$ the excited state wave function is a superposition state:

$$\psi(t) = a\, e^{-i\bar{\omega}_1 t}\, |e_1\rangle + b e^{-i\bar{\omega}_2 t}\, |e_2\rangle \qquad (1)$$

where $\bar{\omega}_j = \omega_j - i\Gamma_j^e/2$ with Γ_j^e being the damping constant of $|e_j\rangle$. and ω_j being E_j divided by $\hbar$. Now, consider the emission of this state to a single ground state level. The intensity as a function time is

$$I(t) = |\mu|^2\, e^{-\Gamma^e \cdot t}\, (1 + \cos \omega_{12} t); \qquad (2)$$

where, for simplicity, we have assumed $\Gamma^e = \Gamma_1^e = \Gamma_2^e$, $a = b = 1/\sqrt{2}$, $\mu = \mu_1 = \mu_2 \equiv$ the value of the dipole moment matrix elements connecting the excited eigenstates to the ground state level, and $\omega_{12} = \omega_2 - \omega_1$. Clearly, the signal exhibits a modulation in time (quantum beats) at an angular frequency of ω_{12}. A measure of the modulation frequency is a measure of excited state energy splittings.

The physical situation assumed in order to derive Eq.(2) illustrates some of the conditions necessary to the observation of intramolecular quantum beats: 1) the coherent excitation of a quantum mechanical superposition state followed by 2) emission to a common ground state level. We emphasize that the phenomenon involves individual molecules, not the collective coherent excitation of a number of molecules (coherence effects of an intermolecular nature are sensitive to ground state or final state splittings).[1]

Physical situations more general than that considered above (e.g., situations where $\Gamma_1^e \neq \Gamma_2^e$, or $a \neq b$, or $\mu_1 \neq \mu_2$) result in beat signals somewhat different than Eq.(2). These differences can be useful diagnostic tools for studying excited state dynamical processes. For instance, for a given beat-modulated decay, the decay of the quantum beat envelope may differ from the decay of the pure exponential term. A determination of the different decay constants can give information concerning pure dephasing, and T_1 and T_2 type rates.[16] Deviations from Eq.(2) may also reflect the nature of excited state coupling interactions, or the presence of more than just two coherently prepared eigenstates. In the section that follows we consider the form of beat-modulated decays expected given various conditions concerning the couplings within a molecule.

2. Quantum Beats and Molecular Couplings

Consider a situation wherein a picosecond pulse coherently excites a molecule from an initial ground state $|g\rangle$ to excited states $|e_1\rangle$ and $|e_2\rangle$, which can emit to the manifold of ground state levels $|f\rangle$. Then one can show[17] that the emission as a function of time after excitation is

$$I(t) = K \sum_{f} \sum_{i,j=1}^{2} \mu_{ig}\mu_{gj}\mu_{jf}\mu_{fi} \, e^{-\Gamma^e t} \, e^{-E_{ij}t/\hbar} \qquad (3)$$

where the μ terms represent absorption or emission dipole matrix elements, K is a constant, $E_{ij} = E_i - E_j$, and Γ^e is as above. Now several cases of Eq.(3) may be discussed.[4] (We shall use the nomenclature "case 1," "case 2," and "case 3" that has appeared in our earlier publications.[4,5]) These cases depend on the form of the $|e_i\rangle$ states. Given that these eigenstates result from the coupling of zero-order states $|a\rangle$ and $|b\rangle$, the quantum beat signal observed will depend on the nature of the zero-order states and on the degree of coupling between them. We have considered the details of this problem elsewhere[4,5] and have outlined three important cases:

a. Case 1 corresponds to interelectronic state coupling, e.g., singlet-triplet coupling. Due to the fact that only the singlet zero-order state has any absorption or emission strength to the ground state, beats are characterized by + cos $\omega_{12}t$ terms, there is no spectral dependence to the beats, and beat modulation depths are dependent on the coupling between $|a\rangle$ and $|b\rangle$.

b. Case 2 corresponds to intraelectronic state coupling, e.g., vibrational coupling within a singlet manifold. For this case one assumes that only one zero-order state $|a\rangle$ has absorption strength from $|g\rangle$ and that the detection is such that only a fluorescence band deriving its strength from $|a\rangle$ is detected. The form of the beating decay for this case is exactly the same as for case 1 (i.e., + cos $\omega_{12}t$ terms with coupling dependent modulation depth).

c. Case 3 also corresponds to vibrational coupling within a single manifold (intraelectronic state coupling). Similar to case 2, one assumes that only $|a\rangle$ has absorption strength from $|g\rangle$. In contrast, however, one assumes that the only fluorescence band that is detected derives all its strength from $|b\rangle$. Case 3 decays have beats that are characterized by - cos $\omega_{12}t$ (phase-shifted) terms. Furthermore, the modulation depth of such decays is always 100%, independent of the coupling between $|a\rangle$ and $|b\rangle$. Both case 2 and case 3 decays can occur in the same fluorescence spectrum. The spectral positions of the various bands provide information concerning the nature of the coupled excited state vibrational levels.

Clearly, the temporal and spectral characteristics of beat-modulated decays can be powerful probes of excited state couplings. In the next sections we review experimental results which further illustrate this point.

3. Anthracene Beats and Energy Redistribution

Since our first report[2] of quantum beat-modulated decays for anthracene excited to S_1 + 1380 cm^{-1}, we have suspected that vibrational coupling within the S_1 manifold is involved with the phenomenon. There are several reasons[5] for this. Firstly, the observation of beats depends on very specific excitation and detection conditions. Secondly, beat modulation depths depend on the fluorescence band detected and on the carrier gas and carrier gas pressure. Thirdly, the application of a static magnetic field to the jet-cooled sample produces no observable effect on the beats, indicating the absence of singlet-triplet coupling in the phenomenon. Finally, the beats occur for excitation to a region of the S_1 manifold in which dispersed fluorescence spectra show features consistent with limited vibrational coupling.

A breakthrough in our understanding of beats in anthracene came when beats were also observed for excitation to the vibronic level at S_1 + 1420 cm^{-1}. For this excitation energy we were able to clearly observe case 2 and case 3-type (see previous section) modulated decays (Fig. 3) in the fluorescence spectrum. Further measurements[5] on the S_1 + 1380 cm^{-1} beats also revealed phase shifts in the decays of various bands in the fluorescence spectrum. The quantum beat behavior for both these excitations has, therefore, been attributed to vibrational coupling within the S_1 manifold. These observations yield direct information concerning <u>restricted</u> vibrational energy redistribution among certain modes of the molecule.[4,5]

One question of interest that remains to be answered regarding beats in anthracene concerns the nature of the vibrational coupling that gives rise to the beats. Two likely possibilities exist. One is a simple anharmonic coupling between vibrational modes. The second is Coriolis coupling, which results from the interaction of rotation and vibration in nonrigid molecules. In examining the consequences of both these coupling schemes, it has become apparent that rotational level structure cannot be ignored in considering the coupling between two vibrational levels.[5] In fact, rotational level effects on the quantum beats for S_1 + 1380 cm^{-1} excitation of anthracene have been observed and discussed.[5]

4. Fourier Transform Spectroscopy: Pyrazine as an example

As illustrated by anthracene, quantum beat spectroscopy can reveal intramolecular couplings and can also provide high resolution data regarding excited state level structures. A more dramatic example of the latter feature is provided by work which we have performed on pyrazine.[7] In the fluorescence decay of this molecule, and other small molecules, beats have been attributed

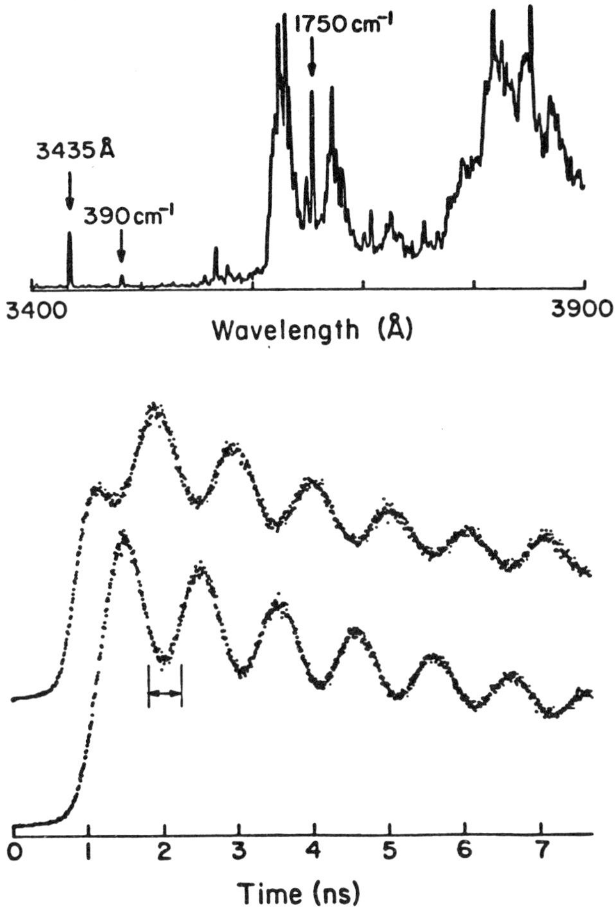

Figure 3. Dispersed fluorescence (top) and beating decays (bottom) for jet-cooled anthracene excited to S_1 + 1420 (see Ref.4). The top decay corresponds to detection of the 390 cm^{-1} band. The lower decay corresponds to detection of the 1750 cm^{-1} band. The 180° phase shift between the modulations of the decays is apparent. The 390 cm^{-1} decay is assigned as "case 2" (plus cosine beats) and the 1750 cm^{-1} decay is assigned as "case 3" (minus cosine beats).

to the singlet-triplet coupling that is known to be characteristic of azaaromatics and dicarbonyls.[18] Our work was concerned with the magnetic field effects on the fluorescence decay and quantum beats of pyrazine excited to its S_1 vibrational level. As is evident from the rich spectrum of Fig. 4, which is a Fourier transform spectrum derived from the beating decay, the fluorescence from many molecular eigenstates is sampled in the detected

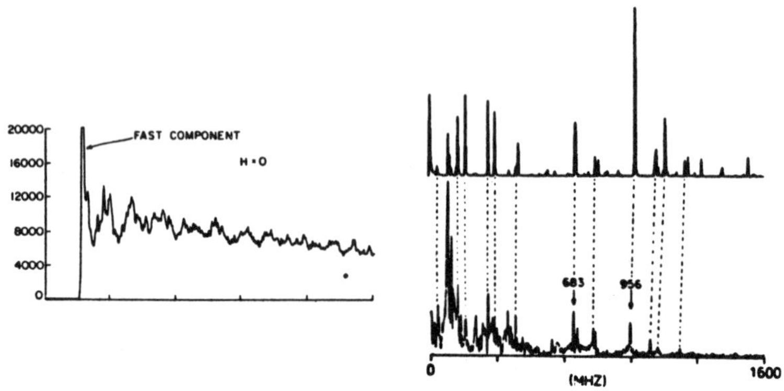

Figure 4. Left: Beat modulated decay of jet-cooled pyrazine excited to S_1 + 0 cm^{-1} in zero magnetic field (see Ref. 7). Right: Molecular eigenstate spectra for pyrazine. The top spectrum represents a simulation of a Fourier beat spectrum using results from high resolution excitation spectra (Ref. 19). The bottom is the Fourier spectrum that is derived from the modulated portion of the decay at left.

fluorescence. Unlike the anthracene case, the magnetic field effect[7] on the beats clearly demonstrates the involvement of S_1 coupling to a triplet manifold. By observing the shifts of Fourier peaks as a function of field strength we could infer the importance of second-order Zeeman effects in influencing singlet-triplet coupling.

The Fourier transform spectra[7] that we have obtained for pyrazine correlate very well with high-resolution fluorescence excitation spectra that have recently been obtained by Kommandeur's group.[19] This correlation of molecular eigenstate spectra from time domain measurements with such spectra from measurements in the frequency domain, is the first of its kind. It clearly shows the utility of using quantum beats for doing ultrahigh resolution spectroscopy without ultranarrow band lasers; a feature that has been extensively exploited in atomic spectroscopy.

IV. PHOTOCHEMISTRY IN BEAMS

Picosecond-jet spectroscopy holds tremendous potential in its application to the study of photochemical processes since unimolecular reaction rates can be measured as a function of well-defined molecular energy and vibrational state. Below, we review results from this laboratory involving this, still novel, area of research.

1. Chromophore-Selective Processes and Photochemistry of Intramolecular Exciplex Formation

A large number of molecules exist which contain several chromophores. Upon excitation of one of these chromophores, the possibility exists that energy may flow intramolecularly into others. The dynamics of such a process, which may involve intramolecular vibrational energy redistribution and molecular geometrical changes, as well as electronic energy transfer, have been studied in this laboratory.[11] Measurements were made on the following system (I):

$A^* - CH_2 - CH_2 - CH_2 - \phi$, where A^* represents an excited anthracene moiety and ϕ represents an N,N-dimethylaniline moiety. This molecule was chosen since we had already done a large amount of work on the spectroscopy of anthracene, and also because this type of molecule has been extensively studied in the solution phase.[20] After picosecond excitation in the jet of single vibronic levels of the anthracene moiety, both anthracene-like fluorescence and red-shifted exciplex fluorescence (emission due to the "product" species formed by the interaction of the anthracene and aniline moieties) were observed. At a given excitation energy the rise time of the product emission matched the decay time of the anthracene-like emission and provided a measure of the rate of energy transfer in the molecule. The rates were found to be very sensitive to the initially deposited excess vibrational energy in the anthracene moiety. An energy threshold to the product formation rate was found at ca. 1000 cm^{-1} (2.9 Kcal/mol). Beyond this threshold the rate increases steeply with energy in such a way that at $\sim$ 3000 cm^{-1} excess vibrational energy the decay constant ($\sim$ 400 ps) of the anthracene-like emission is very much different than that of anthracene (5.7 ns) at an analogous energy. The measurements represent the first observations in the time domain of product formation due to energy flow in an isolated large molecule.[11]

The detailed observations of energy flow in I may be expected to reveal much concerning IVR processes and alkyl chain dynamics. Since after excitation a conformational change involving the propyl chain must occur to bring the aniline moiety in proximity to the excited anthracene group, energy that is initially deposited in anthracene-localized vibrations must redistribute in such a way as to excite propyl chain motions. The rate of product formation is thus expected to be dependent on the rate of IVR in the molecule and the ease with which propyl chain geometrical changes can be made. The experimentally deduced energy threshold to product formation is consistent with rate limiting steps involving rotations about C—C type bonds.

If one calculates a thermally weighted rate constant based

on our jet data and compares this value with solution data, it is found that the calculated constant is greater than the measured solution constant. This is consistent with the hindrance by solvent molecules of the propyl chain geometrical changes needed to bring about product formation. A kinetic analysis involving consideration of alkyl chain dynamics and IVR processes appears in Ref. 11.

2. Isomerization and Internal Rotation

Several large molecules have been studied by the picosecond-jet technique with the aim of characterizing processes involving rotations about double and single bonds. These molecules include trans-stilbene,[6] diphenyl butadiene,[8] styrene,[9] and the composite molecule, bianthryl.[13] Below, we highlight the new findings from these studies. For further details, the reader is referred to the original references.

The photoisomerization process of t-stilbene,

is a classic problem in chemistry. Much work has been done in solution, and more recently, under gas phase, isolated molecule conditions.[21] Our picosecond-jet studies of the process have uncovered several new results. Firstly, the rates for isomerization (inferred from fluorescence lifetime data) show a clear energy threshold at $\sim 1200\ cm^{-1}$ (see Fig. 5). Above this threshold k varies smoothly and monotonically, and there is no dependence on the form of the vibrational mode excited. The whole curve of k vs. E correlates well with quantum yield measurements.[22] Secondly, spectral broadening in the dispersed fluorescence spectra gives indications of the degree of IVR taking place at a given energy. Spectra are very congested for excitation above the threshold. Thirdly, quantum beat-modulated fluorescence decays have been observed. Finally, fluorescence excitation and dispersed fluorescence spectra display low frequency intervals, some of which, in light of calculated results,[23] can be attributed to torsional motions in the molecule. Knowledge of these motions may be the key to understanding the isomerization process.

The energy value for the threshold in stilbene gives a barrier height that is in good agreement with values derived from solution phase work.[21] The characteristics of the threshold and its dependence on intramolecular energy redistribution have been discussed

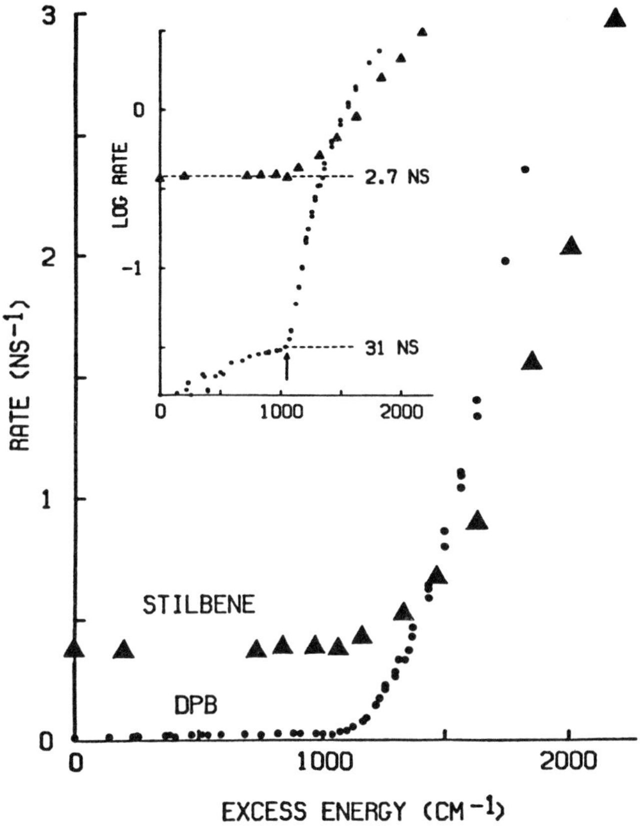

Figure 5. The fluorescence decay rates of jet-cooled trans-stilbene and 1,4-diphenyl-1,2-butadiene as a function of excess vibrational energy in their S_1 manifolds (see Ref. 8). The inset shows the log of the rate vs. excess energy.

elsewhere, and rates have been calculated using the RRKM method.[3,24] When a simple unimolecular k is calculated, the values do not agree well with the experimental ones. On the other hand, if the reaction at a certain excess energy is assumed to involve a reversible process between the trans isomer and an intermediate species, then good agreement is found between RRKM and experimental rates.[24] Of course, RRKM, being a statistical theory, applies only to molecules in which rapid and extensive IVR occurs on the timescale of the photochemical process. Our spectral and quantum beat data indicate that such energy redistribution does occur in t-stilbene above the 1200 cm^{-1} threshold. Furthermore, recent multiphoton ionization experiments[15] on t-stilbene indicate that IVR above the threshold occurs on a timescale of ca. 2 ps.

Experiments similar to those performed on trans-stilbene have also been performed on 1,4-diphenyl-1,3-butadiene. Shepanski et al.[8] have obtained fluorescence lifetimes vs. energy in the jet-cooled molecule (Fig. 5). Again, a threshold is present (this one at ca. 1000 cm^{-1}). The slope of the rate vs. E curve above this threshold is greater than that of the t-stilbene curve. In a statistical rate theory (such as RRKM) this is understandable in terms of the larger density of states in 1,4-diphenyl-1,3-butadiene.

Another difference between t-stilbene and 1,4-diphenyl-1,3-butadiene concerns their low-lying electronic states. The lowest excited singlet state in the larger molecule is A_g. A slightly higher-lying ^1B_u state also exists. The rate threshold in 1,4-diphenyl-1,3-butadiene appears to be due to a crossing of the potential surfaces of these two electronic states. This indicates the important role which excited state electronic interactions can have in determining photoisomerization and other nonradiative processes.

A third molecule, which we have studied in an effort to understand the dynamics of rotations about bonds, is styrene. In this molecule, where the spacing of low-lying electronic states is expected to be larger than for t-stilbene or diphenyl butadiene, we see no evidence for rotations about the alkene double bond.[9] It is concluded that the barrier to such a rotation is comparable to or larger than that of t-stilbene.

9,9'-Bianthryl is the latest molecule we have undertaken to study.[13] For this molecule, one expects significant torsional motions about only one bond. An analysis of fluorescence excitation and dispersed fluorescence spectra, in addition to time-resolved studies, may be expected to give information pertinent to the potential surfaces and energy flow involving this C—C single bond. Initial spectral results have revealed low frequency intervals superposed upon anthracene-like features. Time-resolved studies have shown that at higher energies a saturation in fluorescence lifetime occurs. This phenomenon may reflect the influence of IVR processes. The energy at which this saturation occurs is much less than the analogous energy in anthracene.[5] This probably reflects the influence of the low frequency vibrations associated with the anthracene-anthracene bond in the IVR process.[13]

3. Intramolecular Proton Transfer

Photo-induced proton transfer reactions are a class of reactions, the study of which could give detailed information concerning the dynamics of bond breakage and formation. This is especially true for picosecond-jet studies of such reactions since molecular energies can be well-defined and dynamical behavior may be observed directly in the time domain. In this lab we have studied[10] intramolecular excited state proton transfer in methyl salicylate:

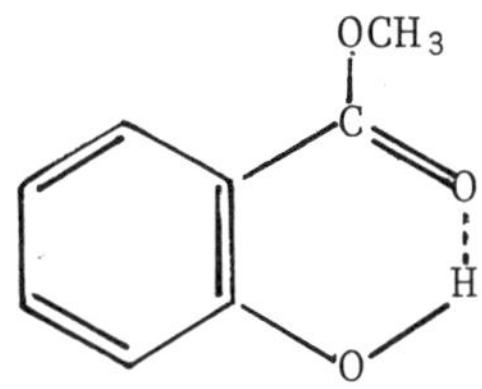

This molecule, upon excitation in the near UV, displays an emission component with a maximum that is shifted ca. 7000 cm^{-1} to the red of the excitation energy. This shifted emission has been attributed[25] to fluorescence from the molecule after it has undergone an intramolecular proton transfer. From our studies in the jet[10] we were able to conclude that the timescale for this process in the isolated molecule is unresolvably short. In fact, spectral evidence combined with lifetime data indicate that an excited state distorted along the proton transfer coordinate is created simultaneously with the absorption of a photon in the isolated molecule. Interestingly enough, methyl salicylate was also found to exhibit threshold behavior in its fluorescence decay rate vs. excitation energy curve. This threshold occurs at 1200-1300 cm^{-1} excess vibrational energy. It may be that such behavior is the norm for molecules which are not rigidly bound in one geometrical configuration.

4. Stepwise Solvation and Photodissociation of Complexes in the Jet

Last year we extended the application of the picosecond-jet technique to the study of the dynamics of molecules in various stages of solvation with various solvents.[12] The idea was to study this "controlled" solvation and its effect on decay processes in the molecule. In addition, we wanted to examine processes involving the photodissociation of solvent-solute complexes. The molecule ("solute") that we studied was isoquinoline (IQ). This particular choice was made because isoquinoline is a more efficient emitter when it is bound to protic solvents than when it is not. It thus displays some interesting photophysics and is much more detectable in a solvated state than when free of any intermolecular interaction.

From the dependence of excitation spectra on solvent concentration, several different solvated species have been identified in helium expansions involving isoquinoline with water, methanol, or acetone. As expected these solvated species have much longer fluorescence lifetimes than their bare molecule counterpart. Dispersed fluorescence spectra reveal the presence of low frequency intervals (≤ 170 cm^{-1}) that can be attributed to vibrational modes arising from complexation. Fluorescence decay rate vs. energy results (Fig. 6) for species assigned as IQ-methanol and IQ-water

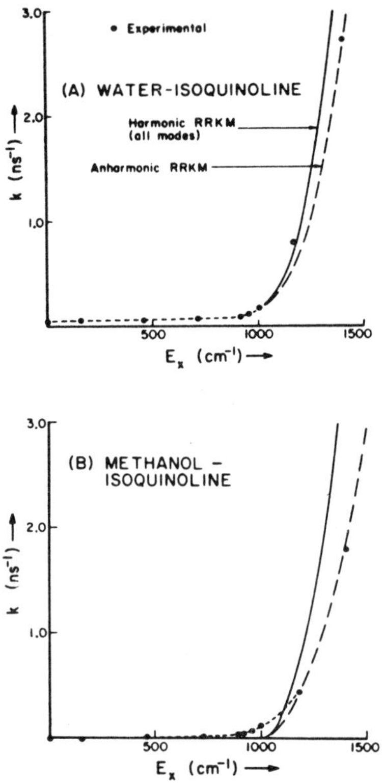

Figure 6. (A) The measured fluorescence decay rate (points) of jet-cooled IQ-H_2O as a function of vibrational energy in the excited state (see Ref. 12). The lines represent results obtained by assuming that solvent-solute dissociation determines the fluorescence decay rate and by using RRKM theory to calculate these dissociation rates (for details see Ref. 24). (B) A plot analogous to (A) for jet-cooled IQ-methanol.

one-to-one complexes display threshold behavior which has been interpreted in terms of hydrogen bond breakage in the excited complexes. The value for the threshold (1000 cm^{-1} or ca. 3 Kcal/mol) is consistent with measured values of hydrogen bond energies.

The breakage of the hydrogen bond in the solvated species that we have studied probes two dynamical processes; the photodissociation process, and vibrational energy redistribution from IQ-localized motions to the dissociation coordinate. The existence of low frequency vibrations in the IQ-solvent complexes might be expected to give rise to rapid and extensive vibrational energy redistribution for energies at and beyond the 1000 cm^{-1} threshold. Consistent with this expectation, results calculated using RRKM theory fit experimental results quite well. Currently, we are examining the effects of cluster size and solvent identity on the photodissociation of IQ-(solvent) clusters.

V. CONCLUSION

In this article we have provided a description of the "picosecond-jet" technique, in which time-resolved spectroscopy with picosecond resolution is applied to molecules cooled via free jet expansion. Work done in this laboratory using the technique was reviewed. Included in the review were studies involving quantum beats and/or photochemical processes in large molecules or molecular complexes. All of these studies focus on the characterization of the dynamics of energy flow and photochemistry in such species. Quantum beats can be used to probe restricted energy redistribution processes. At the other extreme, many of the photochemical processes, which we have studied and have found to exhibit threshold effects, seem to be best characterized as processes which involve rapid energy flow. Although much work is ahead, the results reviewed here indicate that picosecond studies in jets can be very fruitful in probing rates and redistribution processes in molecular systems.

ACKNOWLEDGMENTS

The major part of this article was written when one of us (AHZ) was a guest at Ecole Normale Supérieure; I greatly appreciate the hospitality and efforts of Professor Jean-Claude Lehmann. Also, many thanks are due to Professor Jean Brossel and Dr. Jack Vigué for their help during this time. This work was supported by grants from the National Science Foundation (CHE8211356 and DMR8105034).

REFERENCES

a) IBM Research Fellow
b) Camille & Henry Dreyfus Foundation Teacher-Scholar
c) Contribution No. 6952
1. See, e.g., Zewail, A. H., 1980, Physics Today, 33, p. 27.

2. Lambert, Wm. R, Felker, P. M. and Zewail, A. H. 1981, J. Chem. Phys. 75, p. 5958.
3. Zewail, A. H. 1983, Disc. Farad. Soc. 75, (19).
4. Felker, P. M. and Zewail, A. H. 1983, Chem. Phys. Lett 102, p. 113.
5. Lambert, W. R., Felker, P. M., Syage, J. A. and Zewail, A. H., J. Chem. Phys., submitted; Lambert, W. R., Felker, P. M. and Zewail, A. H., J. Chem. Phys., submitted; Lambert, W. R., Felker, P. M. and Zewail, A. H., J. Chem. Phys., submitted.
6. Syage, J. A., Lambert, Wm. R., Felker, P. M., Zewail, A. H. and Hochstrasser, R. M. 1982, Chem. Phys. Lett. 88, p. 266.
7. Felker, P. M., Lambert, Wm. R. and Zewail, A. H. 1982, Chem. Phys. Lett. 89, p. 309.
8. Shepanski, J., Keelan, B. and Zewail, A. H. 1983, Chem. Phys. Lett. in press.
9. Syage, J. A., Al Adel, F. and Zewail, A. H. 1983, Chem. Phys. Lett. in press.
10. Felker, P. M., Lambert, Wm. R. and Zewail, A. H. 1982, J. Chem. Phys. 77, p. 1603.
11. Felker, P. M., Syage, J. A., Lambert, Wm. R. and Zewail, A. H. 1982, Chem. Phys. Lett. 92, p. 1; Syage, J. A., Felker, P. M. and Zewail, A. H., J. Chem. Phys., to be submitted.
12. Felker, P. M. and Zewail, A. H. 1983, Chem. Phys. Lett. 94, p. 448; ibid. p. 454; Felker, P. M. and Zewail, A. H. 1983, J. Chem. Phys. 78, p. 5266.
13. Khundkar, L. R., Felker, P. M. and Zewail, A. H., work in progress.
14. Millar, D. M. and Zewail, A. H. 1982, Chem. Phys. 72, p. 381.
15. Perry, J. W., Scherer, N. F. and Zewail, A. H. 1983, Chem. Phys. Lett. in press.
16. Zewail, A. H. 1980, Acc. Chem. Res. 13, p. 360; Burns, M. J., Liu, W.-K. and Zewail, A. H. in *Spectroscopy and Excitation Dynamics of Condensed Molecular Systems*, eds. Agranovich, V. M. and Hochstrasser, R. M., Modern Problems in Condensed Matter Sciences, Vol. 4 (North-Holland Publishing Company, Amsterdam-New York-Oxford, 1983) p. 301-435.
17. Haroche, S. in *High Resolution Laser Spectroscopy*, ed. Shimoda, K. (Springer, Berlin, 1976) pp. 254-313.
18. (a) Chaiken, J., Gurnick, M. and McDonald, J. D. 1981, J. Chem. Phys. 74, p. 106; (b) van der Meer, B. J., Jonkman, H. Th., ter Horst, G. M. and Kommandeur, J. 1982, J. Chem. Phys. 76, p. 2099; (c) Okajima, S., Sargusa, H. and Lim, E. C. 1982, J. Chem. Phys. 76, p. 2096.
19. van der Meer, B. J., Jonkman, H. Th., Kommandeur, J., Meerts, W. L. and Majewski, W. A. 1982, Chem. Phys. Lett. 92, p.565.
20. See, for example, Crawford, M., Wang, Y. and Eisenthal, K. B. 1981, Chem. Phys. Lett. 79, p. 529; Okada, T., Fujita, T., Kubota, M., Masaki, S., Mataga, N., Ide, R., Sakata, Y. and Misumi, S. 1972, ibid. 14, p. 563.
21. For a review see Hochstrasser, R. M. 1980, Pure Appl. Chem. 52, p. 2683.

22. Amirav, A. and Jortner, J. 1983, Chem. Phys. Lett. 95, p. 295; Zwier, T. S., Carrasquillo and Levy, D. H., 1983, J. Chem. Phys. 78, p. 5493.
23. Warshel, A. 1975, J. Chem. Phys. 62, p. 214.
24. Khundkar, L. R., Marcus, R. A. and Zewail, A. H. 1983, J. Phys. Chem. 87, p. 2473.
25. Weller, A. 1956, Z. Electrochem. 60, p. 1144; Sandros, K. 1976, Acta Chem. Scand. Ser. A30, p. 761; Klöpffer, W. and Naundorf, G. 1974, J. Lumin. 8, p. 457.

PHOTODISSOCIATION OF S-TETRAZINE AND VIBRATIONAL PREDISSOCIATION PROCESSES OF ITS ARGON VAN DER WAALS COMPLEXES.

Jan Langelaar, Dick Bebelaar and Rudolf P.H. Rettschnick.
Laboratory for Physical Chemistry,
University of Amsterdam,
The Netherlands.

In s-tetrazine (T) vapour short fluorescence decay times (1.5ns-100ps) and high photochemical quantum yields ($\phi_p \approx 1.30$) have been found. The decaytime measurements were carried out with the single photon couting technique, which is a powerful method to perform subnanosecond rise- and decaytime measurements at extremely low light levels. The van der Waals complex T.Ar, prepared in a supersonic jet, dissociates when excited to single vibronic levels of the $S_1(^1B_{3u})$state of tetrazine. Time-resolved information about the vibrational state distribution of the dissociation fragment T yields predissociation rates up to $3.3 \times 10^9 s^{-1}$, depending on the internal energy content of the complex.

1. INTRODUCTION

The photophysics and photochemistry of s-tetrazine has challenged many researchers in the past decades to study this molecule in solution, in matrices and in the gas phase. Since the first spectroscopic observation by Curtius et.al.[1], who indicated the absorption spectrum as a "new kind of molecular spectrum" due to the sharp line spectrum observed, only in the last years a good understanding of the photophysical processes of the molecule has been obtained in great detail. Mason [2] already indicated that the line structure in the visible range (460-560nm) was due to vibronic transitions in the first

K. B. Eisenthal (ed.), Applications of Picosecond Spectroscopy to Chemistry, 293–313.

$n \rightarrow \pi^*$ singlet transition, while the structureless band around 250 nm was assigned to a singlet $\pi \rightarrow \pi^*$ transition (see fig. 1).
Theoretical values for the transition energies were obtained from SCF-CI calculations by Ha [3]. A review of the spectroscopic observations is given in this paper [3], where the experimental results are compared with calculated data.
With the pioneering work of Innes and co-workers [4] a reasonably sophisticated knowledge is now available concerning ground and excited state vibrations and molecular geometry. Some of the eighteen normal vibrations of s-tetrazine and the geometry in ground and excited state are represented in fig. 2.

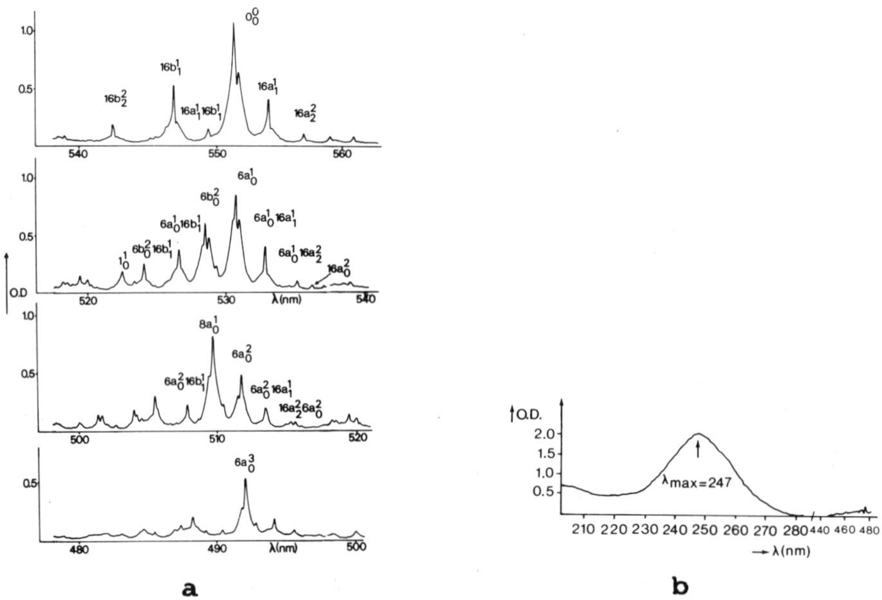

Figure 1. Absorption spectrum of s-tetrazine vapour (20°C). Optical Pathlength: 5 cm. a) $n \rightarrow \pi^*$ transition, b) $\pi \rightarrow \pi^*$ transition.

Extensive single vibronic level gas phase fluorescence studies by Vemulapalli [5], Levy and co-workers [6] and our group [7] completed the photophysical knowledge of s-tetrazine in the low-energy excited states. With these spectroscopic data now a reference book of single vibronic level band assignments is available, an essential tool in studies of energy conversion processes.

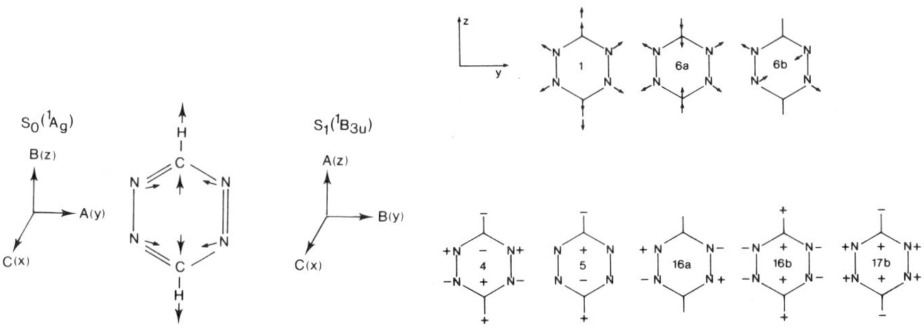

Figure 2. Geometry of s-tetrazine in ground and excited state and some of the normal modes.

A sophisticated understanding of the dynamics of the s-tetrazine molecule, however, is still developing. Fluorescence lifetimes were for the first time estimated by Meyling et al. [8]. From spectroscopic data they assumed an excited singlet state lifetime of about 1.5 ns. Hochstrasser and co-workers [9] measured a decay time of 400 ps for s-tetrazine dissolved in benzene after excitation with a Nd^{3+}-glass laser at 530 nm. In our group [10, 11], we obtained gas phase SVL lifetimes for the excited vibronic levels with an excess energy up to ≈1200 cm^{-1} varying from 1.5 ns down to ≈50 ps depending on the vibrational mode excited (fig. 3).

It has been indicated that the short decaytimes of s-tetrazine should be related to the photochemical dissociation of the molecule into HCN and N_2. The correct pathway of the photochemical decomposition and the dynamical properties, however, are as yet not understood.

Paczkowski et al. [12] recently measured the quantum yields for fluorescence and photodissociation for a number of vibrational excess energies in the first excited singlet state of dimethyl-s-tetrazine vapour (DMT). They concluded that the photochemical decomposition of DMT does not compete directly with the fluorescence process, and therefore had to assume that the photodissociation involves a bottleneck in the non-radiative singlet decay. In the third section of this paper we will present the results of a detailed study concerning photochemical and fluorescence quantum yields of s-tetrazine in the gas phase as a function of the vapour pressure.

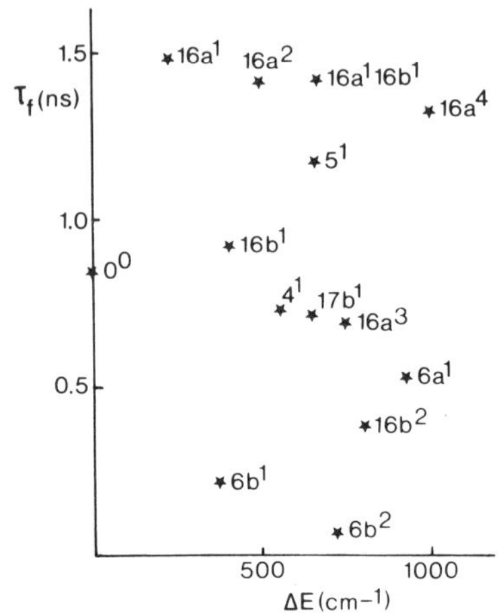

Figure 3. SVL decaytimes of s-tetrazine for the lowest vibronic states (<1200 cm^{-1}).

In a photofragmentation process one ideally wants to study for each molecular state the fragment yield as a function of all the internal quantum numbers and the pathways along which the initial internal energy flows from the initially prepared state to the dissociative bond and into the photofragments.
An interesting technique that is capable of providing some of the desired detailed information, is the dynamical study of predissociation processes of the s-tetrazine-argon van der Waals complex in a supersonic jet. The results will be presented in the last section of this paper. First, however, we will describe the experimental technique used to measure fast fluorescence decay times at extremely low light levels.

2. EXPERIMENTAL TECHNIQUE.

Since the development of picosecond and subpicosecond laser pulse techniques, a δ-pulse excitation source is available for most fluorescence studies to prepare a molecule in its excited state, in order to study the emission as a function of time.

The measurement of the temporal intensity profile I(t) with picosecond time resolution and high sensitivity is the key to the study of dynamical processes in matter. The difficulty in obtaining high time resolution is often due to the low light level of the signal to be detected. High intensity ps laser pulses can be measured with measuring equipment having a short response time like silicon photodiodes and streak cameras (see table I, first column).
The measurement of a weak fluorescence decay has to make use of a fast photodetector with high internal amplification and can be performed
(i) by measuring the waveform of the detector current pulse by sampling or averaging techniques
(ii) by measuring the time position of the single photon response of the detector with respect to the excitation pulse and then accumulate many of such events in a histogram.

Detector	Response time with method a) (FWHM)	Response time with SPC method b) (FWHM in ps)	Sensitive detector area (diameter in mm)	detection limit [1]
Si-photodiode	50 ps	-	0.050	100
Avalanche photodiode	≥ 20 ns	90	0.035	10^{-4}
Photomultipliers				
model XP 2020	2.5 ns	220	5	3.10^{-7}
Cross-field	150 ps	230[2]	5	10^{-6}
Micro-channelplate	900 ps	70	18	10^{-6}
Streak camera	(2 ps)	-	0.025x5	10
Synchroscan	10 ps	-	0.025x5	10^{-1}

[1] Amount of emission photons per excitation pulse with a laser repetition frequency of 88 MHz
[2] Risetime (10-90%) ≤ 80 ps, decaytime limited by electronics.

Table 1. Detector response time and detection limits.

The resolving time for averaging techniques is dependent on the delay spread in the arrival time of the electrons at the output of a fast detector and on the variation of this transit time from pulse to pulse. Hence

$$\Delta t_{aver.} \propto \left[\sqrt{t_\delta^2 + t_s^2}\right]^{1/2}$$

where t_δ=responsetime of the detector amplifier
t_s=transit time spread

For time correlated single photon counting only the transit time spread contributes i.e.

$$\Delta t_{SPC} \simeq t_s$$

So, the time resolution of the single photon counting method in combination with a fast photomultiplier is superior to that of averaging and sampling methods.

As may be seen from table 1, streak cameras provide the shortest resolving time of all electronic methods. This camera, however, has the following drawbacks

- the resolving time is proportional to the time range scan,
- the linearity is poor,
- the sensitive area is small ($\simeq$ 0,12 mm^2, see table 1).
- the minimum sensitivity lies around 1 photon per event, when using a synchronously pumped dye laser.

Hence for weak signals the SPC method has to be preferred. For this technique the application of a synchronously pumped cw dye laser system as a light source is the best choice. A description of the ps laser system used in our equipment has been given elsewhere (10). The only difference is, that with the system nowadays the linewidth of the tunable ps pulses can be chosen between 30 cm^{-1} ($\simeq$ 1 ps) and 0.03 cm^{-1} ($\simeq$800ps) depending on the spectroscopic requirements.

2.1. Single Photon Counting Photochronoscopy

Fluorescence signals have to be observed via a high resolution monochromator (preferably with two exit slits) if high spectral resolution and time resolution has to be combined. This system then also enables one to perform absolute calibration of the dye laser wavelength and linewidth before and after each decay experiment. Most of the time resolved experiments in this research were carried out with a Philips XP 2020 photomultiplier and a single photon counting system (SPC). The time resolution of this combination allows to measure decay times down to 150 ps without the need of deconvolution techniques. The usefulness of the SPC is due to its linearity, high dynamic range and good signal to noise ratio in combination with an extremely high sensitivity. It allows decay time measurements of signals as low as 25 counts per second; the pulse repetition rate of the laser being 85 MHz. These are realistic signal levels encountered in our study on single level decay time measurements of s-tetrazine vapour. A block diagram of the excitation and the detection system is shown in fig. 4. The single photon counting electronics consist of standard Ortec modules. The Constant Fraction Discrimitnator has been selected to give lowest jitter (about 20 ps FWHM) and walk (25 ps peak to peak over a 20 dB range). For short decaytimes ≤ 2ns the apparatus is used with the start and stop pulse reversed. In this configuration the start pulse going to the Time to Amplitude Converter (TAC) is delivered by the PMT and the stop pulse is derived from the laser pulse repetition rate. The TAC produces an output pulse with a height proportional to the time interval between a start and stop pulse. A Multi Channel Analyser (MCA) accumulates the pulses from the TAC. The display of the MCA will be a histogram of the detected photons in time which is equivalent to the fluorescence decay curve.

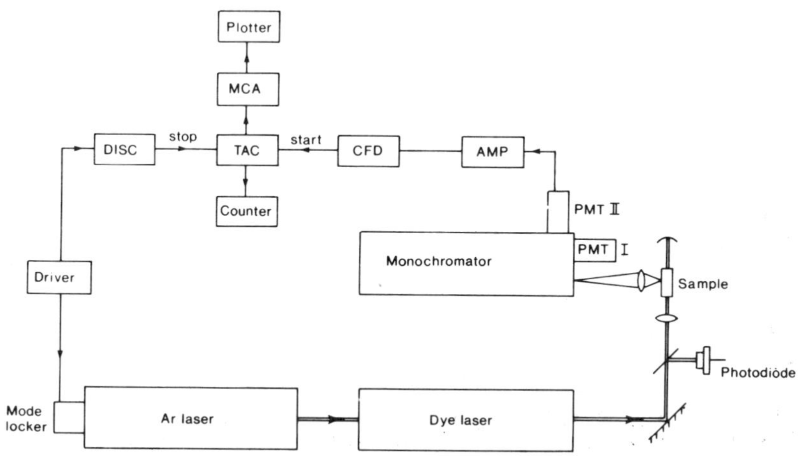

Fig. 4. Schematic representation of the experimental set-up. PMT; photomultiplier (see table 1); CFD; constant fraction discriminator; TAC, time to amplitude converter; MCA, multichannel analyser.

The maximum time interval during which the fluorescence decay can be viewed is the period between the laser pulses and is in our case 10 ns. In order to obtain an accurate interpretation of the decay curve over at least three decades, we accumulate in the MCA to 10^4-counts at the top of the decay curve. For SVL studies on s-tetrazine a measurement normally takes

less than 1 minute counting time; for weak emission studies with a signal level of 25 cps the operating time increases to 1 or 2 hours.
For decay times > 2 ns a low voltage electro-optic modulator is employed to decrease the repetition rate of the dye laser. This enables one to record fluorescence decays with a time window larger than 10ns. The modulator power supply and the start pulses to the TAC are synchronized by dividing the modelocker drive frequency of the ion laser. The PMT-signal then delivers the stop pulses. The start pulses start the time-to-amplitude converter (TAC). If a stop pulse arrives within a preset time after the start, the TAC supplies an output pulse with a height proportional to the time interval between start and stop. The pulse serves as input to the multichannel pulseheight analyser and in this way a photon counting histogram of accumulated counts (ordinate) versus time (abscissa) is obtained. Conversion to absolute time is done by calibration using a pulsegenerator which gives correlated start and stop pulses with a known time difference (Ortec Model 462 Time Calibrator).

2.2. The Photomultiplier Tubes (PMT).

The response of the apparatus to a delta pulse has been measured as a function of wavelength for several photomulitplier tubes. The results are shown in fig.5 for a Philips 56 DUVP and a XP 2020 tube. A half-width of 620 is observed with the 56 DUVP, a widely used tube in single photon counting systems in the past. The response, however, changes considerably by going to shorter wavelengths. The best results have been obtained so far with the XP 2020 and are only slightly dependent on wavelength (curve b, fig. 5).
The contribution of the first dynode to the response can be decreased by changing the focussing voltage and selecting the area on the photocathode from where the fastest area on the first dynode is reached by the photo-electrons. This gives an overall instrumental response of 230 ps (FWHM) at 580 nm. The results are reviewed in table 2. By using a special model Cross Field Photomultiplier we expected a further improvement of the instrumental resolution to 20-50 ps FWHM. Due to one of the electronic components an overall response function of only 230 ps is observed. The leading edge of the response function, however, shows a fast risetime of 80 ps. (10 to 90%). The risetime experiments described in section 4 have therefore been performed with the Cross-Field Photomultiplier.

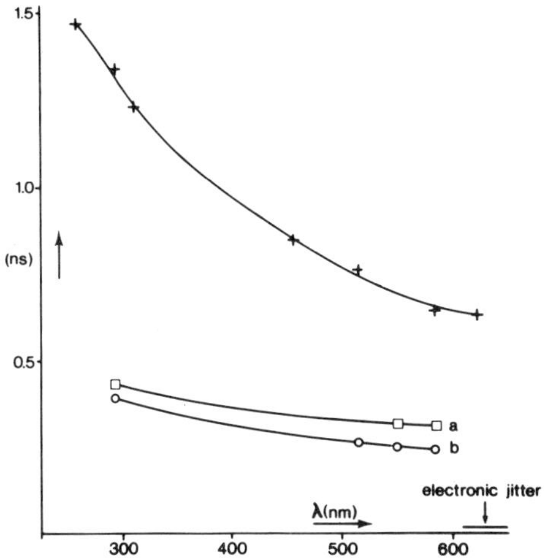

Figure 5. Wavelength dependence of the photomultiplier response (FWHM) to a delta pulse: +, Philips 56 DUVP; □, Philips XP 2020(a); O. Philips XP 2020, after selecting the fastest area of the photocathode (b). The total electronic jitter is 15 ps.

		time resolution FWHM (ps)			
Type	photo-cathode	field-strength at cathode (V/cm)	measured	estimated contribution photo-cathode	estimated contribution first dynode
56 DUVP	bialkali (K_2CsSb)	40	620	540	290
XP 2020	"	180	a[2] 320	120	290
			b[2] 220	120	190
X-field PM	GaAs	2000	230[1]	20-50	

[1] *preliminary; probably limited by CFD electronics.* [2] *see fig. 5.*

Table 2. Photomultiplier response.

The response function of the XP 2020 PMT to a delta pulse at λ=585 nm is given in fig. 6.
Note that the ordinate is a logarithmic scale. The "afterpulses" are at least 3 orders of magnitude weaker than the main signal. The response function shows a signal with a half-width of 235 ps. and a 1/e-decay of 98ps. The afterpulses around t ≃ 14 and 22 ns, are due to processes in the photomultiplier; the weak pulses around 20 and 30 ns are the next pulses from the dye laser indicating an extinction ratio of the modulator of at least 10^4.
The PMT response for the SPC system when the start and stop pulses are reversed is given in figure 7 for λ=546nm. This response function is used to analyse the fluorescence decay time of Erythrosin B at this wavelength. The measurements are described to demonstrate the time resolution of the equipment using the XP 2020 PMT as a detector.

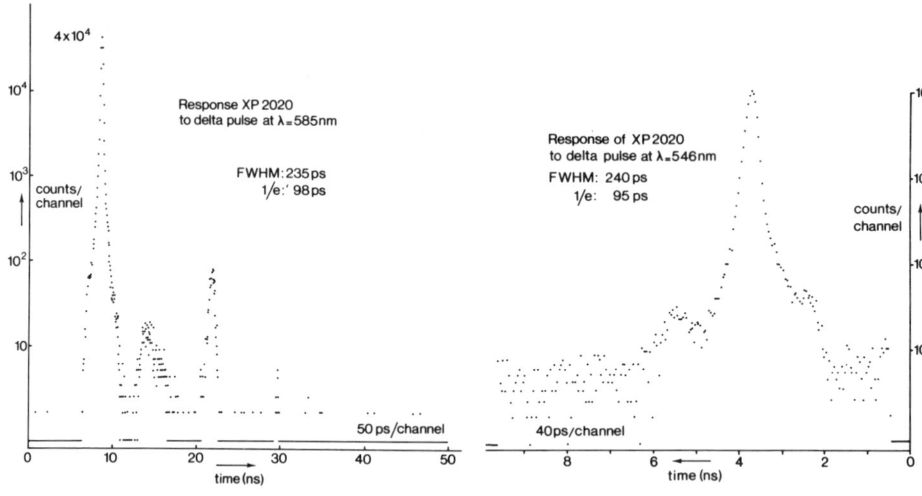

Figure 6. Response function of the XP 2020 photomultiplier to a δ-pulse of λ=585nm (left hand figure).
Figure 7. Response function of XP 2020 to δ-pulse at λ=546 nm. Start and stop pulses are reversed. FWHM=360ps; 1/e decay=95ps.

2.3. Fluorescence decaytime of Erythrosin B.

Erythrosin B is a iodine substituted fluorescein dye It is known that its fluorescence yield is drastically reduced by going from acetone (0.5) to water (0.02) as a solvent. In acetone we observed a decaytime of 2.28 ns, which indicates a quantum yield of approximately 0.57 using a radiative lifetime of 4×10^{-9}s.
The results obtained for Erythrosin B in methanol and water are given in fig. 8. In methanol the decaytime τ=460ps, can directly be obtained from the logarithmic plot as given in fig. 8a. Erythrosin B solved in water, however, gives a fluorescence intensity profile which slightly differs from the response function of the detector as is presented in fig. 7.
The 1/e-decay indicates a decaytime of τ<125ps. A full convolution of the fluorescence intensity profile versus time of Erythrosin B in water with the response function of the equipment gives a decaytime of τ=80±5ps. The convoluted profile, with a decaytime of 80ps, of the experimental intensity profile of Erythrosin B in water as obtained in fig. 8, is presented in fig. 9. The result is in excellent agreement with the observed decaytime (τ=78ps) measured with a streakcamera (13).

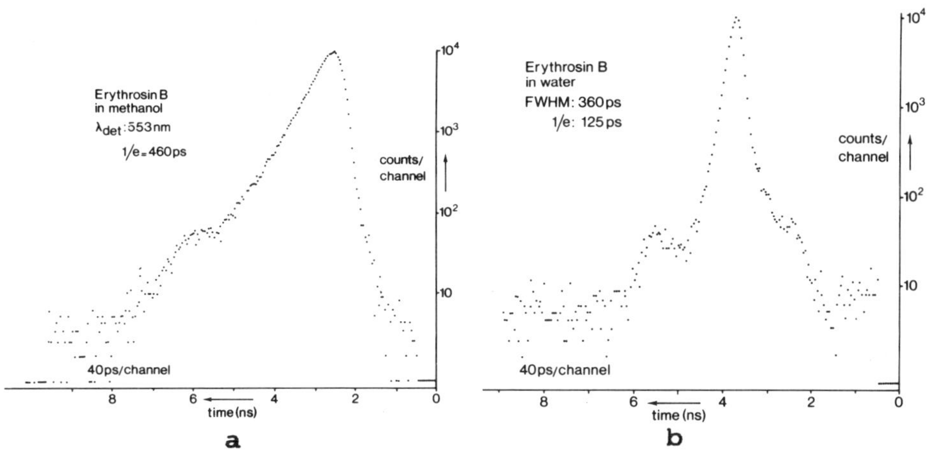

Figure 8. Fluorescence decay curves of Erythrosin B in methanol (a) and water (b) measured by the single photon counting technique; τ_f(methanol)≈460ps; τ_f(water)<125ps.

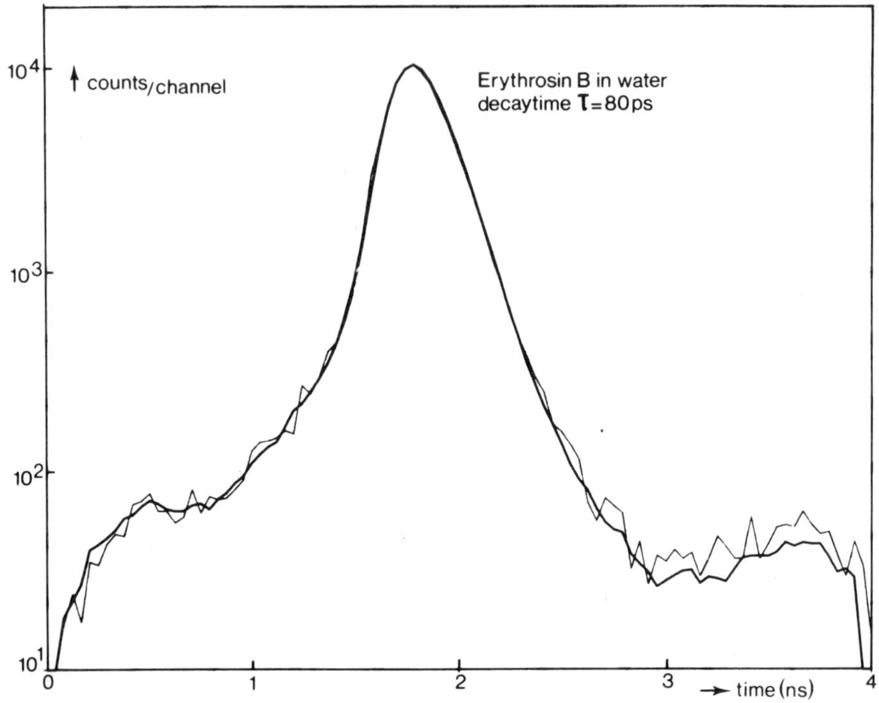

Figure 9. Fatline: fluorescence temporal intensity profile of Erythrosin B in water convoluted with a decaytime of 80ps and the detector response function given in fig. 7. Thin line: experimental curve of fig.8 given for comparison (the abcis is shifted ≈ 2ns compared to fig. 7 and 8).

3. PHOTOCHEMICAL PROCESSES

The photochemical decomposition can be represented by the overall reaction $C_2H_2N_4 \rightarrow 2\ HCN + N_2$. The quantumyield of this process lies around 1, in good agreement with the low value for the fluorescence quantum yield ϕ_f. We already mentioned in the introduction a strong variation of the decaytimes of the vibronic levels. The question then arises whether the strong variations of the dynamics of the processes also indicates a strong variation of the photochemical yield. As we will see further this is not the case, so that we have to conclude that the photochemistry does not take place from the optically prepared level. Furthermore, only two other dark processes are capable to reduce the level decay time under the experimental isolated conditions: internal conversion to a hot groundstate or intersystem crossing to the triplet system. Any significant phosphorescence has not been observed for s-tetrazine vapour. On the other hand a transient absorption is observed [14], which has been attributed to hot ground state molecules. This gives rise to the conclusion that the iso-energetic transition from the optically prepared level in the excited electronic state to a hot ground state level determines the lifetime of the prepared vibronic level.

3.1. The Photochemical Quantumyields.

Absolute photochemical quantumyields (ϕ_p) were obtained for the levels 0^0, $16a^1$, $16b^1$, $6a^1$, $6b^2$ and $6a^1 16a^1$ in the first excited electronic state of s-tetrazine. For these levels we also measured the decaytimes and fluorescence quantumyields (ϕ_f).
The photochemical quantumyield was obtained by exploring in a vapour sample cell, laser photolysis and absorption measurements at the same time. The photolysis was carried out with a cw dye laser ($\Delta\lambda$ = 0.025 nm) by tuning into the top of the absorption band of the vibronic transition under consideration. The optical density of the UV transition around 245 nm (see fig. 1) was measured with a standard spectrometer during the photolysis process.
From the thus obtained data the photochemical quantumyield could be calculated for the levels mentioned above. The mean values of ϕ_p were obtained from two independent runs each consisting of eight measurements at different photolysis periods. The results are given in table 3 together with the vibrational energies, fluorescence quantum yields and decaytimes. The most striking feature is that ϕ_p values are obtained larger than 1, indicating that as a result of the absorption of one photon more than one molecule can dissociate. Therefore one has to assume that bimolecular processes are involved, but due to the extreme experimental conditions (vapour pressure ≃0.7 torr, excited state lifetime ≃1 ns); collisions are only probable with molecules in the electronic ground state. Moreover, introduction of 100 torr argon into the sample cell reduces ϕ_p drastically, but does not influence the excited state decaytime of the prepared level. Reduction of the vapour pressure of s-tetrazine hardly changes the lifetime but it gives rise to a drastic decrease of ϕ_p (fig. 10).

Vibronic level	0^0	$16a^1$	$16b^1$	$6a^1$	$6b^2$	$6a^1 16a^1$
ϕ_p (± 0.05)	1.30	1.32	1.28	1.15	1.20	1.19
τ_f (ns)	0.83	1.45	0.92	0.52	<0.10	0.88
ΔE (cm^{-1})	0	255	403	703	784	958

Table 3. Photochemical quantumyield (ϕ_p) and decaytime (τ_f) for several vibronic levels of s-tetrazine at 1 mbar (≈20°C).

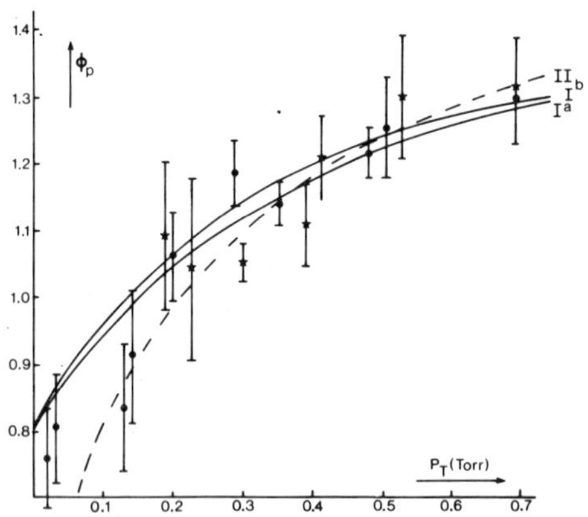

Figure 10. The photochemical quantumyield (ϕ_p) of s-tetrazine as a function of the vapour pressure Δ=16a¹ level; 0=0⁰ level.

A reduction of the vapour pressure from 0,7 torr to 0,07 torr decreases the photochemical yield from 1.3 to 0.8.

From these results we tentatively conclude:

a. At the experimental conditions (p < 0.7 torr) s-tetrazine molecules are isolated during their short stay in the excited electronic state.

b. Internal conversion of the excited level to an electronic ground state level with the same energy content most probably determines the observed fluorescence decaytimes.

c. Photodissociation does not take place from the optically prepared level, but from hot ground state levels.

d. The hot ground state molecules formed are living long enough to encounter a second s-tetrazine molecule and then both molecules dissociate into the photoproducts HCN and N_2.
The accuracy of the photochemical quantum yields at low pressure unfortunately is not sufficient to obtain an accurate value at zero pressure (fig. 10) by extrapolation. A good fit for the different reaction rates (model I in fig. 10) however is obtained proposing the following scheme with an extra bimolecular reactionpath:

$$M_1^{*} \longrightarrow M_0^{**} \longrightarrow 2HCN + N_2$$

$$M_0^{**} \xrightarrow{+\, M_0} [M_0M_0]^{**} \longrightarrow 4HCN + 2N_2$$

The dotted curve, (model II, fig. 10) is obtained when a reaction path is assumed with two collision processes. At this stage the information is not sufficient to make a clear choice between the models in order to decide along which channels the photochemical decomposition takes place.
A better defined photodecomposition process is the predissociation of the s-tetrazine-argon (T.Ar) van der Waals complex. The study of the redistribution of the internal energy as well as the dynamics of the photofragmentation of this system can be performed in great detail for different reaction channels. This will be described in the next chapter.

4. SPECTROSCOPY AND PREDISSOCIATION OF T.Ar.

In a supersonic expansion of argon seeded with s-tetrazine van der Waals complexes are formed during the cooling process. As a result of the low collision energy in the area behind the nozzle the weakly bound complexes are stable. Laserspectroscopic techniques were used for high resolution studies of these complexes. The described experiments were performed in a supersonic jet with a nozzle diameter of 50 or 100μm with an argon stagnation pressure of 1.3 bar. The argon was seeded with 0.03 % s-tetrazine. The ps dyelaser is focused into the jet on a diameter of 50 μm. From this area the emission is collected and focused on the slit of the high resolution monochromator with a dispersion of 0,24 nm/mm.

From the molecular spectra of s-tetrazine in an area between 0,05 and 5 mm behind the nozzle we learn that vibrational relaxation processes do not occur. Under the same conditions however in the T.Ar van der Waals complexes vibrational redistribution and predissociation processes do take place.
The complex transitions show a redshift with respect to the corresponding molecular transition which indicates, that the intermolecular potential well is somewhat deeper in the excited electronic state than in the ground-state. It turns out that the redshift is not equal for all vibronic levels but varies between 7 and 24 cm^{-1}[7]. As a result of the spectral shifts selective excitation of the complex and the free tetrazine molecule can be performed in the jet.
We have measured the fluorescence spectra of T.Ar for different vibronic levels and compared with the molecular spectra under the same conditions.
In fig. 11a part of the emission spectrum of the $\overline{6b^2}$ complex is presented (the complex transitions are indicated by a bar above the notation). If we compare this spectrum with the corresponding emission spectrum of the molecule (fig. 11b) we observe, besides the redshift some additional emissionbands. From our knowledge of the resonant emission spectra we can attribute these extra bands (1 - 5, fig. 11a) resp. to emission from the levels $16b^1$, $\overline{16a^1 16b^1}$, $16a^1$, $\overline{16a^2}$ and $\overline{5^1}$. The complex emission bands indicate the occurence of redistribution of vibrational energy. The two extra molecular fragment transitions ($16a^1$, $16b^1$) must be a result of a predissociation process giving rise to rupture of the van der Waals bond so that the argon atom is ejected
In the next section we will try to evaluate this process on the basis of time resolved experiments.

4.1. Dynamics of the Predissociation Process.

From the spectra of the T.Ar van der Waals complex we have learned that vibrational redistribution and predissociation processes are taking place after excitation of (for instance) the $\overline{6b_0^2}$ transition. Time-resolved experiments might answer the question whether these processes occur parallel or consecutive. For the $6b^2$ complex a decaytime $\tau \leq 100$ ps is measured, for the $6a^1$ level 520 ps and for the 0^0 level 830 ps (see fig. 3 and table 3 for the decaytimes of the molecular levels). These decaytimes are (within experimental error) identical with the decaytimes of the corresponding molecular levels. In fig. 12a we see

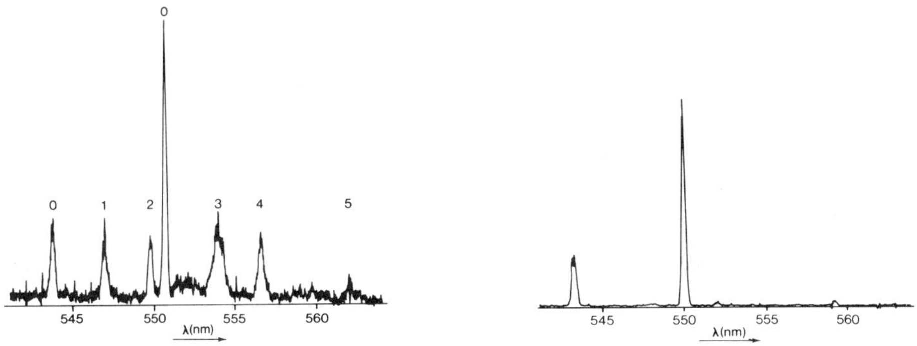

a b

Fig. 11a. Part of the fluorescence spectrum after excitation of the $6b_0^2$ transition of the T.Ar complex (18.888 cm^{-1}, $\Delta\sigma$=1cm^{-1}). The bands are emissions from the levels: $\overline{6b^2}(0)$; $16b^1(1)$; $\overline{16a^1 16b^1}(2)$; $16a^1(3)$; $\overline{16a^2}(4)$; $\overline{5^1}(5)$.
b. Part of the fluorescence spectrum of s-tetrazine after excitation of the $6b_0^2$ transition (18.912 cm^{-1}, $\Delta\sigma$=1cm^{-1}). Both spectra are measured in a supersonic jet at 0.03 mm from the nozzle (50 μ). $\Delta\sigma$ = 10 cm^{-1}, p_0 = 1,2 bar.

that after optical excitation of the $6b^2$ level the vibrational energy is redistributed over three different levels of the complex numbered 2, 4 and 5. For each of these emissionbands it is possible to measure the intensity as a function of time. From the rise- and decay of the temporal intensity profile of each band in principle one might decide in which order and from which level these bands are formed. For the $6b^2$ level the following processes have been observed to take plsace in parallel:

$$\overline{6b^2} \rightarrow \overline{5^1} \quad (1)$$

$$\overline{6b^2} \rightarrow \overline{16a^1 16b^1} \rightarrow 16b^1 + Ar \quad (2)$$

$$\overline{6b^2} \rightarrow \overline{16a^2} \rightarrow 16a^1 + Ar \quad (3)$$

We will follow process (3) in more detail. From the temporal intensity profiles of each of the transitions of this process we could decide that the vibrational redistribution process $\overline{6b^2} \rightarrow \overline{16a^2}$ takes place first,

followed by a vibrational predissociation process $16a^2 \rightarrow 16a^1$ + Ar in which the molecular fragment is formed in the $16a^1$ vibrational state.
The alternative route for predissociation i.e. $6b^2 \rightarrow 16a^1$ + Ar turnes out to be closed. The rise and decay curve of the emission of the optically prepared level $6b^2$ is illustrated in fig. 12a.
The profile of the curve is fully determined by the response function of the equipment (electronics and cross-field photomultiplier). The result indicates a decay-time for the $6b^2$ level of $\tau \leq 100$ ps. The temporal intensity profile of the $16a^2$ emission is presented in fig. 12b. If this level is formed directly from the prepared $6b^2$ level than we must expect that the experimental curve (data points) will be equal to the convaluated curve (drawn line) of the temporal profile of the experimental $6b^2$ emission (fig. 12a) and the decaytime of the $16a^2$ level. A best fit is obtained with a decaytime for the $16a^2$ emission of $\tau(16a^2)$=750ps. The considerable reduction of the decaytime of the $16a^2$ state of the complex as compared to the lifetime of the $16a^2$ state of the free s-tetrazine molecule (1450 ps, see table 3) must be due to an additional deactivation channel i.e. the predissociation process. The temporal intensity profile of the $16a^1$ (fig. 12c) emission of the fragment (data points) finally agrees with the convoluted signal (drawn curve) when we assume that it is prepared from the $16a^2$ state. No good fit is obtained when we assume a formation of a $16a^1$ fragment directly from the optically prepared state $6b^2$ (dashed line in fig 12c). Similar results are observed when the T.Ar complex is prepared in the $6a^1$ level [15], where we also reported a surplus of rotational energy in the $16a^1$ vibrational state of the molecular fragment. In the case of a direct excitation of the $16a^2$ complex level we observed that the dissociation fragment can be formed in the 0^0 or $16a^1$ level depending on the amount of vibrational energy (E_W) of the van der Waals stretch mode which is absorbed as a result of the optical excitation process. The amount of energy E_W can be varied by tuning the laser successively to different wavelengths within the absorption band of the T.Ar complex. Considerations whether dissociation channels are open or closed indicated an upper and lower level for the dissociation energy of the complex of respectively 275 cm^{-1} and 380 cm^{-1} [7].

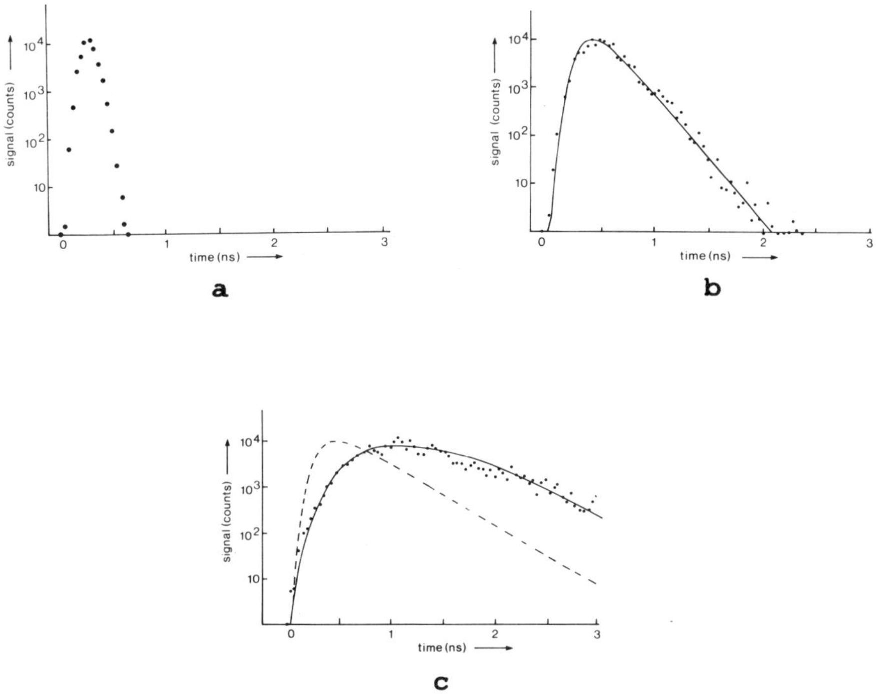

Fig. 12. Data points: Experimental rise and decay curves after $\overline{6b_0}^2$ excitation. Drawn curves: best fit computer simulations using the experimental curve of the foregoing level.
a. $\overline{6b^2}$ resonance emission
b. $\overline{16a^2}$ emission, process: $\overline{6b^2} \rightarrow \overline{16a^2}$, $\tau(16a^2)=750$ ps
c. $16a^1$ emission, process: $\overline{16a^2} \rightarrow 16a^1$, $\tau(16a^1)=1450$ps. Dashed curve: calculated for process $\overline{6b^2} \rightarrow 16a^1$ with $\tau(16a^1)=1450$ps.

In the vibrational predissociation process the 16a vibration turns out to be the active mode. Energy transfer from this molecular vibration to the van der Waals stretch vibration leads to the rupture of the van der Waals bond. The 16a mode is an in plane ring vibration. If we excite an out of plane vibration (6a or 6b) we do not observe a direct predissociation from the prepared mode. First a redistribution has to take place. A scheme for the intermolecular potential of the $\overline{16a^2}$ and $\overline{16a^1}$ complex states is given in fig. 9 where the van der Waals stretch vibration ($42cm^{-1}$) is also indicated. From this diagram it can be seen that depending on the internal energy content of the $\overline{16a^2}$

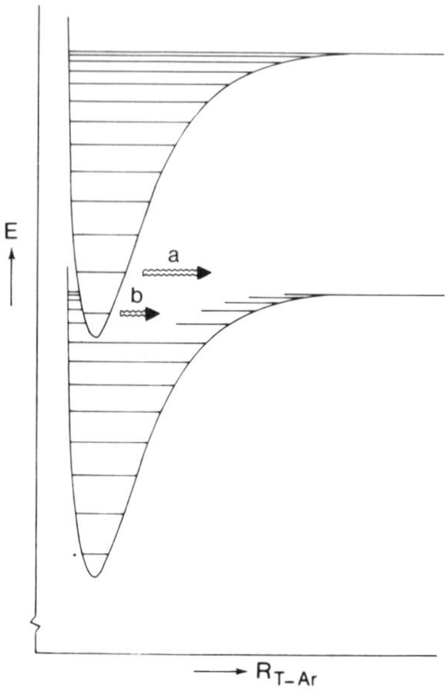

Fig. 13. Schematic diagram of the T.Ar van der Waals intermolecular potential for two vibronic states $16a^2$ and $16a^1$. The levels of the van der Waals stretching mode (τ) are shown as well. Vibrational predissociation of T.Ar ($16a^2\tau^1$) and vibrational relaxation of T.Ar ($16a^2\tau^0$) are indicated by arrows a and b resp.

state the predissociation channel $\overline{16a^2} \rightarrow 16a^1$ + Ar can be open (channel a) or closed (channel b). In case b vibrational redistribution takes place, an inefficient process which can be concluded from the lifetime of this level. Vibrational predissociation (channel a) of the T.Ar complex takes place with a rate $> 10^9 s^{-1}$. The rate turns out to be dependent on the internal energy content, in qualitative agreement with the momentum gap law [16].

ACKNOWLEDGEMENTS

These investigations were supported by the Netherlands Foundation for Chemical Research (SON) and were made possible by financial support from the Netherlands Organization for Pure Research (ZWO).

REFERENCES

1. T. Curtius, A.Darapsky and F. Muller, Ber. Deut. Chem. Ges. 40 ,(1907), 89.
2. S.F. Mason, J. Chem. Soc., (1959) 1240, 1263, 1269.
3. T.K. Ha, Mol. Phys. 29 (1975) 1829.
4. D.V. Brumbaugh and K.K. Innes, Chem. Phys. 59 (1981) 413.
5. G.K. Vemulapalli and T. Cassen, J. Chem. Phys. 56 (1972) 5120.
6. a. D.H. Levy, C.A. Hayman and D.V. Brumbaugh Faraday Discuss. Chem. Soc. 73 (1982) 427
   b. J.E. Kenney, D.V. Brumbaugh and D.H. Levy, J. Chem. Phys. 71 (1979) 4757
7. J.J.F. Ramaekers, H.K. van Dijk, J. Langelaar, R.P.H. Rettschnick, Faraday Discuss. Chem. Soc. 75 (1983) 183.
8. J.H. Meyling, R.P. van der Werf and D.A. Wiersma, Chem. Phys. Lett. 28 (1974) 364
9. R.M. Hochstrasser, D.S. King and A.C. Nelson, Chem. Phys. Lett. 42 (1976) 8
10 J. de Vries, D. Bebelaar and J. Langelaar, Opt. Comm. 18 (1976) 24.
11. J. Langelaar et al., Picosecond Phenomena II ed R.M. Hochstrasser et al. Springer Verlag Berlin (1980) pp. 171.
12. M. Paczkowski et al. Chem.Phys.Lett. 72 (1980) 5
13. W. Yu et al., J. Chem. Phys. 67 (1977) 1766.
14. R.M. Hochstrasser, private communication.
15. J.J.F. Ramaekers et.al., Laser Chem. 2 (1983) 125.

PART V SPECTROSCOPIC TECHNIQUES

SPECTROSCOPY WITH FEMTOSECOND OPTICAL PULSES

C. V. Shank

Bell Laboratories, Holmdel, NJ 07733 USA

Abstract

This paper reviews the recent advances in femtosecond pulse generation and measurement techniques.

Steady progress has continued in the development of ultra-short pulse measurement techniques. (1) For nearly the last 15 years, the shortest reported optical pulse width has continued to drop at nearly an exponential rate (See Figure 1). Only 5 years following generation of the first optical pulses (2) in the picosecond range, the first subpicosecond optical pulses were generated. (3) It was not until 1981 with the advent of the colliding pulse modelocked dye laser (4) that optical pulse widths dipped below 0.1 ps into the femtosecond regime. And most recently pulse compression techniques have further reduced the shortest reported optical pulse width to 30 fs. Further advances appear over the horizon. (5)

The optical configuration of the colliding pulse modelocked laser has been shown to be a significant advance over other passively modelocked dye laser configurations. The laser cavity is in a ring configuration containing two elements: an optically pumped saturable gain medium and a saturable absorber. The amazing simplicity of this laser, in part, accounts for its ability to generate femtosecond optical pulses. When the amount of material in the cavity is minimized the effects of group velocity dispersion are reduced allowing the cavity to sustain the broad bandwidth of oscillating frequencies necessary to form a short pulse. The mechanism (6) for pulse shortening in the colliding pulse laser is similar to that for the passively modelocked dye laser. Optical pulse shaping takes place by

K. B. Eisenthal (ed.), Applications of Picosecond Spectroscopy to Chemistry, 317–324.

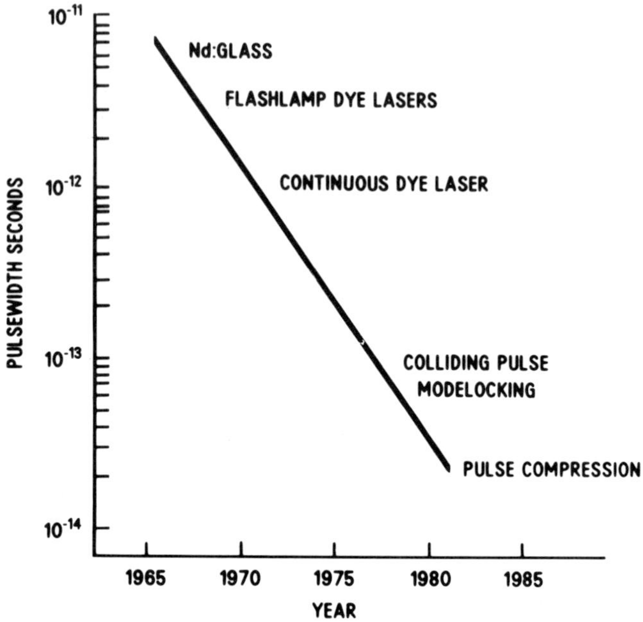

Fig. 1 Plot of optical pulsewidths achieved versus year.

preferential absorption of the leading edge of the pulse by the saturable absorber and preferential amplification on the leading edge of the optical pulse by the gain dye which, combined with the cavity loss, effectively shortens the pulse with each pass around the cavity until the limiting pulse width is achieved. The additional mechanism operative in the colliding pulse configuration is the interaction of the two oppositely directed pulse streams present in the ring laser cavity. These two pulse streams "collide" in a thin 10-μm ethylene glycol stream containing the saturable absorber 3,3'-diethyloxydicarbocyanine iodide. Synchronization of the two counterpropagating pulse streams occurs because the minimum energy is lost in the saturable absorber when the pulses meet in the saturable absorber. The pulses interfere and set up a standing wave pattern in the absorber stream. The standing wave field minimizes the energy lost because the field is most intense where the absorption is saturated whereas the electric field is weakest in the unsaturated field minima. The shortest optical pulses are produced with a thin saturable absorber which confines the standing wave field. With this configuration optical pulses in the range 50 to 60 femtoseconds have been obtained.

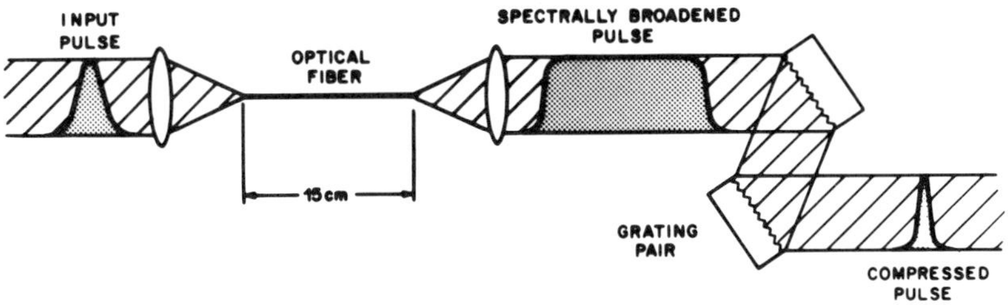

Fig. 2 Schematic of femtosecond optical pulse compression experimental arrangement.

Even shorter optical pulses have been generated by using pulse compression techniques. More than a decade ago Gires and Tournois and Giordmaine, et.al. (7,8) proposed shortening of optical pulses using compression techniques analogous to those used at microwave frequencies. Optical compression is achieved by passing the pulse to be compressed through an optical Kerr medium followed by a dispersive delay line. When an intense optical pulse passes through a nonlinear medium the refractive index is modified by the electric field E

$$n = n_o + n_2(E^2) + \qquad (1)$$

A phase change is impressed on the pulse given by

$$\delta\phi = n_2(E^2)\omega z/c \qquad (2)$$

where ω is the frequency and z is the distance traveled in the Kerr medium. Since the pulse intensity rises very steeply a frequency sweep or "chirp" is impressed on the pulse carrier given by

$$\delta\omega = (\omega z n_2/c)d/dt(E^2(t)) \qquad (3)$$

One problem with this scheme is that other nonlinear processes such as self-focusing can interfere with the pulse compression process. These problems can be overcome by using a single-mode optical fiber as a Kerr medium.

To reassemble the chirped optical pulse a dispersive delay line is required. A nearly ideal linearly dispersive delay line is a parallel pair of gratings. (9) Each wavelength passing through the pair is diffracted at a different angle by the first grating of the pair and recollimated by the second grating. The end result is a wavelength-dependent optical path delay.

In Figure 2 we have schematically drawn the optical setup for femtosecond pulse compression. Optical pulses of 90 fs in duration were focused into a 15-cm long polarization preserving single-mode optical fiber. When a few microjoules of pulse energy was coupled into the fiber, the frequency spectrum was seen to broaden significantly. At about 0.5-nJ energy coupled into the fiber the frequency spectrum was broadened by about a factor of three. The spectral half-width broadened from 60 to 200Å. Optical compression was achieved by the grating pair. The grating pair also served the useful purpose of compensating the group dispersion in the pulse measurement optics. In Figure 3 we have plotted the measured autocorrelation function of the 30-fs compressed optical pulse.

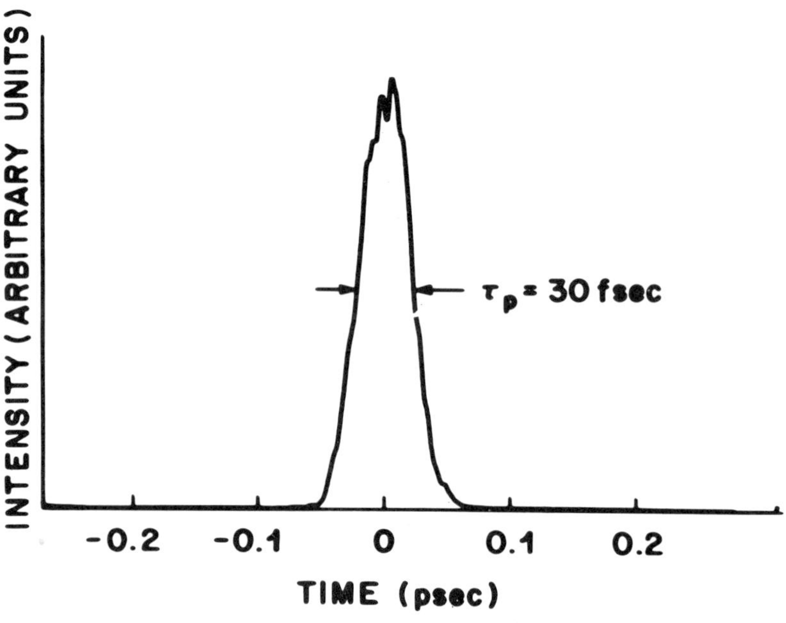

Fig. 3 Experimentally measured autocorrelation function of a compressed 30 femtosecond optical pulse.

Advances have also taken place in the amplification of ultrashort optical pulses. With a four-stage dye amplifier pumped by a frequency doubled Nd:Yag laser, 70-fs optical pulses have been amplified to peak pulse powers of several gigawatts while preserving the optical pulse width. (10) These pulses have been used to generate a picosecond continuum in a 400-μm stream of ethylene glycol. (11) By using such a thin generating medium, it has been possible to minimize the frequency sweep due to group velocity dispersion. With this technique, the generated continuum has a frequency sweep comparable to the pulse width over a wide frequency range. Frequency components have been observed from 190 to 1600 nm.

Applications of these short pulse optical techniques have been numerous. In the next section of this paper I am going to describe a new technique for performing time resolved spectroscopy on surfaces. The particular problem that will be addressed will be the structure of crystalline Si during a phase transition to the melted state following excitation with a femtosecond optical pulse.

Second harmonic generation from the surface of a medium with inversion symmetry, such as silicon, was first reported by Bloembergen and Chang. (12) In general, the nonlinear response of a medium with a center of inversion symmetry requires the consideration of higher order nonlocal contributions to the susceptibility. For a medium with cubic symmetry we can write the i-th component of the nonlinear source polarization as

$$P^{NLS}(2\omega) = (\delta - \beta)[E(\omega)\cdot\nabla]E_i(\omega) + \beta E_i[\nabla\cdot E(\omega)] + (2i\omega/c)\gamma[E(\omega) \times B(\omega)]_i + \zeta E_i(\omega)\nabla_i E_i(\omega) \qquad (4)$$

where δ,β,ζ and ζ are frequency dependent constants. The first three terms give rise to an isotropic contribution to the second harmonic polarization and the last term provides an anisotropic contribution which is determined by the crystal symmetry.

In the experiments described here we direct our attention to the <111> surface of silicon. The projection of the bulk symmetry onto the <111> surface is three fold symmetric. We can describe the second harmonic radiation from this surface with p polarized fundamental and second harmonic by

$$I(2\omega) = KI^2(\omega)[\cos 3\phi + A(\Theta)]^2 \qquad (5)$$

where ϕ is the angle of rotation about the surface normal, K is a constant, and $A(\Theta)$ is the isotropic contribution to the nonlinear polarization which is a function of angle of incidence, Θ. For $A(\Theta)=0$ the generated second harmonic is six fold symmetric and for $A(\Theta)=1$ a three fold symmetry is expected. In the experiments to be described here the angle of incidence is 45° and the second harmonic is observed to have nearly three fold symmetry indicating $A(\Theta) \sim 1$.

In our experiments we use a pulse at 620 nm with a duration of 90 femtoseconds to excite the <111> surface of a polished silicon wafer and measure the second harmonic generated by a second much weaker pulse at delayed time. The probing beam sampled the central 10 percent of the sample area and the probe intensity was at least an order of magnitude weaker than the weakest pump intensity. The pumping pulse was incident normal to the semiconductor surface and was 100 μ in diameter. The sample was mounted on a combination rotation and translation stage. A computer controlled the rotation and translation of the wafer so that each pulse of the laser was incident on a new spot on the silicon wafer. This was absolutely necessary because multiple shots from the laser produce cumulative damage to the surface. The laser was operated at 10Hz. The probing pulse was incident on the surface at an angle of 45 degrees with the polarization parallel to the plane of incidence. The primary difficulty with this experiment is isolating the second harmonic radiation from the pumping pulse from the much weaker second harmonic signal generated by the probing pulse. Since the second harmonic generated by the probe pulse is collinear with the reflected fundamental it is possible to spatially filter to prevent unwanted second harmonic light from the pump from entering the detection system. This was easily checked by noting that the detected second harmonic signal from the probing pulse went to zero when the probe pulse was blocked.

The threshold for amorphous surface layer formation to be at $0.1 J/cm^2$ and define this intensity to be E_{th}. As we have discussed, the observed variation of the second harmonic as function of the angle of rotation, ϕ about the <111> axis exhibits a nearly threefold rotation symmetry. Maxima and minima in the second harmonic from the surface are observed every 120 degrees.

In Figure 4 we have plotted the second harmonic as a function of angle of rotation of the crystal, ϕ in polar coordinates for two different power levels.

The electron density at this pumping level is approximately $10^{21}/cm^2$. The polar plot in Figure 4(a) reveals that the rotational dependence of the second harmonic shows little change. These experiments indicate that even at these high densities bonding electrons still dominate the second harmonic generation process. The small secondary peak of the second harmonic radiation at $\phi=60°$ disappears shortly after excitation, possibly indicating that the plasma causes a more perfect cancellation between the isotropic and anisotropic contributions to the second harmonic tensor.

In Figure 4(b) the optical pumping power is set 2 times threshold. The polar plot reveals a large change in the symmetry

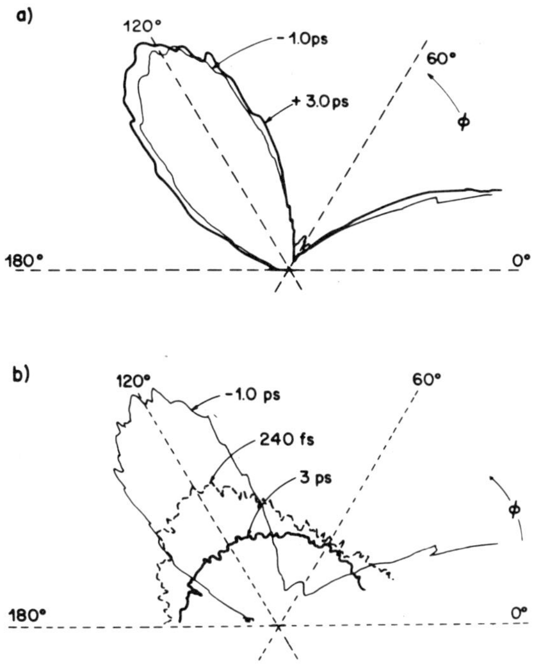

Fig. 4 (a) This is plot of the measured second harmonic signal generated by the probe as a function of rotation about a normal to <111> surface of Si. The pumping power was set at $0.6E_{th}$. (b) Pump power is increased to $2.0E_{th}$.

of the second harmonic following excitation. Before t=0 the second harmonic is clearly anisotropic. After 240 femtoseconds has elapsed, the width of the lobe at $\Theta=120°$ broadens and the minimum becomes less distinct. Apparently the surface is beginning to lose order. Within the first picosecond, the second harmonic coming from the surface becomes nearly isotropic. This is precisely what would be expected if the surface order makes a transition from a crystalline structure to a disordered liquid. At a pumping level of $3.0E_{th}$ a similar process takes place but somewhat more rapidly.

In conclusion, we have been able to observe the transition from crystalline order on the <111> surface to disorder consistent with melting of the surface on a time scale of less than a picosecond using femtosecond pulse excitation. The large decrease in the second harmonic is probably due to the loss of resonant enhancement of the second harmonic following melting. For the crystalline material the second harmonic energy, 4.0 eV, is very close to the sharp peak in absorption coefficient for silicon. These results for the first time clearly demonstrate the utility of nonlinear optical processes to time resolve structural changes at a semiconductor surface.

References

1. C. V. Shank, Science, 219, 1027 (1983).
2. A. J. DeMaria, and Heynau, Appl. Phys. Lett., 38, 671 (1981).
3. C. V. Shank, and E. P. Ippen, Appl. Phys. Lett., 24, 373 (1974).
4. R. L. Fork, B. I. Greene, and C. V. Shank, Appl. Phys. Lett., 38, 671 (1981).
5. C. V. Shank, R. L. Fork, R. Yen, R. H. Stolen, and W. J. Tomlinson, Appl. Phys. Lett., 40, 9 (1982).
6. G. H. C. New, IEEE J. Quantum Electron., QE-10, 115 (1974).
7. F. Gires and P. Tournois, C. R. Acad. Sci., 258 6112 (1964).
8. J. A. Giordmaine, M. A. Duguay, and J. W. Hansen, Quant. Electron., 4, 252 (1968).
9. E. B. Treacy, IEEE J. Quant. Electron., QE-5, 454 (1969).
10. R. L. Fork, C. V. Shank, and R. T. Yen, Appl. Phys. Lett., 41, 223 (1982).
11. R. L. Fork, C. V. Shank, C. Hirlimann, R. Yen, and W. J. Tomlinson, Optics Letters, 8, 1 (1983).
12. N. Bloembergen, R. K. Chang, S. S. Jha, and C. H. Lee, Phys. Rev. 174, 813 (1968).

ALTERNATIVE TECHNIQUES FOR PICOSECOND SPECTROSCOPY

Anthony E. Siegman

Stanford University, Stanford, California

ABSTRACT

Photoacoustic detection methods and tunable laser-induced grating techniques can both provide useful new measurement capabilities for studying ultrafast physical and chemical processes and lifetimes. The photoacoustic technique, which measures the total acoustic impulse produced by two variable-delay picosecond pulses in an absorbing sample, provides a sensitive and effective technique for measuring the absolute cross sections and lifetimes of weakly absorbing excited states, with time resolution limited only by the optical pulsewidths. The transient grating techniques, which measure the probe beam diffraction from the moving excited-state gratings induced in a sample by two dye laser beams with a tunable difference frequency, provide an effective method for measuring picosecond and subpicosecond physical phenomena through measurements made in the frequency domain. Tunable transient grating measurements on organic dyes using both two-laser and three-laser methods will be described.

1. INTRODUCTION

This lecture will be focused primarily (and necessarily, in view of the author's qualifications) on certain useful experimental techniques for studying fast physical and chemical processes using lasers, rather than on the actual chemical results, although a few illustrative results will also be included. I will discuss first the virtues of *photoacoustic detection* as a promising alternative detection method for making picosecond spectroscopic measurements, as demonstrated in recent work by one of my students, Dr. J.M.

K. B. Eisenthal (ed.), Applications of Picosecond Spectroscopy to Chemistry, 325–351.

Heritier (1). I will then describe several tunable laser-induced grating techniques recently studied by another of my students, Dr. R. Trebino (2). By using these grating techniques, fast non-linear phenomena can be studied in the frequency domain, using tunable lasers, rather than in the time domain using picosecond pulses.

2. PHOTOACOUSTIC SPECTROSCOPY

Photoacoustic spectroscopy is a well-established technique in which a periodically modulated or pulsed laser beam passes through an absorbing sample, which may be a solid, liquid or gas. The laser energy deposited in the absorbing sample then creates a localized temperature increase, and hence a pressure rise, leading to a subsequent acoustic impulse which is detected by a sensitive microphone. A major advantage of the photoacoustic technique is that it responds directly to the energy absorbed by the sample, rather than the possibly very small change in intensity of the transmitted beam. As a practical matter, by using a sensitive microphone extremely small amounts of absorbed energy can be detected, even with a noisy or unstable laser source. In addition the technique is sensitive only to that portion of the absorbed energy that goes into nonradiative rather than radiative decay, and it thus provides different information than the two-photon fluorescence detection often used in picosecond spectroscopic experiments.

Conventional photoacoustic spectroscopy has thus become a widely used technique for straightforward spectroscopic investigations of weak absorption effects in a very wide variety of materials (3). In this lecture we will illustrate how this same photoacoustic detection method can be employed to make picosecond time resolved measurements (4,5), by using pairs of laser pulses with variable pulse delay between them.

3. AN ELLIPTICAL PHOTOACOUSTIC CELL FOR PULSED PHOTOACOUSTIC MEASUREMENTS

Conventional photoacoustic measurements are commonly made using either a cylindrical gas cell with a microphone inserted in one side; or a liquid cuvette with a microphone either immersed or clamped against one wall; or a solid sample with a microphone directly contacting the sample. For the pulse experiments to be reported in this lecture, however, we have developed a somewhat more elegant photoacoustic cell (6) with the elliptical geometry shown in Figure 1.

In this cell the laser beam is sent along one focal line of a

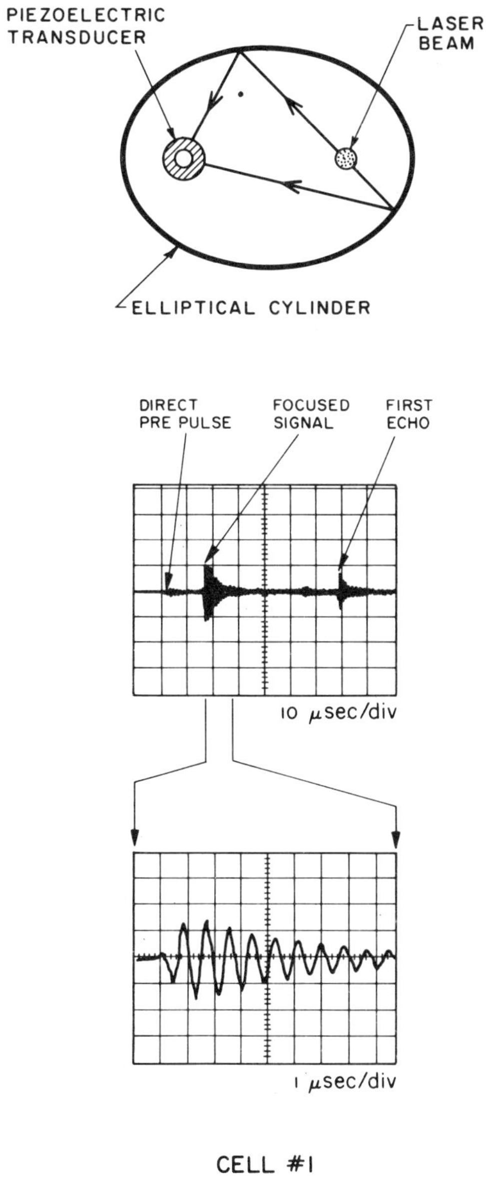

Figure 1. Elliptical Cell Cross Section and Pulsed Photoacoustic Response

cell in the form of an elliptical cylinder (in fact, a converted solid-state laser pump cavity). The cylindrical sound wave radiated from this line source is then focused onto a cylindrical piezoelectric transducer located at the other focus of the elliptical cell. This transducer consisted in some cases of a hollow cylindrical piezoceramic transducer which was poled in the radial direction, with metal film electrodes on the inner and outer cylindrical surfaces; and in other cases of a slender solid rod transducer poled in the axial direction and having electrodes on the ends of the cylinder.

The additional traces in Figure 1 illustrate the type of response produced in this cell by a short Q-switched or single mode-locked laser pulse traveling down the laser beam axis. In the center trace the oscilloscope is triggered at the time of the laser impulse. The first response is then produced by a small amount of acoustic energy which travels across the centerline of this cell and is directly intercepted by the piezoelectric transducer to produce a weak pre-pulse, as seen in the oscilloscope trace. The majority of the acoustic signal, however, is reflected off the cell walls and focused on to the transducer to produce the strong focused signal seen after an acoustic transit time of slightly more than 25 microseconds.

The lower trace in Figure 1 shows an expanded view of this focused signal. The sinusoidal ringing at approximately 1 MHz represents the shock-excited acoustic response and acoustic resonance in the piezoelectric transducer itself.

There is of course a large acoustic mismatch between the liquid in the cell and the piezoelectric transducer. The major portion of the acoustic energy in the liquid is thus reflected from the piezoelectric transducer, travels back to the laser axis, and then returns again to produce an echo after a time interval equal to twice the initial time delay. Many further such echoes can be observed if the sweep rate of the oscilloscope is appropriately reduced.

It is possible to calculate analytically the acoustic wave response produced in the liquid medium inside such a cell by an incident laser pulse that is gaussian in both space and time (7). Figure 2 illustrates the type of thermally induced pressure wave produced in the liquid medium (not the piezoelectric transducer) for different beam diameters and laser pulse widths. Note that the acoustic pulse width is determined by a combination of the laser pulsewidth and the acoustic transmit time across the laser beam spot.

It should be noted that an additional impulse response (not shown) will also be produced in the cell by an electrostrictive

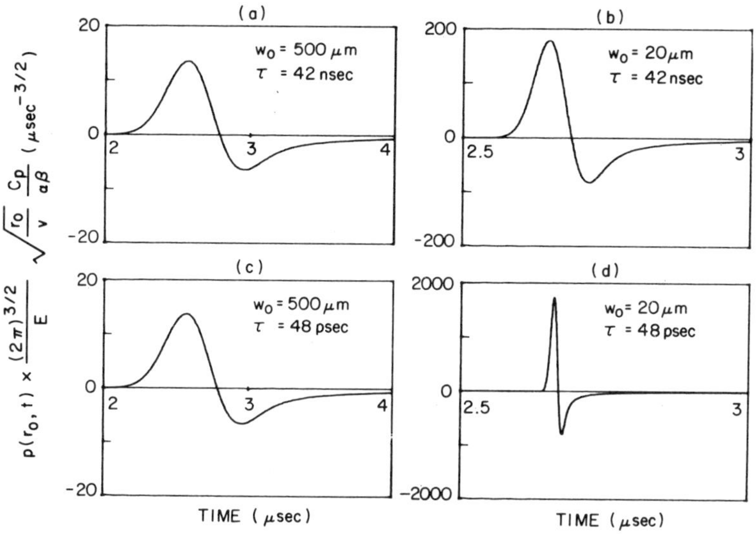

Figure 2. Absorptive Pressure Waveforms for Different Excitations

response which is inevitably present in all liquid solvents. This electrostrictive response represents an ultimate lower limit to the sensitivity for photoacoustic measurements in very weakly absorbing samples.

4. ELLIPTICAL CELL PERFORMANCE

The usual signal detection method in pulsed photoacoustic experiments with weakly absorbing samples is to capture the strongest peak of the initial impulse response with a gated integrator (3,4), although other electronic detection schemes can also be used. The major advantage of the elliptical cell compared to conventional cells is that it collects essentially all of the acoustic energy into a single impulse, rather than dispersing a large amount of the acoustic energy into multiple reflections as in conventional cells. As a consequence it creates a more clean and orderly impulse response than is characteristic of conventional cell designs.

As a simple check on the sensitivity of our cell, we used one particular version of this cell in conjunction with a tunable pulsed dye laser (~10 nsec tunable pulses) to measure the very weak visible absorption line near 610 nm in liquid benzene due to the sixth overtone of the C-H stretch, as measured previously by Patel and Tam (8). Figure 3 shows this absorption line as measured in our cell using different pulse energies and dilutions of

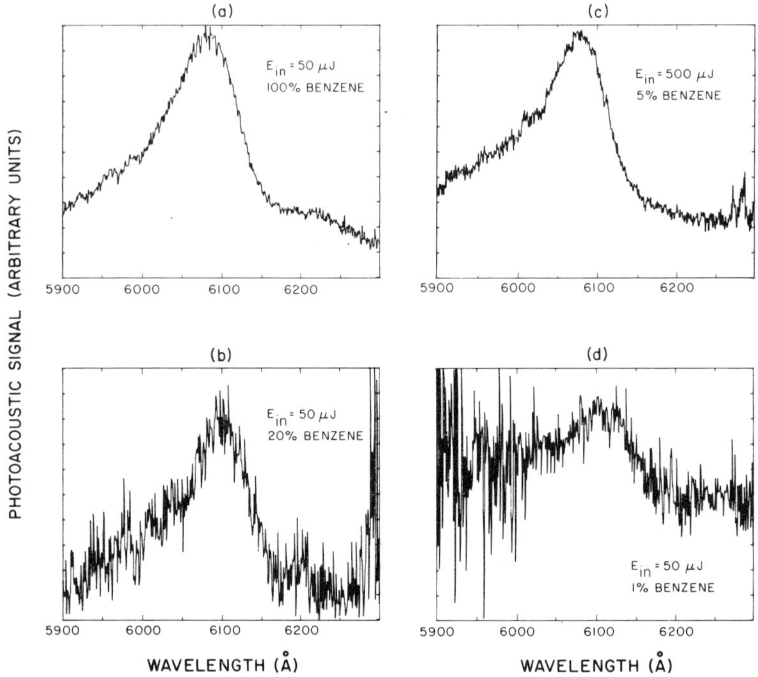

Figure 3. Photoacoustic Spectra of Diluted Benzene

benzene in carbon tetrachloride. The lineshape measured in our cell using 50 μJ pulses compares closely with the results measured using ~1 mJ pulses by Patel and Tam. From the observed signal-to-noise ratio of $S/N \approx 1$ that we obtain using 50 μJ pulses and 1% benzene, and averaging over 100 pulses at each point, we estimate that the absolute sensitivity of our cell comes within at least a factor of 3 of the very carefully engineered cell developed by Patel and Tam.

The occasional large noise bursts in the more sensitive traces in Figure 3 result primarily from the presence of small absorbing particles in the sample cell, a difficulty which is particularly important in pulsed photoacoustic measurements. Elimination of this noise source requires both careful cleaning of the photo-acoustic cell, and possibly the distillation of high purity solvents directly into the sample cell.

5. PICOSECOND PHOTOACOUSTIC MEASUREMENTS

A dual-pulse technique by which picosecond time-resolved measurements of excited state lifetimes can be carried out, using a photoacoustic cell whose intrinsic time response is limited to the microsecond range, has now been demonstrated by us (4) as well as by Bernstein, Rothberg and Peters (5). This technique is illustrated in Figure 4. As shown in the lefthand part of this diagram, we consider a typical situation in which an initial excitation pulse at wavelength λ_0 excites atoms or molecules from a ground state S_0 into an excited state S_1, delivering a certain amount of energy into both radiative and nonradiative decay in the process. This excitation pulse will thus produce a certain amount of photoacoustic response.

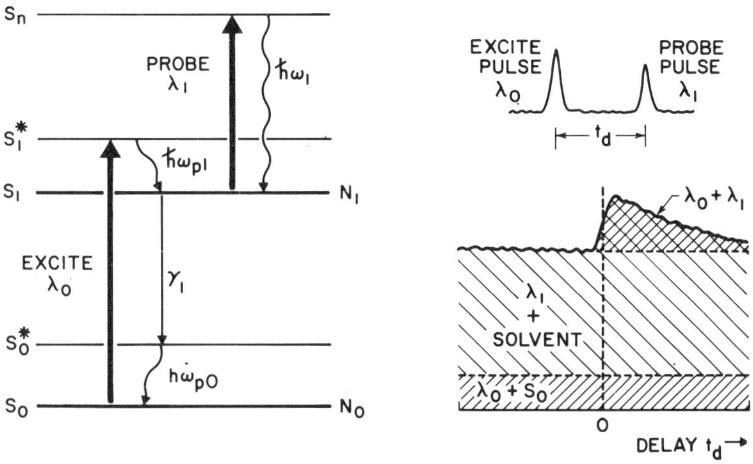

Figure 4. Energy Level Diagram

A probe pulse at a different wavelength λ_1, which is absorbed by the excited but not the ground state, is then sent through the same sample volume with a variable delay relative to the excitation pulse. In many molecular systems essentially all the energy absorbed from this probe pulse will then be converted directly through fast nonradiative decay into heat, and thus into an additional photoacoustic signal, with nearly 100% quantum efficiency.

If the excitation and probe pulses have nanosecond or shorter pulse widths and time delays, the resulting total photoacoustic response (i.e., the net piezoelectric transducer voltage) will be essentially the response to a single acoustic impulse with an amplitude proportional to the total heat dumped into the atomic or molecular system by the combined excitation plus probe pulses. As shown in the right-hand portion of Figure 4, the excitation pulse by itself will produce on each shot a constant or delay-independent background acoustic impulse, independently of the probe pulse.

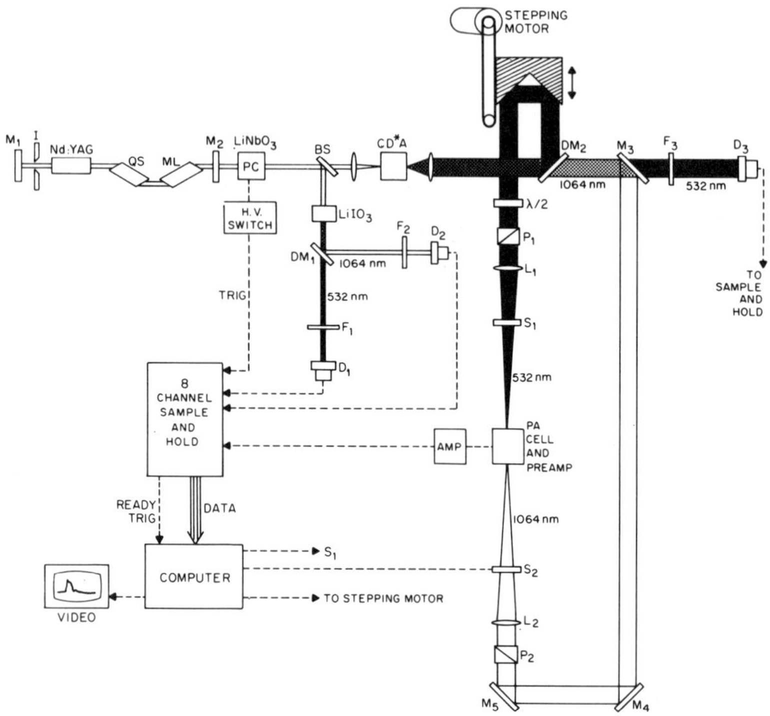

Figure 5. Experimental Set-up for Two-Pulse Photoacoustic Experiments

The amplitude of this portion of the response will depend upon the total fraction of the excitation from the ground state that goes into nonradiative as contrasted to radiative decay.

In the experiments reported in this paper the probe pulse by itself also produced an additional delay-independent background signal due to unavoidable absorption of the infrared probe pulse in the liquid solvents employed. With visible or ultraviolet pulses, this additional background signal can be greatly reduced (5) or eliminated.

The essential point, however, is that when the probe pulse follows the excitation pulse, absorption of the probe pulse by the excited state molecules then produces an additional delay-dependent photoacoustic signal, which decreases with delay in a fashion that directly follows the decay of the excited state populations in state S_1. This delay-dependent portion of the signal can be readily extracted from even quite large background signals by appropriate signal processing techniques, thus providing a direct method for measuring the buildup and decay of the excited state population.

6. EXPERIMENTAL RESULTS

To give a preliminary demonstration of this type of picosecond photoacoustic spectroscopy, we employed the fairly conventional laser setup shown in Figure 5. This system involves a repetitively pumped, Q-switched, and mode-locked Nd:YAG laser, followed by a single-pulse selector, frequency doubler, and an optical delay line for the frequency-doubled pulses. (This system was used because it was available; other types of cw mode locked lasers with higher repetition rates and greater pulse stability would probably be preferable for this type of experiment).

Figure 6 then shows the delay-dependent portions of the photoacoustic signals produced using this system in solutions of Rhodamine 6G and Rhodamine B in ethanol or in heavy water, with the measured signal as a function of pulse delay plotted on linear scales in the upper curves, and the same data plotted on logarithmic scales in the lower curves. The second impulse occurring after about 5 nanoseconds delay in each case comes from a second excitation pulse which was occasionally transmitted by a malfunctioning single pulse selector.

The results for ethanol solutions in the first two examples give excited-state lifetimes that compare very well with the standard values for these dyes. The third plot in Figure 6 illustrates measured signals for Rhodamine 6G in D_2O made with the probe electric field vector oriented either parallel or perpendicular to the excitation pulse, in order to examine molecular orientation effects. Although the experimental data in these curves are rather noisy, it is clear that the dipole moments of the ground state and excited state transitions are parallel rather than perpendicular, and that with some improvement in laser stability and data collection efficiency, both the orientational relaxation and the fluorescent lifetime for the Rhodamine 6G sample could be extracted from these curves.

7. DISCUSSION

This general method of photoacoustic detection is obviously similar in concept to the two-quantum fluorescent detection technique often used for picosecond spectroscopy; but this technique requires only an acoustic detector, and can be implemented in solutions or at wavelengths where no fluorescent emission may be available. The amplitude of the observed acoustic signal is also obviously sensitive to very different properties of the molecules

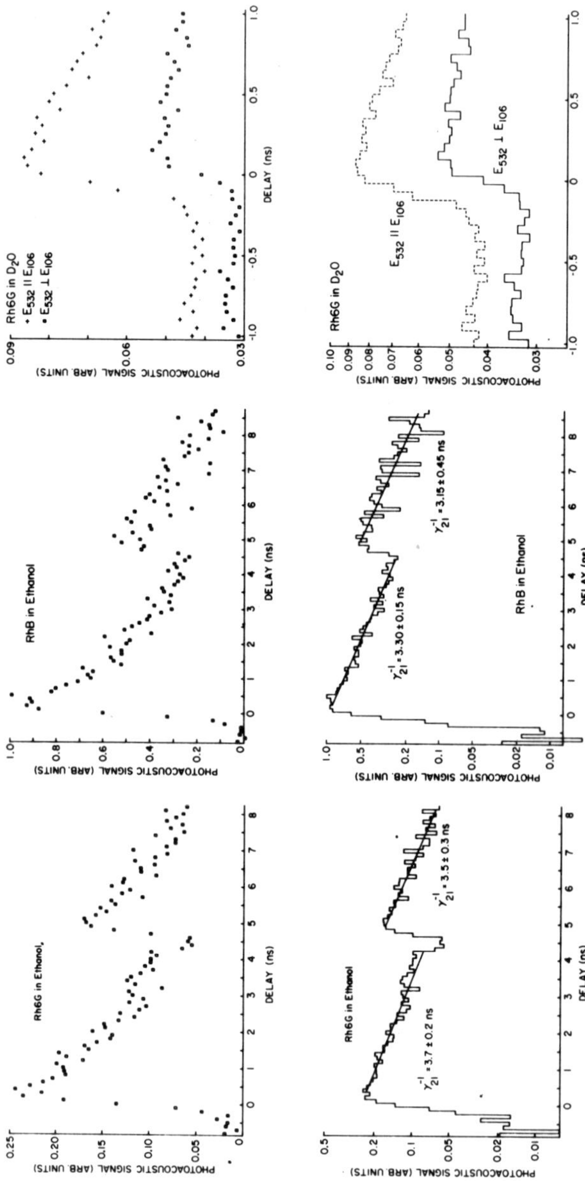

Figure 6. Photoacoustic Measurements of Excited State Lifetimes

involved, and in particular to the nonradiative rather than radiative processes in the molecular system.

In addition to being a new and useful method for measuring excited state dynamics, this type of picosecond photoacoustic spectroscopy also provides a very useful new tool for measuring absolute excited state cross sections in weakly absorbing excited states, as also noted by Bernstein, Rothberg and Peters (5). To see this, suppose that one first extracts out of the measured signal that portion of the delay-independent background due only to ground-state and solvent absorption (as can fairly easily be done). This delay-independent portion of the photoacoustic signal due to the excitation pulse by itself then corresponds to a heat deposition per unit volume given by

$$H_0 = N_1 \hbar\omega_{ph} \tag{1}$$

where N_1 is the density of molecules lifted into state S_1 per shot, and $\hbar\omega_{ph}$ is the nonradiative quantum of energy decay from this state.

At the same time, the net energy per unit volume absorbed and rapidly converted into heat deposition by probe pulse excitation of these excited molecules is given by

$$H_1 = N_1 \sigma_1 W_1 e^{-t_d/\tau_1} \tag{2}$$

where σ_1 is the excited state absorption cross section; L the absorption length; W_1 the probe pulse energy fluence; t_d the time delay between excitation and probe pulses; and τ_1 the lifetime of the single state (assuming single exponential decay).

The easily measured ratio of the delay-dependent signal at zero delay to the delay-independent background due to the excitation pulse alone is thus given by

$$R = \frac{\sigma_1 W_1}{\hbar\omega_{ph}} \tag{3}$$

All the quantities in this equation except σ_1 can then be fairly readily either directly measured or estimated, while the excited state population N_1 has cancelled out. One can thus determine the excited state cross section σ_1 *without needing to know the total number of atoms* N_1 *lifted into the upper level.* Since the measurement of the excited state population N_1 is usually the most difficult aspect of any other technique for measuring excited state cross sections, the photoacoustic technique thus provides a very convenient new technique for measuring absolute excited state cross sections. Very weak excited state absorptions can also be observed in this fashion, especially since the probe pulse energy fluence W_1 can be made quite large if necessary.

We thus suggest that the combined picosecond pulse plus photoacoustic detection technique described here can become a useful and effective method for measuring lifetimes and absolute cross sections of excited states in liquids, solids, gases and surface layers. This approach should be particularly applicable for measurements on weakly absorbing systems, or systems where excited state fluorescence is not available as a detection mechanism. The elliptical cell design is convenient for certain types of samples, but is by no means essential. Much shorter laser pulses than used in our experiment are of course now available, and can equally well be used to gain greater time resolution. The use of more stable cw mode-locked lasers with greater pulse uniformity and higher repetition rate would substantially improve the signal averaging experiments, and further improvements in cell design and electronics are also possible.

8. TUNABLE TRANSIENT GRATING MEASUREMENTS

We turn now to the measurement of very fast atomic or molecular lifetimes or recovery rates using transient laser-induced grating techniques (2,9,10).

As an elementary model for this kind of experiment we can first consider the situation shown in Figure 7, in which two excitation beams at frequencies ω_1 and ω_2 -- privided perhaps by long-pulse dye lasers -- impinge at an angle in an experimental sample. These two excitation beams are made nearly coincident in frequency but one or the other is made tunable about their common frequency to give a difference frequency $\Delta\omega = \omega_2 - \omega_1$. A third or probing beam at a possibly quite different frequency ω_3 is also passed through the sample.

From a nonlinear optics viewpoint (9), these three beams can then mix or interfere with each other as a result of any nonlinear response in the sample medium. For example, the two excitation beams at ω_1 and ω_2 can interfere with each other to produce a molecular excitation at each point in the sample which is sinusoidally modulated at the tunable difference frequency $\Delta\omega = \omega_2 - \omega_1$. The probe beam can then be modulated by this time-varying molecular excitation to produce frequency sidebands at $\omega_3 \pm \Delta\omega$, as well as other higher-order sum and difference frequencies.

From an induced grating viewpoint, however (10), the two beams at ω_1 and ω_2 can be viewed as interfering inside the sample to

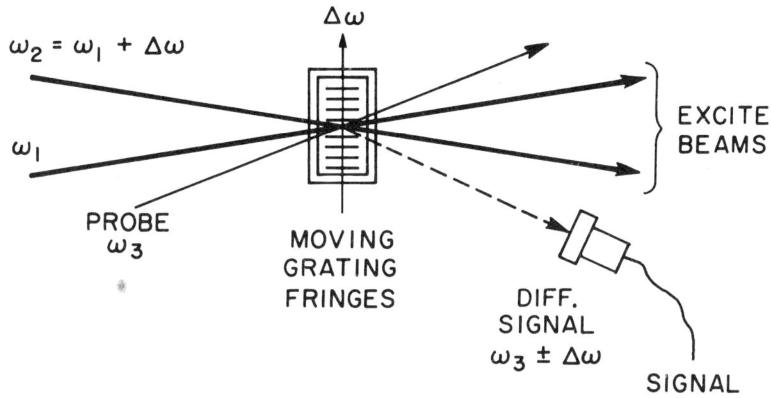

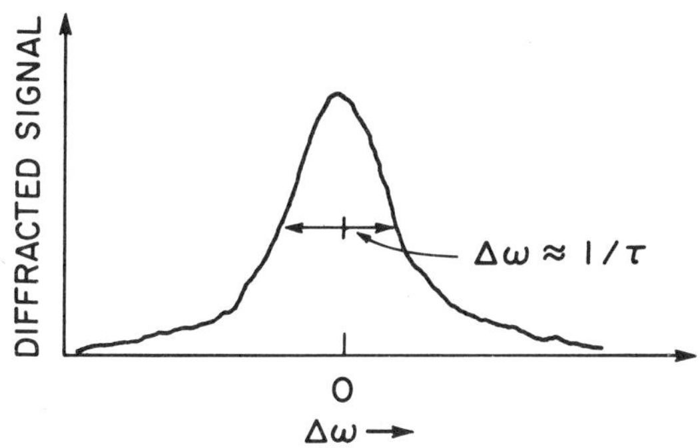

Figure 7. Three-Frequency Transient Gratings

produce a moving interference pattern, whose fringes sweep with frequency $\Delta\omega$ past any point in the sample. If these moving interference fringes produce any type of excited-state or ground-state population density fringes, or any type of nonlinear index fringes in the sample, and if the probe beam then arrives at the appropriate Bragg angle for this grating pattern, the existence of this moving grating may then be detected by diffraction of the probe beam at ω_3 from this moving grating. Such a grating will cause a diffracted signal to emerge from the sample in an appropriate direction determined by the phase matching condition.

The diffracted intensity that will be measured in such an experiment will depend, however, on the tunable difference frequency between the two excitation beams, in a manner that it determined by the sample relaxation rate $1/\tau$. If the absolute value of the frequency difference $\Delta\omega$ is small (much less than $1/\tau$), then the optical intensity fringes will move slowly and will appear nearly stationary to the material. That is, each atom or molecule in the experimental sample will see the local intensity oscillating up and down at a frequency slow compared to its response rate. The excitation of excited states in the sample can then easily follow the intensity modulation in the interference fringes, resulting in maximum visibility and maximum diffraction efficiency for the induced grating.

When the absolute value of $\Delta\omega$ becomes much greater than $1/\tau$, however, the intensity fringes will sweep past each point in the sample too quickly for the molecules to follow the sinusoidal variation. Although a space- and time-averaged "smear" of excited state molecules may still be produced, the visibility of the induced grating fringes will be washed out, so that the sample will diffract very little probe light.

We can thus expect to see a cutoff or breakpoint between high and low diffraction efficiency regimes to occur for a detuning given by $\Delta\omega \quad 1/\tau$, as illustrated in the lower part of Figure 7. It can be shown, in fact, that the diffracted signal lineshape in this type of experiment will be given by

$$I_{diff}(\Delta\omega) = |H(\Delta\omega)|^2 = |\mathfrak{F}[h(t)]|^2 \qquad (4)$$

where $h(t)$ is the system impulse response. That is, $h(t)$ is the absorption change $\Delta\alpha(t)$ or the index change $\Delta n(t)$ at the probe frequency that would be produced by a delta-function impulse arriving at the excitation frequency, and $H(\Delta\omega)$ is the Fourier transform of this system impulse response. For a system having a single exponential impulse response with decay time τ, the diffracted signal intensity will be the square of a Lorentzian lineshape with a half-width given by $\Delta\omega_h = 1/\tau$.

9. TRANSIENT GRATING GEOMETRIES

The type of geometry under which this sort of experiment is carried out is an important design parameter. Examples of three different geometries under which this type of induced grating experiment can be performed, while still satisfying the necessary Bragg angle conditions, are illustrated in Figure 8 under the assumption that ω_3 is slightly smaller than ω_1 or ω_2. (Recalling that the grating fringes or wavefronts in each case will bisect the angle between the two excitation beam k-vectors).

In each of these three cases the Bragg angle for proper phase matching of the probe beam at ω_3 to the grating produced by ω_1 and ω_2 will change as one tunes either of the excitation frequencies ω_1 or ω_2. To put this in another way, if the direction of ω_3 is fixed at the proper Bragg angle corresponding to the

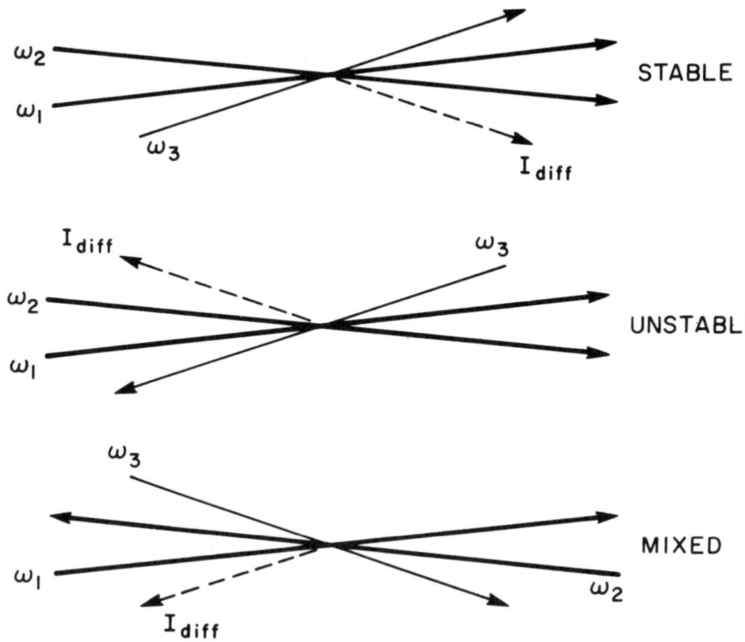

Figure 8. Examples of Transient Grating Geometries

case $\omega_1 = \omega_2$ or $\Delta\omega = 0$, then a phase mismatch will develop as one tunes either ω_1 or ω_2. Detailed examination shows that the resulting phase mismatch will be negligible in some configurations ("stable Bragg angle"), while in other configurations the resulting phase mismatch becomes so large with excitation tuning as to greatly limit the available bandwidth in these geometries.

The top sketch in Figure 8 represents in particular the classic small-angle grating or standard holographic geometry. This situation has a very stable Bragg angle, or a large tuning range before phase mismatch effects become important. However, the diffracted signal beam unavoidably lies close to both of the excitation beams, so that detection of a weak scattered beam becomes more difficult.

In the limit of $\omega_1 = \omega_2 = \omega_3$, the middle sketch corresponds to what is commonly considered to be the standard phase conjugation geometry (11), although the notations for beams 1, 2 and 3 is usually different. This geometry has the diffracted signal beam widely separated in angle from both of the excitation beam directions; but it has a very unstable Bragg angle, which is to say, a very narrow phase-matching bandwidth at fixed angular directions.

Finally, the third case is a mixed case, in that it is stable or has a relatively broad phase-matching bandwidth against tuning of the ω_2 but not the ω_1 excitation beam; while the diffracted signal unavoidably falls near or (at degeneracy) on top the ω_1 excitation beam. This case corresponds to the standard versions of polarization spectroscopy (12) or Lippman (bulk) holography.

10. TWO-FREQUENCY TRANSIENT GRATING EXPERIMENTS

The simplest way in which transient grating experiments can be carried out is to make the probe frequency ω_3 coincident with one of the excitation frequencies ω_1 or ω_2, so that only two lasers (only one of which need be tunable) are required. We then have in essence a degenerate wave mixing case; and the Bragg angle condition then requires that the probe beam run either exactly parallel or antiparallel to one of the excitation beams. This can be accomplished in either of two ways, corresponding to the second or third sketches in Figure 8. (The geometry in the first sketch could also be used, with polarization discrimination between probe and excitation beams, but this is usually avoided because of the increased excitation beam interference problems).

One approach is to use the phase conjugate geometry, corresponding to the middle sketch, as described and illustrated by Kenney-Wallace elsewhere in this volume. This clearly corresponds to a situation in which the diffracted signal travels oppositely to the coparallel excitation signal, thus providing a zero background detection situation; but this also corresponds to a very unstable Bragg angle condition, so that the useful bandwidth is seriously reduced. This bandwidth reduction can only be minimized by using a very short interaction length, as employed by Kenney-Wallace.

The alternative two-laser configuration, corresponding to the bottom sketch, represents exactly the polarization spectroscopy configuration used by Song, Lee, and Levenson and others (12). In this case the phase matching bandwidth is unlimited (against tuning of ω_2), but the diffracted signal inherently emerges exactly parallel to the other excitation laser beam, thus requiring polarization discrimination to separate the diffracted signal.

We have in fact carried out a number of such two-laser polarization spectroscopy experiments, using the experimental configuration shown in Figure 9. In this setup a Q-switched and frequency-doubled neodymium-YAG laser simultaneously pumps two dye lasers, one of which is tuned through the frequency of the other. Figure 10 illustrates the type of experimental results for diffracted signal versus tuning that can be obtained using this technique

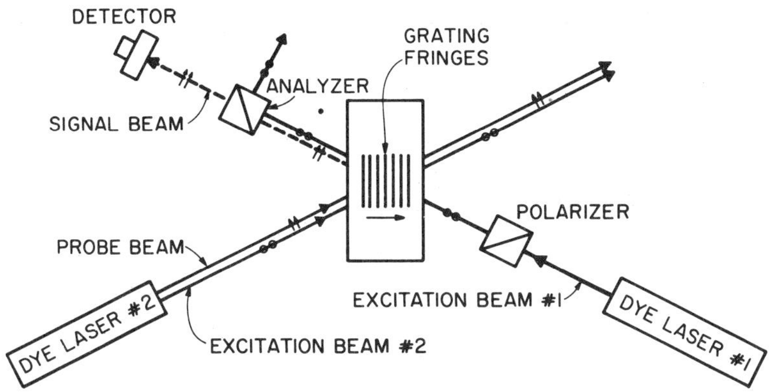

Figure 9. Two-Laser Transient Grating Geometry

both for a saturated absorption grating in a solution of malachite green in water, and for a refractive index grating induced using the optical Kerr effect in a transparent sample of carbon disulfide.

The malachite green data indicate a ground-state recovery time of about .7 psec, which is on the order of, but somewhat shorter than, other measured values; while a rough fit of the CS_2 results indicates a half-width of approximately 3 cm^{-1}, corresponding to an exponential decay time of ~1.6 ps, in generally good agreement with the values for the orientational time constant of CS_2 obtained by other researchers using picosecond pulse techniques. (The asymmetry is due to the polarizer leakage.) The faster exponential decay components (with time constants in the 0.3 ps range) which have been seen in CS_2 by workers using femtosecond pulses are not obtainable from these preliminary results. These faster time components would of course appear in the tails of the observed signal, where the resonant peak merges with the background.

The primary advantages of the two-laser technique, as contrasted with genuine three-laser tunable grating experiments, are that the system is relatively simple and generally quite easy to align. One important disadvantage is that with ω_3 degenerate with ω_1 and ω_2, one has limited freedom to select the specific molecular state (or states) to be probed.

Another serious disadvantage of the two-laser technique is that there is an inevitable background signal in all the results, due to induced dichroism in absorption grating experiments, or to induced birefringence in refractive index experiments, as well as potential coherent background from leakage through imperfect or

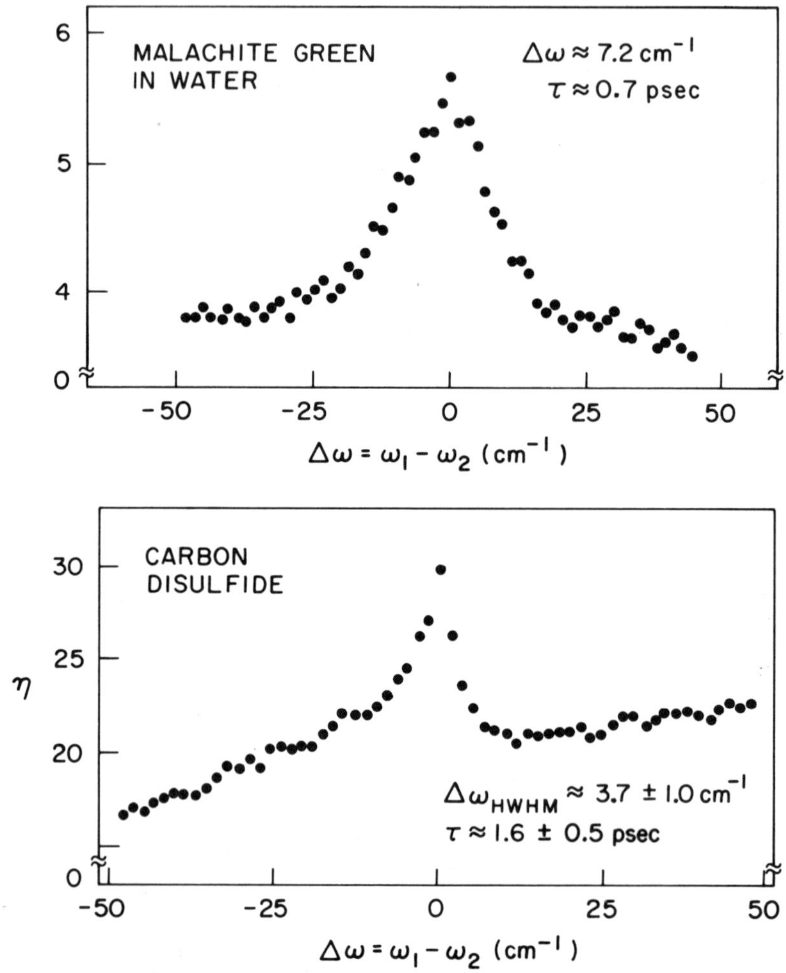

Figure 10. Two-Laser Experimental Results

improperly adjusted polarizers. In two-laser saturable absorption experiments, such as the malachite green results shown in Figure 10, for example, the ratio of peak diffracted signal to induced dichroism background should be exactly 4:1 under ideal conditions.

In our malachite green results shown in Figure 10, as well as in others reported in the literature (12), the observed peak to background ratio has always been less than this, due primarily to saturation effects in the sample. A simple calculation shows in fact that the effects of saturation should be to reduce the peak to background ratio to approximately

$$\frac{\text{Peak diffracted signal}}{\text{Induced dichroism background}} \approx 4 \times [1 - I/I_{sat}] \qquad (5)$$

where I is the excitation intensity and I_{sat} the saturation intensity in the absorbing medium. Under these conditions, for a system with a single exponential decay, the relationship between diffracted signal half width and lifetime should then become

$$\Delta\omega_h \tau \approx 1 + I/I_{sat} \qquad (6)$$

When we apply these correction factors both to our results and to the earlier results of Song, Lee, and Levenson (12), we find that in both cases the apparent response time of the ground state of malachite green in water, interpreted as a single exponential, corresponds to a recovery time of 1.2 ps. This is somewhat shorter than the approximately 2 ps derived from earlier picosecond pulse measurements (13).

11. THREE-LASER GRATING EXPERIMENTS

In order to overcome the limitations of the two-laser approach and demonstrate the virtues of the background-free three-laser grating technique, Trebino has recently set up a complete three-laser tunable transient grating system, using three dye lasers all pumped by the same Q-switched neodymium-YAG laser, as illustrated in Figure 11. The full details of this apparatus and the associated

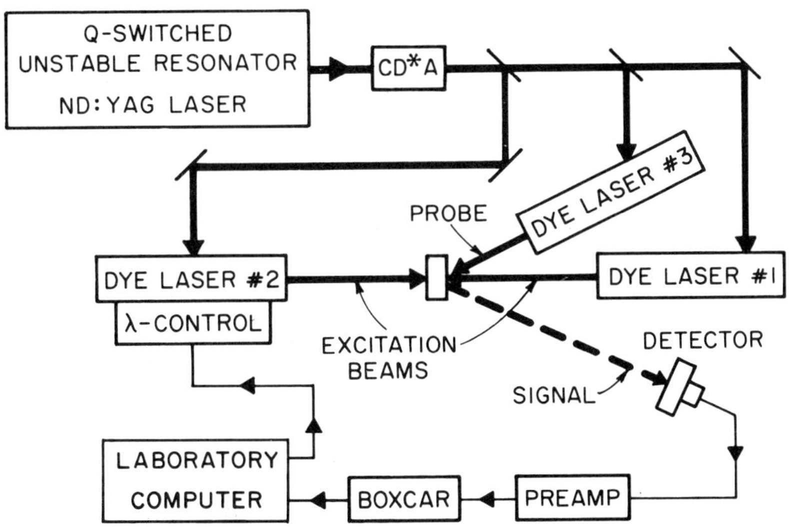

Figure 11. Experimental Apparatus for Three-Laser Experiments

experimental techniques are reported elsewhere (2). In setting up these experiments, one particularly useful technique has been to accomplish the initial alignment of the three beams by replacing the experimental sample with a dummy sample in which very strong thermal gratings (to be discussed below) can be obtained, and then reverting to the regular sample for the final measurements.

Figure 12 illustrates experimental results that were obtained by using this technique to make measurements on a dilute solution of malachite green in water. The probe wavelength of 609 nm in this experiment, because it is slightly shorter than the excitation wavelength, probes primarily the absorption grating of "holes," or locally saturated regions, burnt into the ground-state population of the malachite green molecules; and the observed lineshape thus measures primarily the ground-state population recovery in the malachite green dye.

The slightly asymmetric and clearly non-Lorentzian lineshape observed in this figure was interpreted as follows. The suggested decay laws for triphenylmethane dyes such as malachite green that have been suggested in earlier studies include: (a) a single exponential, $\exp(-t/\tau)$; (b) a cubic exponential (14), $\exp(-kt^3)$; (c) the sum of two exponentials, with time constants τ_1 and τ_2; and (d) a continuous range of exponentials with time constants distributed uniformly over a certain range (15). To test each of these hypotheses, we have fitted the experimental data in Figure 12 to the expression

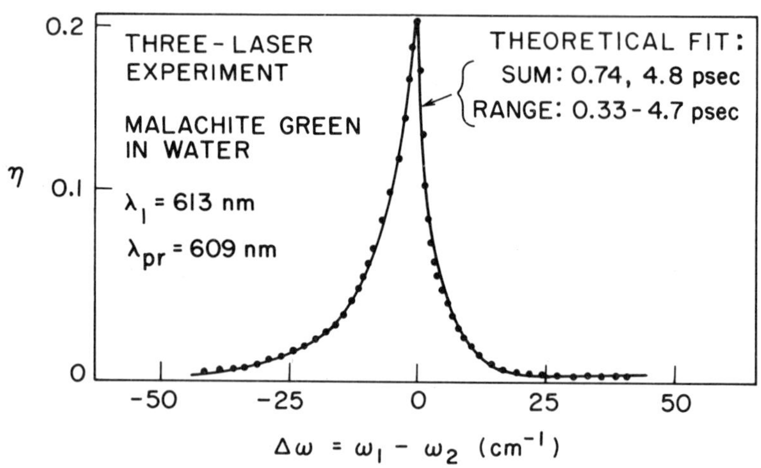

Figure 12. Three-Laser Experiment on Malachite Green in Water

$$I_{diff}(\Delta\omega) = I_0 + (K_1 + K_2\Delta\omega)\left\{\left[H_1(\Delta\omega) + C_1\right]^2 + \right.$$
$$\left. + \left[H_2(\Delta\omega) + C_2\right]^2\right\} \quad (7)$$

where I_0 represents a constant background amplitude; the $K_1 + K_2\Delta\omega$ factor represents a possible linear variation of experimental sensitivity across the laser tuning range (due to changes in detector sensitivity, linear absorption of the dye solution and so forth); the components H_1 and H_2 represent the real and imaginary parts of the Fourier transforms of each of the impulse responses listed above, in turn; and C_1 and C_2 represent possible coherent backgrounds from the solvent.

When this fitting procedure is carried out in turn for each of the four proposed responses listed above, we find that our experimental results cannot be satisfactorily fitted by either a single exponential or a cubic exponential. Instead the results must be fitted either by a dual exponential recovery law of the form

$$h(t) = 0.9 \exp[-t/(0.78 \pm 0.10 \text{ ps})] + 0.1 \exp[-t/(7.4 \pm 3.0 \text{ ps})] \quad (8)$$

or by a continuous range of exponentials extending from 0.34 ps to 7.2 ps. (These two fits are essentially indistinguishable from the experimental data or from each other).

We thus conclude that the dominant component in the ground state recovery of malachite green in water has a recovery time of 0.78 ± 0.10 ps, a result which fits better with extrapolated data versus viscosity from earlier work (16) than do the larger values given by earlier measurements (13,17,18). In particular, we suppose that the somewhat longer value of ~1.2 ps obtained from the two-laser measurements results in part from forcing a single exponential fit onto what is in fact at least a dual-exponential decay.

12. THERMAL GRATINGS

One significant difficulty that does arise with many transient grating experiments is illustrated by the results shown in Figure 13 for a three-laser transient grating experiment carried out on a dilute solution of malachite green in ethanol. Note that the central peak observed in this experiment is at least an order of magnitude stronger and very much narrower than the results for the same dye in water. In agreement with an earlier proposal by Andrews and Hochstrasser (10), we believe the physical explanation for the very strong central peak observed in this and many similar

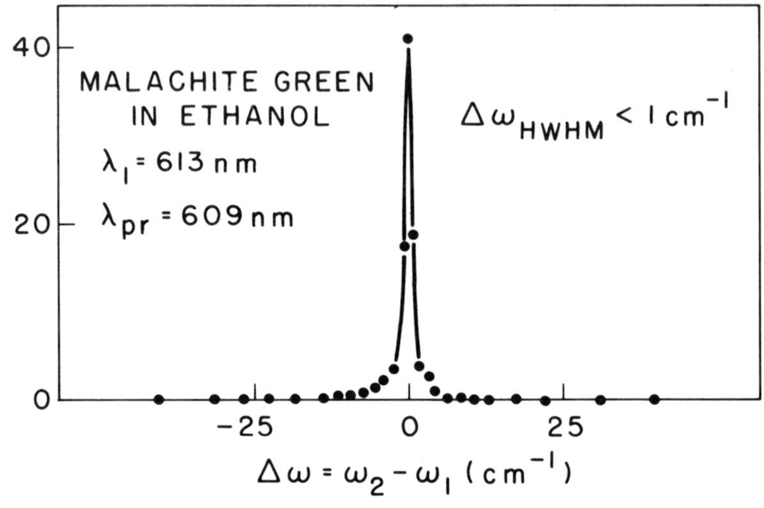

Figure 13. Three-Laser Experiment on Malachite Green in Ethanol

experiments is that it results from a laser-induced thermal grating. That is, even the comparatively small amount of heat dumped into the experimental solution at the peaks of the transient absorption grating is sufficient to cause a slight local heating or temperature grating in the sample; and this in turn is almost instantaneously converted into an index of refraction change through the dn/dT value of the solvent at constant density.

Given the long relaxation time (microseconds or longer) characteristic of a thermal grating in a typical liquid medium, it may seem difficult to conceive of observing thermal grating effects in experiments involving high-speed moving gratings produced by two independent lasers at different frequencies. The essential point however is that thermal gratings, although they decay very slowly, can be very strong in many organic solvents, and can build up quite rapidly (as rapidly as the optical energy delivered to the sample can be converted to local thermal excitation through nonradiative relaxation). The nearly instantaneous rise time of the transient grating thus leads to a potentially very high frequency of response. These strong thermal grating spikes can thus hide the desired population grating signals, which are contained in the much wider but weaker pedestal underneath the central thermal grating signal.

Using an appropriate model for the thermal grating and for a partially coherent laser pulse, we can predict theoretically that the width of the thermal grating spike in this type of experiment should be exactly twice the linewidth of the dye laser for a

Lorentzian laser line. This was confirmed experimentally, as shown in Figure 14, by comparing the thermal grating signal with a simultaneous measurement of the laser linewidth, using an etalon to measure the laser spectrum. The expected factor of two in linewidth is very closely confirmed.

As a particularly simple method of studying these thermal gratings, we have also carried out a number of experiments involving the two-laser grating geometry, but using only the interference between a single partially coherent laser beam and a delayed portion of the same beam. The setup for these experiments is shown in Figure 15. A detailed analysis which will be published elsewhere (2,19) shows that the expected diffraction efficiency versus the time delay τ_d between the two beams in such an experiment (assuming infinitely fast rise time for the thermal grating) should be given by the fourth order coherence function

$$\eta = \frac{1}{\tau_p^2} \int_{-\tau_p/2}^{\tau_p/2} \int_{-\tau_p/2}^{\tau_p/2} \Gamma_{(4)}(t_1, t_2+\tau_D; t_1+\tau_D, t_2)\, dt_1 dt_2 \tag{9}$$

where Γ is the fourth order coherence function as usually defined in coherence theory, and τ_p is the laser pulse length. It would be interesting to add a finite response time for the thermal gratings to this analysis, and then to explore whether this approach could be used to measure the (very short) time scale during which nonradiative relaxation inside a molecule spreads out and is converted into thermal excitation in the immediately surrounding solvent.

Reduction or elimination of the thermal grating effects in laser grating experiments requires the use of a high specific heat and low dn/dT solvent such as water, together with the use of more highly coherent laser signals so that the narrow central thermal grating spike can be separated from the weaker and broader signals in the pedestal which contain the desired lifetime information. Thermal gratings are of course not a problem in nonabsorbing materials, such as optical Kerr materials, so that measurements of lifetimes in these interesting systems are not obscured by thermal grating effects.

13. SUMMARY

The two-laser approach to transient grating spectroscopy, in both its polarization spectroscopy and phase conjugation forms, has already proven to be a very useful technique for studying fast photochemical effects. The three-laser approach adds substantial additional flexibility to this technique, making possible

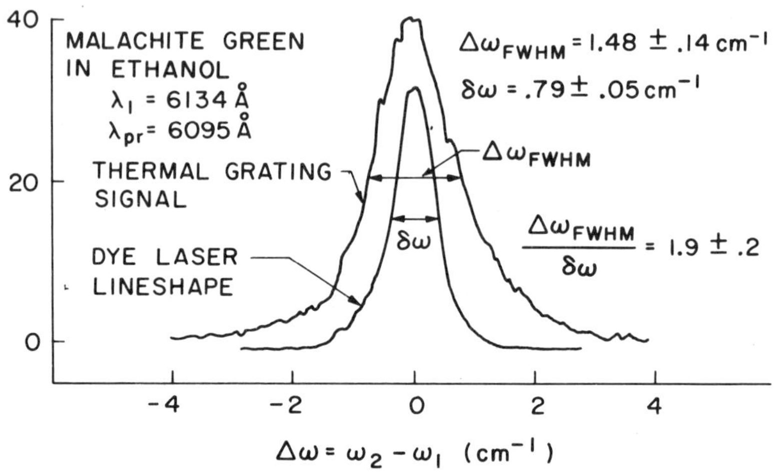

Figure 14. Two-Laser Thermal-Grating Diffraction Efficiency vs. $\Delta\omega$

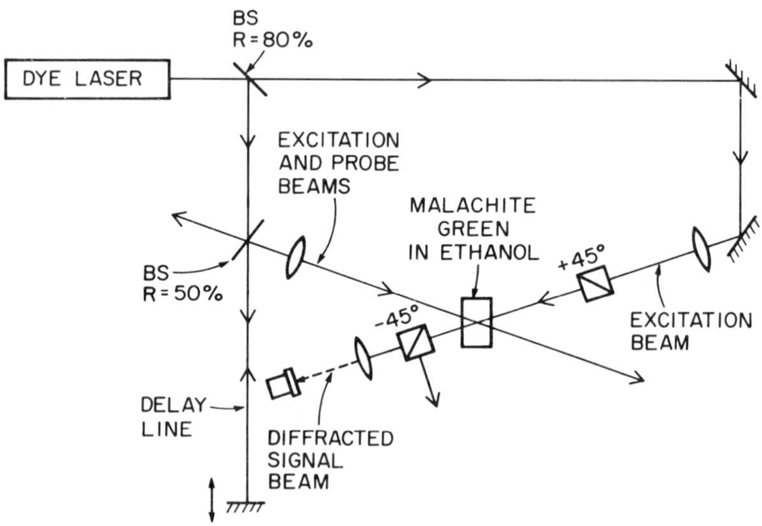

Figure 15. Experimental Setup for Studying Thermal Gratings Using a Single Laser in the Two-Beam Grating Geometry

interesting and important new measurements. The technical difficulties associated with this technique can be overcome, as demonstrated by the results reported here. Thermal gratings must be taken into account in all these induced grating experiments; but the thermal gratings effects also offer new potentials both for studying thermal and phonon phenomena and for measuring coherence properties of the laser beams themselves.

REFERENCES

1. This material comes largely from a Ph.D. dissertation by Jean-Marc Heritier, *Picosecond Spectroscopy Using a Photoacoustic Detector*, Ph.D. dissertation submitted to the Department of Applied Physics, Stanford University, March 1983; plus additional references given below.

2. This material comes largely from a Ph.D. dissertation by Rick Trebino, *Subpicosecond-Relaxation Studies using Tunable-Laser-Induced-Grating Techniques*, Ph.D. dissertation submitted to the Department of Applied Physics, Stanford University, May 1983; plus additional references given below.

3. For introductory reviews of photoacoustic spectroscopy, see Yon-Han Pao, ed., *Optoacoustic Spectroscopy and Detection*, (Academic Press, New York, 1977); or Allen Rosencwaig, *Photoacoustics and Photoacoustic Spectroscopy*, (John Wiley & Sons, Inc., New York, 1980). See also, C.K.N. Patel and A.C. Tam, 1981, *Pulsed Optoacoustic Spectroscopy of Condensed Matter*, Rev. of Mod. Phys. 53, pp. 517-550.

4. J-M. Heritier and A.E. Siegman, October 1983, *Picosecond Measurements Using Photoacoustic Detection*, IEEE J. Quantum Electron. QE-19, in press.

5. M. Bernstein, L.J. Rothberg and K.S. Peters, *Picosecond Time-Resolved Photoacoustic Spectroscopy*, in Picosecond Phenomena III, K.B. Eisenthal, R.M. Hochstrasser, W. Kaiser, and A. Laubereau, eds., Berlin: Springer-Verlag, 1982; pp. 112-115; and *Time-Resolved Photoacoustic Spectroscopy in the Picosecond Regime*, Chem. Phys. Lett. 91, pp. 315-318 (September 1982).

6. J-M. Heritier, J.E. Fouquet, and A.E. Siegman, 1982, *Photoacoustic Cell Using Elliptical Acoustic Focusing*, Appl. Opt. 21, pp. 90-93.

7. Jean-Marc Heritier, January 1983, *Electrostrictive Limit and Focusing Effects in Pulsed Photoacoustic Detection*, Opt. Comm. 44, pp. 267-272.

8. C.K.N. Patel and A.C. Tam, 1979, *Optoacoustic Spectroscopy of Liquids*, Appl. Phys. Lett. 34, p. 467; and A.C. Tam, C.K.N. Patel and R.J. Kerl, 1979, *Measurement of Small Absorption in Liquids*, Opt. Lett. 81.

9. T. Yajima, 1975, *Non-Linear Optical Spectroscopy of an Inhomogeneously Broadened Resonant Transition by Means of Three Wave Mixing*, Opt. Comm. 14, pp. 378-382; T. Yajima and H. Souma, 1978, *Study of Ultrafast Relaxation Processes by Resonant Rayleigh-Type Optical Mixing. I. Theory*, Phys. Rev. A. 17, pp. 309-323; and T. Yajima, H. Souma and Y. Ishida, *Study of Ultrafast Relaxation Processes by Resonant Rayleigh-Type Optical Mixing. II. Experiment on Dye Solutions*, Phys. Rev. A, 17, pp. 324-334; P.L. Decola, R.M. Hochstrasser and H.P. Trommsdorf, 1980, *Vibrational Relaxation in Molecular Crystals by Four-Wave Mixing: Naphthalene*, Chem. Phys. Lett. 72, pp. 1-4; J-L. Oudar and Y.T. Shen, 1980, *Nonlinear Spectroscopy by Multiresonant Four-Wave Mixing*, Phys. Rev. A, 22, p. 1141.

10. A.E. Siegman, 1977, *Proposed Picosecond Excited-State Measurement Method Using a Tunable-Laser-Induced Grating*, Appl. Phys. Lett. 30, pp. 21-23; M. Sargent III, 1976, *Laser Saturation Grating Phenomena*, Appl. Phys. 9, pp. 127-141; J.R. Andrews and R.M. Hochstrasser, 1980, *Transient Grating Studies of Energy Deposition in Radiationless Processes*, Chem. Phys. Lett. 76, pp. 207-212; and J.R. Andrews and R.M. Hochstrasser, 1980, *Transient Grating Effects in Resonant Four-Wave Mixing Experiments*, Chem. Phys. Lett. 76, pp. 213-217.

11. Robert Fisher, ed., *Optical Phase Conjugation*, (Academic Press, 1983).

12. F. Keilmann, 1976, *Infrared Saturation Spectroscopy in p-Type Germanium*, IEEE J. Quantum Electron. QE-12, pp. 592-597; J.J. Song, J.H. Lee, and M.D. Levenson, 1978, *Picosecond Relaxation Measurements by Polarization Spectroscopy in Condensed Phases*, Phys. Rev. A, 17, pp. 1439-1447; F. Keilmann, 1977, *Tunable-Laser-Induced Grating Dip for Measuring Sub-picosecond Relaxation*, Appl. Phys. 14, pp. 29-33; J.R. Andrews and R.M. Hochstrasser, 1980, *Femtosecond Relaxation of an Iron Porphyrin Observed with Polarization Spectroscopy in a Three-Level System*, Proc. Natl. Acad. Sci. USA, 77, pp. 3110-3114.

13. E.P. Ippen, C.V. Shank, and A. Bergman, 1976, *Picosecond Recovery Dynamics of Malachite Green*, Chem. Phys. Lett. 38, pp. 611-614.

14. Th. Förster und G. Hoffman, 1971, *Die Viskositätsabhängigkeit der Fluoreszenzquantenausbeuten einiger Farbstoffsysteme,* Z. Phys. Chem. N.F. Bd. 75, S. pp. 63-76.

15. D.A. Cremers and M.W. Windsor, 1980, *Study of the Viscosity-Dependent Electronic Relaxation of Some Triphenylmethane Dyes Using Picosecond Flash Photolysis,* Chem. Phys. Lett. 71, pp. 27-32.

16. W. Yu, F. Pellegrino, M. Grant and R.R. Alfano, 1977, *Subnanosecond Fluorescence Quenching of Dye Molecules in Solution,* J. Chem. Phys. 64, pp. 1766-1773.

17. M. Stavola, G. Mourou and W. Knox, 1980, *Picosecond Time-Delay Fluorimetry Using a Jitter-Free Streak Camera,* Opt. Comm. 34, pp. 404-408.

18. P. Wirth, S. Schneider and F. Dörr, 1977, *S_1-Lifetimes of Triphenylmethane and Indigo Dyes Determined by Two-Photon Fluorescence Technique*, Opt. Comm. 20, pp. 155-158.

19. Rick Trebino, Eric K. Gustafson and A.E. Siegman, *Thermal-Grating Formation with Partially Coherent Radiation from a Single Source,* to be published.

LIST OF PARTICIPANTS

Professor F. Dorr, Institut fur Physikalische und Theoretische Chelmie, Technische Universitat Munchen, G-8056, Lightemberg-strasse 4, Garching, Germany.

Professor Kenneth B. Eisenthal, Box 616 Havemeyer Hall, Department of Chemistry, Columbia University, New York, NY 10027 USA.

Professor J. Faure, Lab. Physico-Chemie, Batiment 350-Centre d'Orsay 91405 Orsay Cedex, France.

Professor G.R. Fleming, Department of Chemistry, University of Chicago, 5735 South Ellis Avenue, Chicago, Illinois 60637 USA.

Professor C. Flytzanis, Laboratoire d'Optique Quantique, Ecole Polytechnique, Palaiseau, 91128, Cedex France.

Professor C. Harris, Department of Chemistry, University of California, Berkeley, California 94720 USA.

Professor R. Hochstrasser, Department of Chemistry, University of Pennsylvania, Philadelphia, Pennsylvania 19104 USA.

Professor G.A. Kenney-Wallace, Department of Chemistry, University of Toronto, Canada M5S 1A1.

Professor J. Langelaar, Laboratorium voor Fysische Chemie Nieuwe Achtergracht 127 1018WS Amsterdam, The Netherlands.

Professor Alfred Laubereau, Lehrstuhl fur Experimentalphysik III der Universitat Bayreuth, 8580 Bayreuth Universitatsstrasse 30, Postfach 3008, Bayreuth, Germany.

Professor Noboru Mataga, Department of Chemistry, Faculty of Engineering Sciences, Osaka University, Tokyonaka, Osaka, Japan 560.

Dr. Paolo Mataloni, University of Rome, Italy.

Professor Kevin S. Peters, Department of Chemistry, Harvard University, 12 Oxford Street, Cambridge, Massachusetts 02138 USA.

Sir George Porter, The Royal Institution, 21 Albermarle Street, London, W1X 4BS England.

Dr. C.V. Shank, Bell Laboratories, Crawford Corner Road, Holmdel, New Jersey 07733 USA.

Professor Anthony E. Siegman, Edward L. Ginzton Laboratory, Stanford University, Stanford, California 94305 USA.

Professor Villy Sundstrom, Department of Physical Chemistry, University of Umea, 901 87 Umea, Sweden.

Professor Michael R. Topp, Department of Chemistry, University of Pennsylvania, Philadelphia, Pennsylvania 19104 USA.

Mr. John A.S. Walker, North Atlantic Treaty Organization, Scientific Affairs Division, B-1110 Brussels.

Professor Douwe Wiersma, Department of Physical Chemistry, University of Groningen, Nijenborgh 16, Groningen 9747AG, The Netherlands.

Professor Keitaro Yoshihara, Institute for Molecular Science, Myodaiji, Okazaki 444, Japan.

Professor Ahmed Zewail, California Institute of Technology, Division of Chemistry and Chemical Engineering, Pasadena, California 91125 USA.

INDEX